Ronald M. Miller

May 18, 1977

D1218945

Protochordates

SYMPOSIA OF THE ZOOLOGICAL SOCIETY OF LONDON
NUMBER 36

Protochordates

(*The Proceedings of a Symposium held at The Zoological
Society of London on 17 and 18 January, 1974*)

Edited by

E. J. W. BARRINGTON

*School of Biological Sciences, University of Nottingham,
Nottingham, England*

and

R. P. S. JEFFERIES

*British Museum (Natural History), Department of Palaeontology,
London, England*

Published for

THE ZOOLOGICAL SOCIETY OF LONDON

BY

ACADEMIC PRESS

1975

ACADEMIC PRESS INC. (LONDON) LTD.

24/28 Oval Road

London NW1

U.S. Edition published by

ACADEMIC PRESS INC.

111 Fifth Avenue,

New York, New York 10003

Library of Congress Catalog Card Number: 74-18517

ISBN: 0-12-613336-0

PRINTED IN GREAT BRITAIN BY
J. W. ARROWSMITH LTD., BRISTOL

CONTRIBUTORS

*BARRINGTON, E. J. W., FRS, *School of Biological Sciences, University of Nottingham, University Park, Nottingham NG7 2RD, England* (p. 129)

COURTNEY, W. A. M., *Department of Zoology, Westfield College, Hampstead, London NW3 7ST, England* (p. 213)

DILLY, P. N., *Department of Anatomy, University College London, Gower Street, London WC1E 6BT, England* (p. 1)

FLOOD, P. R., *Institute of Anatomy, University of Bergen, Arstadvollen 19, Bergen, Norway* (p. 81)

‡GUTHRIE, D. M., *Department of Zoology, University of Aberdeen, Aberdeen AB9 2TN, Scotland* (p. 43)

JEFFERIES, R. P. S., *British Museum (Natural History), Department of Palaeontology, Cromwell Road, London SW7, England* (p. 253)

SOUTHWARD, E. C., *Marine Biological Association of the United Kingdom, The Laboratory, Citadel Hill, Plymouth PL1 2PB, England* (p. 235)

¶THORNDYKE, M. C., *Institute of Urology, Department of Pathology, St. Paul's Hospital, Endell Street, London* (p. 159)

THORPE, A., *Department of Zoology and Comparative Physiology, Queen Mary College, Mile End Road, London E1 4NS, England* (p. 159)

WATTS, D. C., *Department of Biochemistry and Chemistry, Guy's Hospital Medical School, London Bridge, London SE1, England* (p. 105)

WEBB, J. E., *Department of Zoology, Westfield College, Hampstead, London NW3 7ST, England* (p. 179)

WELSCH, U., *Anatomisches Institut der Universität Kiel, 2300 Kiel, Germany* (p. 17)

WILLMER, E. N., FRS, *Clare College, Cambridge, England* (p. 319)

* Present address: Cornerways, 2 St. Margarets Drive, Alderton, Tewkesbury, Gloucestershire, GL20 8NY.

‡ Present address: Department of Zoology, University of Manchester, Manchester, M13 9PL.

¶ Present address: Department of Zoology, Bedford College, Regent's Park, London, NW1 4N5.

ORGANIZERS AND CHAIRMEN

ORGANIZERS

E. J. W. BARRINGTON and R. P. S. JEFFERIES, *on behalf of The Zoological Society of London*

CHAIRMEN OF SESSIONS

E. J. W. BARRINGTON, FRS, *School of Biological Sciences, University of Nottingham, University Park, Nottingham NG7 2RD, England*
D. M. GUTHRIE, *Department of Zoology, University of Aberdeen, Aberdeen AB9 2TN, Scotland*
J. E. WEBB, *Department of Zoology, Westfield College, Hampstead, London NW3 7ST, England*
J. Z. YOUNG, FRS, *Department of Anatomy, University College London, Gower Street, London WC1E 6BT, England*

CONTENTS

The Pterobranch *Rhabdopleura Compacta:* its Nervous System and Phylogenetic Position

P. N DILLY

The Fine Structure of the Pharynx, Cyrtopodocytes and Digestive Caecum of Amphioxus (*Branchiostoma lanceolatum*)

U. WELSCH

The Physiology and Structure of the Nervous System of Amphioxus (the Lancelet) *Branchiostoma Lanceolatum* Pallas

D. M. GUTHRIE

Fine Structure of the Notochord of Amphioxus

P. R. FLOOD

Evolution of Phosphagen Kinases in the Chordate Line

D. C. WATTS

Problems of Iodine Binding in Ascidians

E. J. W. BARRINGTON

The Endostyle in Relation to Iodine Binding

A. THORPE and M. C. THORNDYKE

The Distribution of Amphioxus

J. E. WEBB

The Temperature Relationships and Age-structure of North Sea and Mediterranean Populations of *Branchiostoma Lanceolatum*

W. A. M. COURTNEY

Fine Structure and Phylogeny of the Pogonophora

EVE C. SOUTHWARD

Fossil Evidence Concerning the Origin of the Chordates

R. P. S. JEFFERIES

The Possible Contribution of the Nemertines to the Problem of the Phylogeny of the Protochordates

E. N. WILLMER

Symp. zool. Soc. Lond. (1975) No. 36, 1–16.

THE PTEROBRANCH *RHABDOPLEURA COMPACTA:* ITS NERVOUS SYSTEM AND PHYLOGENETIC POSITION

P. N. DILLY

Department of Anatomy, University College, London, England

SYNOPSIS

Rhabdopleura has a very primitive nervous system. The cell bodies and their fibres are confined within the epithelial cell layer. Bundles of fibres radiate from a collection of nerve cell bodies in the collar region of the animal. The collar ganglion consists of a peripheral rind of neurons enclosing a tangled neuropil. There are many synapses within the neuropil, occurring between axons and axons, axons and dendrites, and between axons and cell bodies. The neuropil synapses are remarkable for the differing sizes and contents of the synaptic vesicles.

There are some synapses within the neuropil that contain a pair of membranes which have not previously been described in association with synapses. These membranes lie within the cell cytoplasm in close association with the synaptic thickening, usually on the post-synaptic side. Their function is unknown. Most of the synapses are polarized.

The fibres within the neuropil vary in diameter but there are no especially large giant fibres. It is possible to identify axons and dendrites, since many of the dendrites have a large number of tubules in their cytoplasm. There are fibrils as well as tubules in the cell processes. The nerve cells giving rise to these processes are uniform in size and appearance. Amongst them are other epithelial cells that have a great number of glycogen granules in their cytoplasm, a feature typical of glial cells in higher vertebrates.

Peripherally the neuromuscular junctions, or motor end plates, are remarkable for their dense contents of mitochondria. Neurociliary synapses between nerve fibres and cells bearing cilia are common. Neurociliary synapses can be polarized in either direction or they can have synaptic vesicles on both sides of the membrane junction.

Rhabdopleura is considered to be a primitive hemichordate which evolved from a stock close to the origin of deuterostomes; this view seems to exclude the possibility of a direct origin of the chordates from the echinoderms.

INTRODUCTION

The protochordates and hemichordates are animals between the vertebrates and the invertebrates. While there is no doubt that the protochordates are closely related to the vertebrates, the hemichordates are in a much more equivocal position. There is convincing evidence of a phylogenetic relationship between hemichordates and both echinoderms and chordates. Enteropneusts and pterobranchs make up the two major groups of hemichordates, the pterobranchs having remained the more primitive; the lophophore-like arms and tentacles are thought to represent a primitive feature of the phylum that has been lost in the enteropneusts.

For abbreviations used in the figures see p. 16.

The pterobranchs have many features in common with the echinoderms and the lophophorates, but the presence of gill clefts in all but *Rhabdopleura*, and the dorsal nerve collar, which is sometimes hollow in other hemichordates, points their chordate affinities. However, the lack of a notochord and the differences in general body structure exclude the hemichordates from the phylum Chordata. It is generally agreed that the hemichordates are in some real sense primitive, and it is of great interest to compare their nervous systems with those of the more advanced Urochordata and Cephalochordata. Recent studies on these groups (Flood, 1968; Dilly, 1969 a, b, c; Dilly, Welsch & Storch, 1970; Guthrie, 1975; Lane, 1968), have made a study of the pterobranch nervous system desirable, especially since in most groups of the Phylum Chordata the type of nervous system is more primitive than that of the other large Metazoa, with the exception of the Coelenterata.

The nervous system of *Rhabdopleura* has previously been described by Schepotieff (1907a). He described in the base of the dorsal collar epidermis a collar ganglion which gave several branches: a median anterior branch to the cephalic shield, a posterior dorsal nerve trunk, and a pair of connectives, as well as branches that run along the dorsal and ventral sides of the arms. He described ganglion cells in the dorsal nerves and circumenteric connectives, as well as branches that run along the dorsal and ventral sides of the arms. Similar organisation has been found in *Cephalodiscus* (Anderson, 1907; Schepotieff, 1907b), and Komai (1949) described the "cerebral" ganglion of *Atubaria* as sending off nerve trunks from both its anterior and posterior ends. The nervous system itself consists of some nerve cells and a layer of nerve fibres lying within the epidermis immediately above the basement membrane, in the primitive ectodermal position. Such a primitive system may be a good place to hunt for clues for the beginnings of those features that are typical of more advanced brains.

MATERIAL AND METHODS

Colonies of *Rhabdopleura compacta* were obtained by dredging. They were found attached to the concave inner surfaces of the separated shells of dead *Glycymeris* (a lamellibranch). For histological investigations the whole colony was removed intact from the shell using a cornea knife, and then immersed in the chosen fixative, usually 10% neutral formol in sea water. The colonies were embedded in paraffin wax and sectioned serially at 5 μm intervals. Staining was usually by Mallory's triple stain or the Nonidez Cajal techniques.

For electronmicroscopy it was necessary to dissect the animals free from the coenecium, usually by breaking the erect tube of the coenium and cutting the muscular stalk with iridectomy scissors, and then using watchmakers' forceps to pull on the muscular stalk to remove the zooid from the tube. The animal was then transferred to a cavity slide by means of a bacteriological pipette, and the cavity flooded with fixative, which was either 4% glutaraldehyde or 1% osmic acid. In each case the fixative was made with freshly filtered sea water, and buffered with either a veronal acetate or cacodylate system. The pH in each case was adjusted to between 7·4 and 7·6. The specimens were fixed for two hours, and then dehydrated in graded ethanol solutions. They were stained for 8 h in 6% uranyl acetate solution made up in absolute ethanol and were then washed in absolute ethanol and soaked in epoxy-propane for 30 mins. Embedding was in Araldite using the multiple changes technique of Gray (1964). Sections were cut with glass knives on a Porter Blum ultramicrotome, and stained on the grids using lead citrate (Reynolds, 1963). Despite much care it is difficult to prepare such tiny specimens for electron microscopy.

RESULTS

Collar ganglion nerve cells and glia

The collar ganglion of *Rhabdopleura* lies in the epidermis. It consists of a mass of nerve cells separated by nerve fibres from the subepithelial basement membrane. Some of the epithelial cells that lie near the nerve cells have masses of glycogen-like particles in their cytoplasm, a feature that is found in glial cells of many vertebrates (Fig. 6). Although glial processes can be seen extending for short distances into the neuropil, glial or epithelial cell processes have not so far been seen traversing the ganglion and extending to the basement membrane (a feature common in *Saccoglossus* (Dilly, 1969b)). Usually the neurons lie peripherally and the fibres are deep to them, but below the neuropil there are some nerve cells between the fibres and the basement membrane. The nervous system maintains its intraepidermal position throughout the zooid. The neurons send processes into the fibre mass; some of these processes have dense tubular contents and others are relatively empty (Figs 1, 4, 5). Many of the neurons in the collar ganglion are unipolar, with a single process leaving the cell body and penetrating the mass of the neuropil before dividing into its axon and dendrite-like branches (Fig. 1). This initial part of the cell process has features in common with both axons and dendrites of vertebrates. It has a dendrite-like content of neurotubules and an axon-like content of vesicles.

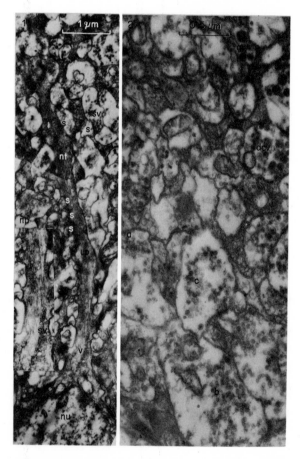

FIG. 1. Montage of a unipolar neuron with the unipolar cell process penetrating the neuropil. In structure it has features of both an axon and a dendrite, see inset for details. Several presumed synaptic sites occur along its length.

FIG. 2. An area of neuropil from the collar ganglion. It contains several different sorts of membrane bound vesicles as well as some electron dense granules.

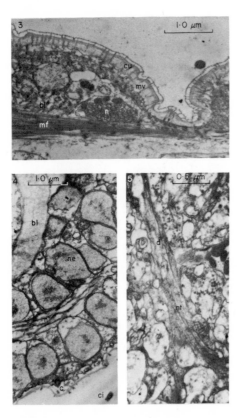

FIG. 3. Section of the body of *Rhabdopleura* showing a presumed motor nerve in close contact with a cell containing a mass of myofibrils:

FIG. 4. Section of the edge of the collar ganglion where it becomes continuous with the collar connective. The uniform structure of the neurons is apparent as well as their intraepidermal position.

FIG. 5. Longitudinal section of part of a dendrite in the collar ganglion neuropil. Several large electron dense droplets are seen in this field.

Fibres

Within the neuropil the fibres with large numbers of tubules are identified as dendritic; they usually have many synapses along their surfaces (Fig. 1). If it is assumed that the clearer profiles are axons, then most of these synapses are axo-dendritic. Most of the synapses appear polarised; that is the vesicles appear only on one side of the pair of increased membrane densities usually considered the site of synaptic transmission. The side containing the vesicles is presumably presynaptic, and the relatively empty profiles are post-synaptic.

Synapses and vesicles

The mass of nerve fibres deep to the nerve cell bodies is a neuropil containing many synapses. These synapses are of several different types (Figs 6, 7, 8, 9, 10, 11). They may be morphologically polarised, with the synaptic vesicles crowded against only one

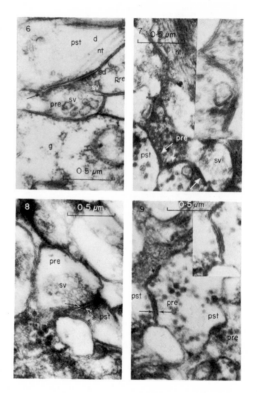

FIG. 6. Synapse within the neuropil of the collar ganglion; part of a large glial cell occupies the lower part of the plate. Two separate presynaptic processes abut against a common post-synaptic dendrite. The vesicle populations in the presynaptic processes are different. In one of them the vesicles appear to be underlined by an increase in synaptic density whereas in the other a vesicle appears to be opening into the synaptic cleft (arrow).

FIG. 7. Synapse in the neuropil showing "extra" membranes, see inset, associated with the presumed areas of transmission (arrows). The process of a unipolar neuron crosses the field and contains both neurotubules and vesicles as well as a mitochondrion.

FIG. 8. Synapse in the neuropil showing vesicles fused with the presynaptic membrane (→), and an increase in thickness of the postsynaptic membrane probably caused by the fusion of microtubules with it (* →).

FIG. 9. Synapse in the neuropil showing "extra" membranes on both sides of the synaptic junction (see inset) and a possible serial synapse.

side of the synaptic membranes, or they may occur on both sides of the increases in membrane densities. Some of the presynaptic profiles have homogeneous vesicle contents, while others may have a mixture of vesicle sizes. Many of the vesicles have dense cores, which may occupy the whole vesicle or just part of it (Figs 2, 11, 13).

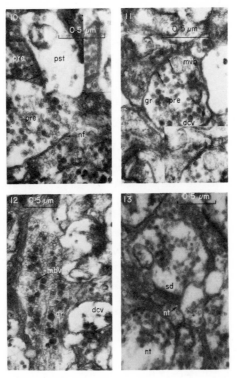

FIG. 10. Synapse in the neuropil showing "extra" membranes that in this case are continuous with the cell membranes of a presynaptic process (Inset).

FIG. 11. Synapse in the neuropil showing an "extra" pair of presynaptic membranes with electron dense granules in the presynaptic process (Inset).

FIG. 12. A large ending in the neuropil showing a wide range of vesicles from multivesicular bodies to large granules, which are apparently not membrane bound.

FIG. 13. Oblique section of a synapse in the neuropil. The synaptic density appears to consist of several pairs of membranes lying more or less parallel to one another.

Little is known of the physiological significance of these morphological differences but it seems reasonable to suppose that the appearances indicate substances of differing neurophysiological

activity. In some of the presynaptic endings, besides this wide
spectrum of vesicles, there are also multivesicular bodies. These
bodies, which are around 250 nm in diameter, contain clear
vesicles, usually of uniform size around 30–50 nm in diameter (Figs
11, 12).

The most unusual feature of these synapses is the presence of
an extra pair of membranes close to the synaptic junction. These
membranes usually have an increased electron density similar to
that of the adjacent synaptic membranes. These "extra" membranes usually run parallel to the cell membranes (Figs 7, 9, 10, 11).
Sometimes the extra membrane appears to be part of a membrane-
bound process that is inserted peg-like into the cell, near the
synaptic junction (Fig. 10). Sometimes these extra membranes can
appear on both sides of the synapse. The separation of the
electron-dense cell membranes of the pre- and post-synaptic
processes is often increased at the presumed synaptic junction.
Sometimes it remains more or less the same as the separation at
other presumed non-synaptic sites (Figs cf 8, 9). So far no "tight"
junctions with the membranes fusing have been observed.

Besides these increases in electron density of the synaptic
membranes, the membranes are sometimes increased in thickness.
Usually this increase in thickness affects both membranes. Thickening of the presynaptic membrane seems to involve a spread of
the increased density to adjacent vesicles and tubules, while
post-synaptic thickening leads to a spread of the increased density
to the "extra" membranes or associated neurotubules (Figs 7, 8).
These thickenings do not occupy the whole length of the adjacent
cell membranes between two profiles; they are usually between
$0 \cdot 25 \ \mu$m and $0 \cdot 35 \ \mu$m long, and several may occur along the length
of contact between two cells. Synapses have been identified be-
tween axon and axon, axon and dendrite and axon and cell body.
Occasional sequential synapses, where a process is post-synaptic to
one fibre and apparently pre-synaptic to the next profile, have
been seen (Fig. 9).

The synaptic vesicles, besides being crowded against the dense
region of the synaptic membrane, are occasionally fused with it
(Figs 6, 8), and sometimes this increase in density extends around
part of the membrane bounding the vesicle (Fig. 8). Perhaps these
are clues to the processes involved in the release of some or all of
the vesicle contents into the synaptic cleft. Where the vesicle
content of the profile is mixed, it is usual for the different sorts of
vesicles to be randomly spread throughout the profile. In some

regions, profiles with the same sort of vesicle contents seem to be vaguely grouped and not spread out randomly throughout the field. Besides vesicles, these synaptic endings contain mitochondria. Mitochondria are especially numerous in the motor end plates.

Nerve fibres, neurofibrils and neurotubules

The presumed motor nerves that lie in close association with the cells containing muscle fibres occur in bundles, usually of uniformly small fibres about 70 nm in diameter (Fig. 3). The muscle fibres themselves appear to be inserted into the cuticle of the animal, as in the pogonophores, where they are inserted into the basement membrane below the epidermis and then via tonofibrils into the very similar cuticle (Southward, 1975). Neurotubules and neurofilaments are often found in the synaptic terminals, as well as in the processes leading to them. In electronmicrographs the neurotubules are hollow, about 12–15 nm in diameter, with a central cavity 8 nm in diameter. They can extend for considerable distances through the cell cytoplasm and often have vesicles associated with their surfaces. In the fibres they are often grouped together in bundles, or they may occupy most of the fibre cytoplasm (Figs 1, 5, 6, 13). The neurofibrils are solid and smaller, being about 8 nm in diameter. They, too, are often crowded together into bundles that may become densely stained with lead citrate.

Most of the fibres appear to be passing either horizontally or vertically within the neuropil. Those that do so are to some extent grouped into oval bundles by glial cell processes. Some of the glial cells have lipid droplet contents, perhaps the result of phagocytosis. There are often several glial cells associated with one nerve cell body and it is my impression that there are more glial cells than neurons in the collar ganglion. Where glial cells abut against one another, there are often desmosome-like structures between them.

Within the neuropil the nerve fibre diameters vary considerably from 0·1 μm to 1·5 μm (Fig. 2). Their outlines are often irregular, but where they are cut in longitudinal section the diameter is more or less constant. Occasionally there are fibres of greater diameter than those that are found in the peripheral arm nerves (Dilly, 1972).

The nerve fibres that enter and leave the cerebral ganglion do so in discrete bundles, in contrast to those in *Saccoglossus*, where they are spread throughout the epidermal layer (Dilly, 1969a). These bundles of fibres maintain their primitive epidermal position throughout the animal. Peripherally, besides motor end plates there are many neurociliary synapses, but other specialized receptor organelles have not been seen.

Nerve connections within the colony

It is known (Stebbing & Dilly, 1972) that if one zooid in a colony is mechanically stimulated then the other zooids may respond as well. The zooids are joined together by a stalk that passes through a thick-walled black stolon. Sections of the muscular part of the stalk have revealed easily identifiable nerve fibres, but so far sections of the stolon have not done so. The black stolon, however, is not empty; it contains cells surrounded by a basement lamella, as well as a mass of extracellular material and debris. None of the profiles within the sections of the stolon can be identified with confidence as nerve fibres, although some of them have a dense granular content similar to that found within the neuropil.

The dormant buds of *Rhabdopleura* consist of a mass of yolk surrounded by two layers of cells that are separated by a basement membrane, the whole being surrounded by a thick-walled capsule. There are nerve fibres between the outermost layer of cells and the basement lamella. It is not known if these fibres are connected with the animals in the rest of the colony.

DISCUSSION

General position and morphology of the nervous system

The collar ganglion of *Rhabdopleura* is very primitive. It is intraepidermal in position and consists of a simple layer of neurons and a neuropil, and is not hollow. As such it is even more primitive than the enteropneust nervous system and, of course, the dorsal tubular nervous system in the cephalochordates (Guthrie, 1975) is far advanced from this condition. The hollow nervous system of the larva of the tunicate *Ciona* is also far advanced beyond that of *Rhabdopleura*.

This considerable difference between the nervous systems of the pterobranchs and the enteropneusts, which are grouped together as hemichordates, suggests that the pterobranchs may have separated early from a common hemichordate ancestor of the two groups. This view is strengthened by the many striking similarities between the *Rhabdopleura* nervous system and that of the echinoderms, for this suggests that the pterobranchs are still close to the common ancestor, while the enteropneusts have evolved further towards the urochordates and the cephalochordates. Indeed in many ways *Rhabdopleura* is very similar to the lophophorates. However, it may be that the remarkably small size of the adult *Rhabdopleura* is a sign of neoteny and that this primitive nervous system is to some extent a neotenous feature.

Synaptic considerations

The morphology of the presumed synaptic junctions is similar throughout the hemichordates and the protochordates. In these groups there is an increase in electron density of the adjacent cell membranes. Actual increase in thickness of the synaptic membranes is rare throughout the hemichordates but it occurs more often in *Rhabdopleura* than in the other groups. This increase in thickness in *Rhabdopleura* may be some form of staining artefact, but it seems more probable that it is caused by the staining of organelles associated with the synaptic membranes. Post-synaptic thickenings are well known in vertebrate synapses and those in *Rhabdopleura* may be a clue to early evolutionary stages of such post-synaptic thickenings. Within the synaptic cleft between the cells there is little or no extracellular material deposited in *Rhabdopleura*, such as is found in the cerebral ganglion of adult *Ciona* (Dilly, 1969c). Synaptic cleft material seems to be a feature of synapses in more advanced vertebrates.

Synaptic vesicles

The wide range of vesicle sizes and contents within single profiles in the neuropil of *Rhabdopleura* contrasts with the much more uniform vesicle populations in the enteropneust *Saccoglossus* (Dilly, 1969b) and in the urochordate *Ciona* (Dilly, 1969c) and in the cephalochordate amphioxus (Flood, 1968). This wide range of vesicle forms is presumably a primitive condition, from which certain vesicles have become selected during evolution to provide the much more circumscribed populations found in advanced nervous systems. The stage at which the vesicle populations are homogeneous probably represents a considerable advance in the evolution of the nervous system. Variations in vesicle size may be related to the amount of material that they contain, and small vesicles may be partially discharged large vesicles. Amongst the wide population of vesicles in *Rhabdopleura* it is not possible to single out any specific population that might for any reason be designated neurosecretory granules. Yet such granules can be found in the enteropneusts and urochordates. There are some granules within the neuropil of *Rhabdopleura* that are not membrane-bound (Figs 11, 12). Similar granules are found within the cells of the stolon connective, suggesting that there is a flow of material between the zooids of the colony. If these granules are identical, then perhaps this is evidence for neurosecretion. But as yet the only evidence of their similarity of function is a similar shape and size, and to ascribe a neurosecretory function to them is little more than speculation.

Neurons and fibres

The neurons are arranged around the neuropil of the collar ganglion in a similar mode to those in the cerebral ganglion of adult *Ciona* (Dilly, 1969c). The nerve cell bodies in *Rhabdopleura* are remarkably uniform in appearance; so far no giant neurons similar to those in *Saccoglossus* have been found, and there are no giant fibres. Close proximity of the nerve fibres to one another, without any intervening glial processes or cytoplasm, is found throughout the hemichordates and the protochordates, and raises the problem how impulses within a single fibre remain discrete. As yet there is no satisfactory answer. Pickens (1970) has suggested that ephaptic conduction is a feature between the tiny axons in the abdominal nerve cord in the enteropneust, *Ptychodera*. It may be, therefore, that when several fibres are stimulated in a bundle of tiny fibres in intimate contact, as in *Rhabdopleura* and most of the hemichordates and protochordates, then the volley will tend to synchronize by interaction between them. Such synchrony has been observed in amphioxus (Guthrie, 1975). This effect may serve to speed up impulses in the small fibres in situations of considerable stimulus to the animal, and may be effective in evoking rapid responses to them.

The intraepithelial position of the nervous system in the adult *Rhabdopleura* corresponds with its position in the larva (Dilly, 1973). This may be a primitive feature that persists in an adult which, as already suggested, has neotenous features (Hyman, 1959).

The nerve fibres that radiate to and from the collar ganglion in *Rhabdopleura* are arranged in bundles, in contrast to those in *Saccoglossus*, where they are spread throughout the epidermis. The organization of the nerve fibres into bundles in *Rhabdopleura* is a more advanced condition than in the enteropneusts, where the only evidence for grouping of fibres into bundles is the thickening in certain regions of the generalized subepithelial layer (Dilly, 1969b). *Rhabdopleura* is not as advanced in nerve fibre separation as is amphioxus, where it may be possible to separate motor and sensory bundles of fibres.

Motor end-plates

In the peripheral part of the nervous system of *Rhabdopleura* there are large motor end-plates against cells that contain masses of muscle fibres. These motor end-plates are characterized by large numbers of mitochondria, more than are found in *Ciona* (Panko, unpublished observations). This concentration of mitochondria perhaps indicates a much higher metabolic activity in the end-plates of *Rhabdopleura*.

Neurociliary synapses

Neurociliary synapses are common in the tentacles of *Rhabdopleura* zooids. They can be polarized in either direction, suggesting that some cilia are sensory whereas others are motor, but most of them must surely have both effector and receptor functions. There are many more cells bearing cilia than there are nerve fibres in the tentacles of *Rhabdopleura*. There must therefore be either some means of non-neuronal conduction between ciliated cells (Dilly, 1972) or else direct response. There is certainly good coordination between movements of the body, arms, tentacles and cilia (Stebbing & Dilly, 1972), however this is achieved.

Apart from cilia, there are no obvious receptor organelles in *Rhabdopleura*. This is in contrast to the more advanced urochordates, where (in *Ciona*) both the larva and the adult have special organelles (Dilly, 1961, 1962, 1964, 1969a; Dilly & Wolken, 1973).

Phylogenetic affinities

The nervous system of *Rhabdopleura*, a pterobranch, gives little evidence for its chordate affinities. The animal does not have a recognizable notochord nor a tubular nervous system. The hemichordates have an extremely equivocal position between the invertebrates and the chordates, and *Rhabdopleura*, if judged by its nervous system, is the most primitive amongst them and closest to the ancestor that hemichordates are believed to share with the echinoderms. The pterobranchs are surely hemichordates, being related to enteropneusts especially by the presence of gill slits in all but *Rhabdopleura*. Relation to echinoderms is indicated by the early embryological development of the hemichordates, for the mode of formation of the gastrula and coelom is much like that of the echinoderms, while the early tornaria larva is very similar to the bipinnaria of the asteroids. Yet an affinity between hemichordates and chordates is also indicated, for the dorsal nerve collar cord of hemichordates, which is sometimes hollow, is similar to the dorsal nerve cord of chordates. However, the lack of a notochord and the differences in general body structure exclude the hemichordates from the phylum Chordata, and make it necessary to regard them as a separate phylum.

Of the two major classes of the hemichordates, the pterobranchs, as already suggested, may be considered the more primitive. The tripartite division of the coelom seems a general character, in that it is present not only in echinoderms but also in lophophorates as well. There is, moreover, a considerable similarity of body plan between the lophophorates and the pterobranchs, and of their mesocoels in particular. There are, of course, differences in detail, but it is very difficult to describe the tentaculated mesosome of the pterobranchs or lophophorates so that one or other group is excluded. The complete crown of tentacles around the mouth of lophophorates could, of course, be used, but the presence of a good sized protosome in the pterobranchs would explain it. This difference between the lophophorates and the pterobranchs is probably less significant when it is remembered that the enteropneusts completely lack tentacles.

The evidence suggests that one has to place the phylum Hemichordata, including the pterobranchs, near to the origin of the chordates. However, the striking parallels between the pterobranchs and the lophophorates suggest that they are very closely related, that they probably arose from a common ancestor, and very near to the origin of deuterostomes. These seem to exclude the possibility of a direct origin of the chordates from the echinoderms, although a contrary view is held by Jefferies (1975).

Acknowledgements

I wish to thank Professor J. Z. Young FRS for much advice and encouragement throughout this study. Dr A. Stebbing helped me obtain the specimens. Mr R. Moss and Miss D. Bailey gave much valuable technical assistance.

REFERENCES

Andersson, K. A. (1907). Die Pterobranchier der Schwedischen Südpolar-Expedition 1901–1903. *Wiss. Ergebn. schwed. Südpolarexped.* **5**: 1–122.

Dilly, P. N. (1961). Electron microscope observations of the receptors in the sensory vesicle of the ascidian tadpole. *Nature, Lond.* **191**: 786–787.

Dilly, P. N. (1962). Studies on the receptors in the cerebral vesicle of the ascidian tadpole. 1. The otolith. *Q. Jl microsc. Sci.* **103**: 393–398.

Dilly, P. N. (1964). Studies on the receptors in the cerebral vesicle of the ascidian tadpole. 2. The ocellus. *Q. Jl microsc. Sci.* **105**: 13–20.

Dilly, P. N. (1969a). Studies on the receptors in *Ciona intestinalis*. III. A second type of photoreceptor in the tadpole larva of *Ciona intestinalis. Z. Zellforsch. mikrosk. Anat.* **96**: 63–65.

Dilly, P. N. (1969b). The nerve fibres in the basement membrane and related structures in *Saccoglossus horsti* (Enteropneusta). *Z. Zellforsch. mikrosk. Anat.* **97**: 69–83.

Dilly, P. N. (1969c). Synapses in the cerebral ganglion of adult *Ciona intestinalis. Z. Zellforsch. mikrosk. Anat.* **93**: 142–150.

Dilly, P. N. (1972). The structure of the tentacles of *Rhabdopleura compacta* (Hemichordata) with special reference to neurociliary control. *Z. Zellforsch. mikrosk. Anat.* **129**: 20–39.

Dilly, P. N. (1973). The larva of *Rhabdopleura compacta* (Hemichordata). *Mar. Biol.* **18**: 69–86.

Dilly, P. N., Welsch, U. & Storch, V. (1970). The structure of the nerve fibre layer and neurochord in Enteropneusts. *Z. Zellforsch. mikrosk. Anat.* **103**: 129–148.

Dilly, P. N. & Wolken, J. J. (1973). Studies on the receptors in *Ciona intestinalis*. IV. The ocellus in the adult. *Micron* **4**: 11–29.

Flood, P. R. (1968). Structure of the segmental trunk muscles in *Amphioxus* with notes on the course and "endings" of the so called ventral root fibres. *Z. Zellforsch. mikrosk. Anat.* **84**: 389–416.

Gray, E. G. (1964). Tissue of the central nervous system. In *Electron microscopic anatomy*: 369–417. Kurtz, S. M. (ed.). London: Academic Press.

Guthrie, D. M. (1975). The physiology and structure of the nervous system of amphioxus (the lancelet) *Branchiostoma lanceolatum* Pallas. *Symp. zool. Soc. Lond.* No. 36: 43–80.

Hyman, L. H. (1959). *The invertebrates. 5 Smaller coelomate groups.* New York: McGraw-Hill.

Jefferies, R. P. S. (1975). Fossil evidence concerning the origin of chordates. *Symp. zool. Soc. Lond.* No. 36: 253–318.

Komai, T. (1949). Internal structure of the pterobranch *Atubaria. Proc. Japan Acad.* **25**: 19–24.

Lane, N. J. (1968). Fine structure and phosphatase distribution in the neural ganglion and associated neural gland of tunicates. *J. Cell Biol.* **39**: 171a.

Pickens, P. (1970). Conduction along the ventral nerve cord of a hemichordate worm. *J. exp. Biol.* **53**: 515–528.

Reynolds, E. S. (1963). The use of lead citrate at high pH as an electron opaque stain for electron microscopy. *J. Cell Biol.* **17**: 208–212.

Schepotieff, A. (1907a). Die Anatomie von *Rhabdopleura. Zool. Jb.* (Anat.) **23**: 463–534.

Schepotieff, A. (1970b). Knospungsprozess von *Cephalodiscus*. *Zool. Jb.* (Anat.) **25**: 405–494.

Southward, E. C. (1975). Fine structure and phylogeny of the Pogonophora. *Symp. zool. Soc. Lond.* No. 36: 235–251.

Stebbing, A. R. D. & Dilly, P. N. (1972). Some observations on living *Rhabdopleura compacta* (Hemichordata). *J. mar. biol. Ass. U.K.* **52**: 443–448.

Abbreviations used in figures

a, small empty vesicles, b, small vesicles with electron dense contents, bl, basement lamella, c, mixed small vesicles, ci, cilium, cu, cuticle, d, dendrite, dcv, dense cored vesicles, g, glial cell, gr, dense granules without an apparent bounding membrane, mf, muscle fibrils, mv, microvillus, mvb, multivesicular body, n, nerve bundle, ne, neuron, nf, neurofilaments, np, neuropil, nt, neurotubule, nu, nucleus, pre, presynaptic process, pst, post synaptic process, s, synapse, sd, region of synaptic density, sv, synaptic vesicles, up, unipolar cell process, v, vesicles.

Symp. zool. Soc. Lond. (1975) No. 36, 17–41.

THE FINE STRUCTURE OF THE PHARYNX, CYRTOPODOCYTES AND DIGESTIVE CAECUM OF AMPHIOXUS (*BRANCHIOSTOMA LANCEOLATUM*)

ULRICH WELSCH

Department of Anatomy, University of Kiel, West Germany

SYNOPSIS

Endostyle. The electron microscope confirms the existence of several cell-zones in the endostyle, which correspond to the six zones as described in the light-microscopical literature; zones 4 and 5, however, cannot be clearly separated in the electron microscope. Zones 1, 3 and 6 possess a comparable ultrastructure. The heterochromatin-rich nuclei form columns. The apex of each cell bordering the lumen bears one cilium, the basal part of which is surrounded by a collar of microvilli. Zone 2 consists of slender prismatic cells containing a strikingly euchromatin-rich nucleus at their base; the cytoplasm is characterized by extensive rough ER-cisterns, a well-developed Golgi-apparatus, and various secretion granules, which in general are rather pale. The cells of zone 4/5 contain a basally situated nucleus and electron-opaque lipid inclusions, the frequency of which decreases from medial to lateral. In the supranuclear region rough ER-cisterns and a uniform population of electron-dense secretion granules occur.

Epibranchial groove. The median part of the epibranchial groove consists of one layer of prismatic cells, in the basal half of which a large vacuole is located. The nucleus thus lies in the upper part of the cells, which in addition contain rather large secretion granules of median densities. The apical surface bears loosely arranged slender microvilli.

Digestive caecum. The majority of cells forming the wall of this caecum are of uniform appearance and are characterized by a basally situated nucleus, electron-dense lipid inclusions, numerous mitochondria, large secretion granules and rough ER-cisterns, both of the latter in a supranuclear position. Occasionally cells of presumably endocrine function have been found.

INTRODUCTION

In the anterior part of the alimentary tract of amphioxus two structures occur which are of particular interest to the comparative anatomist since they are homologous with organs of the vertebrates: 1) the endostyle (= hypobranchial groove) which corresponds to the vertebrate thyroid gland, and 2) the digestive caecum (= hepatic caecum) which is considered to be the precursor of the vertebrate liver or both liver and pancreas.

Light microscopical studies have shown that the endostyle forms the ventral floor of the pharynx. Its prismatic epithelium is characterized by a special arrangement of the cellular nuclei. Beside sections with basally located nuclei, others can be found in which the nuclei form columns extending from the base to the upper surface of the epithelium (Krause, 1923). The main function of the endostyle is the secretion of mucous substances (Olsson,

1963). Barrington (1958) has demonstrated its ability to bind
iodine.

The dorsal roof of the pharynx is formed by the epibranchial
groove which is lined by vacuolated and tall ciliated cells (Krause,
1923) and which is rich in various enzymes (Welsch & Storch,
1969).

The lateral parts of the pharynx consist of the pharyngeal bars
(primary and secondary ones) which are covered by different
ciliated and non-ciliated cells and which contain coelomic canals
and blood vessels (Krause, 1923).

The digestive caecum is lined by about 100 μm tall ciliated cells
containing various granules and vacuoles; it is considered to
secrete digestive enzymes.

Because of their close structural and functional relation to the
peribranchial cavity and associated coelomic spaces and blood
vessels, a brief description of the cyrtopodocytes (Kümmel, 1967) is
included in the present communication. These cells, which usually
are called solenocytes, have been shown by Brandenburg & Kümmel (1961) to possess fundamental structural resemblances to the
podocytes of the vertebrate kidney.

In this paper the light-microscopical findings, the literature on
which can be found in the classical handbooks of zoology, will be
only briefly mentioned. They will be extended by the electron-
microscopical observations which have been made in continuation
of our earlier paper (Welsch & Storch, 1969).

MATERIAL AND METHODS

Amphioxus (*Branchiostoma lanceolatum*) were obtained from the
Biological Station, Helgoland (North Sea). They were cut into
1 mm thick slices which were immersed for 2 h in cold glutaraldehyde fixative (buffered in 3·5% phosphate buffer, pH 7·4). After
repeated rinses in cold phosphate buffer (pH 7·4), the slices were
postfixed for 2 h in 4% osmic acid, dehydrated in ethanol, and
embedded in araldite. Electron microscopes: Zeiss EM 9A,
Siemens 101.

FINDINGS

Peribranchial cavity

The pharynx is surrounded by the peribranchial cavity which is
lined by a squamous or cuboidal epithelium exhibiting no particu-

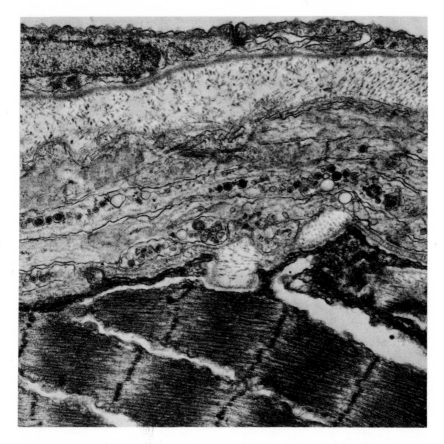

FIG. 1. Epithelium of peribranchial cavity (top of Fig.) and underlying filaments, nerve and muscle cell. Note microtubules as well as granular and vesicular inclusions in the nerve fibres (× 18 000).

lar features. Ventrally—above the pterygeal muscle—it is under-lain by bundles of nerve fibres (Fig. 1).

Endostyle

The endostyle forms the dorsal and most conspicuous part of a tissue complex which is established by the ventral union of the pharyngeal bars (Fig. 2). The various cell-zones of the endostyle will be classified according to the suggestions of Barrington (1958) and Olsson (1963).

The biconcave (in cross-sections) central cell-column (= zone 1) consists of four to six superimposed rows of elongated cells containing heterochromatin-rich nuclei. Their cytoplasm does not

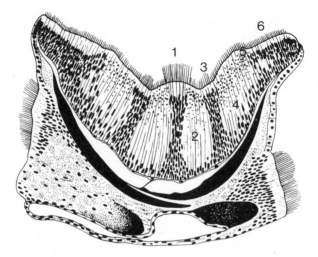

FIG. 2. Drawing of light microscopical section through the endostyle and underlying vessels and epithelia: 1–6, cell-zones in the endostyle. After Krause, 1923.

exhibit any striking characters. It contains bundles of tonofilaments, microtubules, glycogen and predominantly apically located granules and vesicles. The surface cells bear one cilium each, which is surrounded by microvilli which originate from a collar of cytoplasm (Figs 3a and 9b) in a manner similar to those recently described by Lyons (1973) in anthozoans. This author discussed the function of cells bearing such cilia and their distribution in the animal kingdom. In close association to the ciliary rootlet relatively large mitochondria and an accessory centiole can be found (Figs 3a and 9d). Multivesicular bodies also occur regularly in the apical region. Rough ER and ribosomes are infrequent. The small Golgi apparatus, which normally lies in supranuclear position, may give rise to the above-mentioned granules and vesicles. Occasionally cells have been found—mainly basally—which were filled with small vesicular inclusions. A well developed apical terminal web is present (Fig. 3a), the individual filaments of which terminate in macula or zonula adhaerens-like cell junctions. Occasionally tight junctions have been observed apically. Their infrequent detection may be due to inadequate fixation. Also at the base of the epithelium the cells are interconnected by desmosomes.

FIG. 3a–c. Endostyle. (a) Surface of epithelial cells of zone 1; mitochondria are regularly to be found closely attached to the ciliary rootlet (× 15 000). (b) Basal parts of zone 1 (1) and 2 (2). Note difference in chromatin-contents of the nuclei and the well-developed rough ER cisterns in zone 2 (× 3600). (c) Detail of apical part of a cell from zone 2. Note abundance of various secretory granules which resemble mucous droplets (× 18 000).

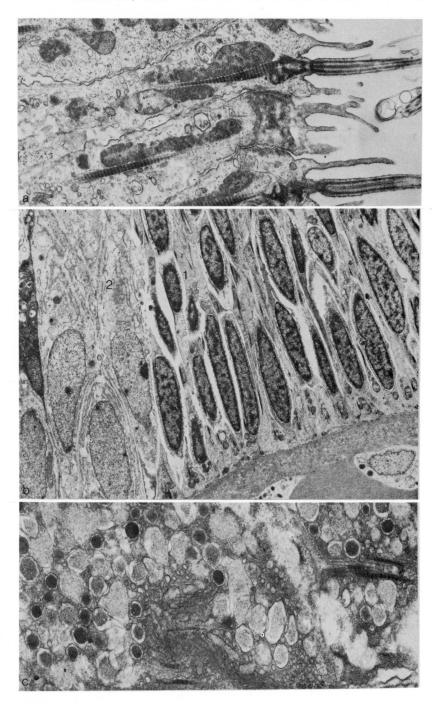

Zone 2 follows laterally to zone 1 (Fig. 3b) without transitional cell-forms. The strikingly euchromatin-rich cellular nuclei are in a basal position and contain one to two nucleoli. The cytoplasm is characterized by extensive rough ER cisterns (mainly basally), a well developed Golgi apparatus which is generally vertically arranged, and numerous granular inclusions of various electron densities (Fig. 3b and 3c). Presumably these granules undergo a process of maturation: initially electron-dense, they loosen up later and become paler (Fig. 3c). Not all of the cells are in the same phase of this process; thus some appear darker than others. Often a number of pale granules appear to fuse. Mitochondria are rare, small fields of glycogen occur regularly (α and β particles). In almost all sections a few necrotic cells with electron-dense cytoplasm have been found (Fig. 3b, left side).

The arrangement and fine structure of the cells of zone 3 correspond to those of zone 1.

In the broad zone 4 the nuclei are again in a basal position, their periphery containing strands or clumps of heterochromatin (Figs 5a and 6a). The main features of these cells are large electron-dense lipid inclusions (Figs 4a and 6a) which occur predominantly in the basal parts of the cells, a rather uniform population of electron-dense secretion granules which are to be found mainly in the apical half of the cells (Figs 5 and 6a), and extensive rough ER cisterns (Fig. 5a). The cisterns of the well-developed Golgi apparatus are generally filled with electron-dense material; they are surrounded by numerous vesicles and granules (Fig. 5b). Fields of glycogen can regularly be found (Fig. 5c), often in close contact with the lipid inclusions. Mitochondria are again infrequent. Beside the above mentioned secretion granules, bigger granular inclusions of varying densities also make their appearance. At the epithelial base, spined vesicles have often been seen (Fig. 4c). The apical surface bears a cilium and microvilli.

An individual zone which can be clearly distinguished as zone 5 has not been detected in the electron microscope (Fig. 6a). However, between zones 4 and 6 there are occasionally to be observed a few cells which contain rather pale mucous droplets, similar to

FIG. 4a–c. Endostyle. (a) Basal parts of zone 3 (3) and 4 (4). Note basally located desmosomes in zone 3, and large electron-dense lipid droplets, glycogen and secretory granules in zone 4. Lower right corner of the picture: blood vessel (B) (× 8000). (b) Detail from blood vessel, which contains amoebocytes (A) and densely packed particles (× 24 000). (c) Basement lamina and filaments under the endostyle (zone 4). Note spined vesicle at the basal plasma membrane of epithelial cell (× 36 000).

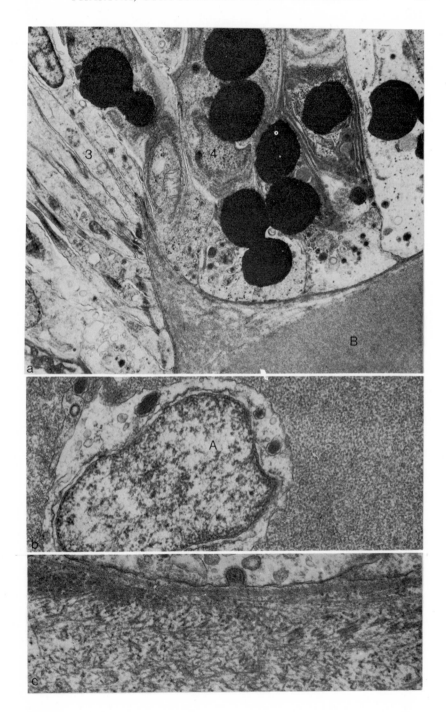

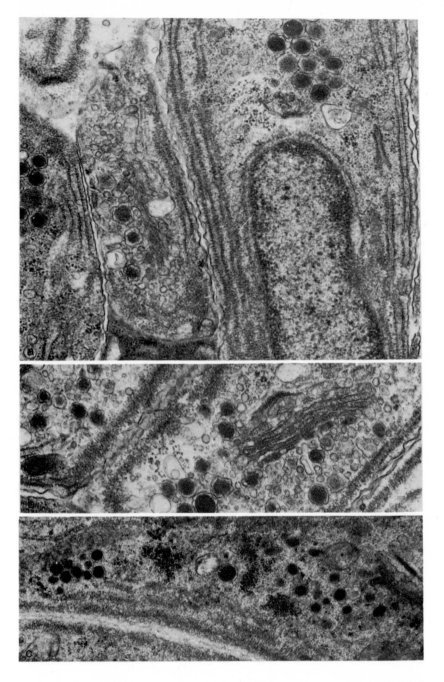

FIG. 5a–c. Endostyle, zone 4; details of the cytoplasmic organization of the epithelial cells. (a) Rough ER cisterns and secretory granules. (b) Golgi apparatus. (c) Secretory granules and glycogen. (a–c) (× 24 000).

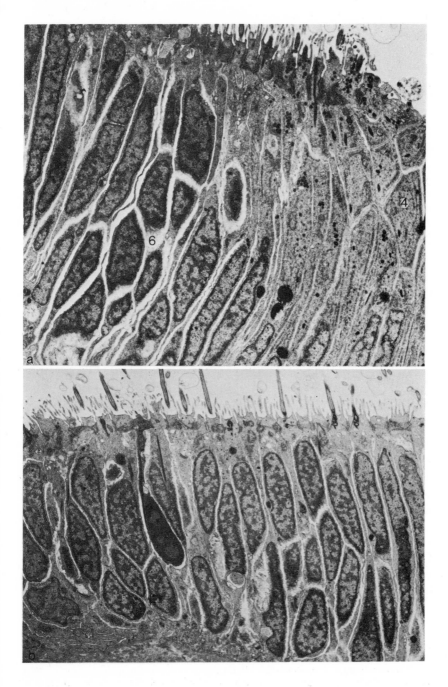

FIG. 6a, b. Endostyle. (a) Zone 4 (4) and 6 (6). Note absence of a clearly definable zone 5, which corresponds to the transitory zone between zone 4 and 6, in which numerous secretory granules occur in the apex of the cells (× 5000). (b) Lateral part of zone 6 (× 5000).

those to be found in zone 2, and clearly different from the dark secretion granules typical for zone 4.

In the last zone, zone 6, the epithelium is relatively low (Fig. 6b). The organization of the epithelial cells resembles that of the cells in zone 1 and 3 (tall cilia, apically located vesicles and multivesicular bodies, tonofilaments, apical and basal desmosomes).

Outside the endostyle the ventral part of the pharynx is covered by tall ciliated cells or flat or cuboidal cells (Fig. 2). Between the ventrally located flat or cuboidal epithelial cells, nerve fibres regularly occur.

The endostylar epithelium rests on a basement lamina below which a dense feltwork of randomly oriented filaments occurs which exhibit periodic cross-striations (Fig. 4c). Large areas of rather electron-dense and homogeneous material correspond to the terminations of the pharyngeal bars (Fig. 7). Within this filamentous or amorphous material two spaces occur.

1. One which is lined by an epithelium, the individual cells of which are squamous or cuboidal. Their pale cytoplasm is above all characterized by basally situated myofilaments (myoepithelial cells). Thin and thick filaments can be distinguished (Fig. 8a). The nucleus is euchromatin-rich (Fig. 8a). This epithelium is surrounded by a basement lamina. This space is interpreted to represent the endostylar coelomic canal.

2. The second space is characterized by its fine-particulate contents and the presence of a few amoebocytes (Fig. 4a and 4b). It is not lined by an epithelium, but there is a clear border between the filaments and the fine-particulate contents which is less electron-dense than its surroundings. This space is interpreted to represent the endostylar blood vessel, the ventral aorta.

Pharyngeal bars

Two types of pharyngeal bars can be distinguished: primary and secondary ones (see e.g. Krause, 1923). One of the main differences which can be seen in the microscope, concerns the

FIG. 7a–d. Pharyngeal bars. (a) Secondary bar with outer blood vessel (B) which—exceptionally—is lined by a cell which might be an endothelial cell (E). L: Lumen of peribranchial cavity ($\times$ 3000). (b) Primary bar with coelomic channel (C) at its outer aspect. This channel is lined by myopepithelial cells (M). L: Lumen of peribranchial cavity ($\times$ 6000). (c) Primary bar with coelomic channel (C) and blood vessel (B), containing profiles of amoebocytes., L: Lumen of peribranchial cavity ($\times$ 2500). (d) Secondary bar with typical blood vessel (B) which is not lined by endothelium; note absence of coelomic space. L: Lumen of peribranchial cavity; arrow points to pigment containing cell ($\times$ 2500).

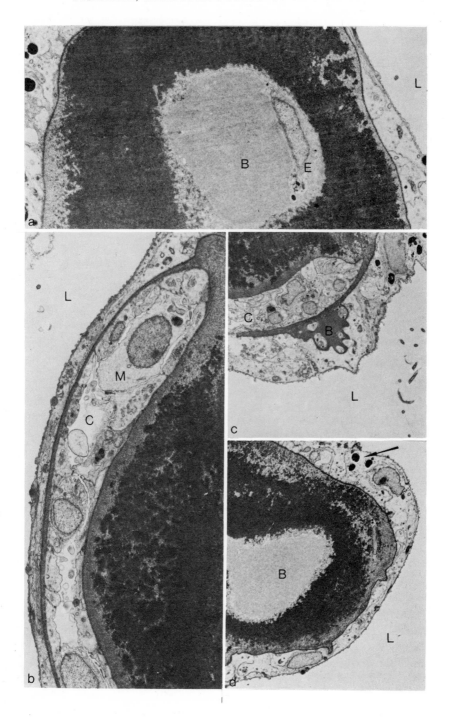

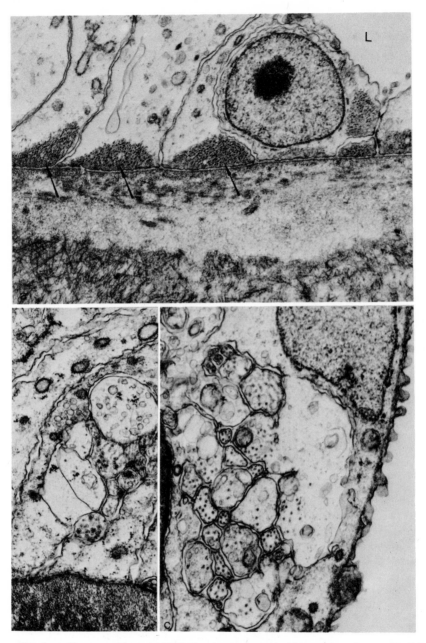

FIG. 8a–c. Pharyngeal bars. (a) Wall of coelomic channel which consists of myoepithelial cells. Arrows point to myofilaments. L: Lumen of coelomic space (× 6000). (b–c) Profiles of nerve fibres between the epithelial cells covering the outer aspect of the pharyngeal bars (× 8000).

presence of a coelomic canal, which is to be found only in the primary bars.

Primary bars. The medial part of these bars—directed towards the lumen of the pharynx—is covered by tall ciliated cells. Each cell bears one tall cilium which is surrounded by a collar of microvilli. In this respect (Fig. 9b and 9d) and in the organization of their cytoplasm these cells closely resemble the cells of zones 1, 3 and 6 of the endostyle. The same applies to the cells covering the dorsal and ventral aspects of the pharyngeal bars, thus lining the passage leading from the pharynx to the peribranchial cavity. Immediately behind the opening into the peribranchial cavity the bars are covered by cells which contain large membrane-bound and electron-dense pigment inclusions (Fig. 7a and 7d), which can also be seen in unstained light-microscopical sections. The nuclei of these cells are pale. Between these cells, nerve fibres are regularly to be seen (Fig. 8b and 8c). The outer aspect of the bars is generally covered by pale flat cells (Fig. 7b, c, d) containing a euchromatin-rich nucleus and between which nerve fibres are also frequently to be found. Occasionally this outer part of the bars is covered by prismatic mucus cells.

Below this outer epithelium there occurs a coelomic canal which is lined by a mesothelium (containing basal myofilaments (Fig. 8a) and pale nuclei), and which is surrounded by a basement lamina and dense material, this being in continuity with the amorphic and electron-dense material forming the skeleton of the bars (Fig. 7a–d). Primary bars are to be seen on Fig. 7b and 7c, which show the outer parts with the coelomic canals.

Inside the skeleton of the bar two blood vessels (outer and inner) occur. They are generally not lined by an endothelium, cells resembling endothelial cells having only rarely been met with (Fig. 7a). They contain amoebocytes and densely packed blood particles. Outside the coelomic canal a similar vessel occurs (Fig. 7c).

The cross-canals ("Querkanäle", Krause, 1923) are generally covered by mucus cells.

Secondary bars. The fine structure of these bars closely resembles that of the primary ones. The outer aspect lacks the coelomic canal (Fig. 7d) and its skeleton here is directly covered by the squamous or cuboidal epithelium (occasionally also by mucus cells), which also contains numerous nerve fibres.

The synapticulae are covered by a flat or cuboidal non-ciliated epithelium with pale nuclei.

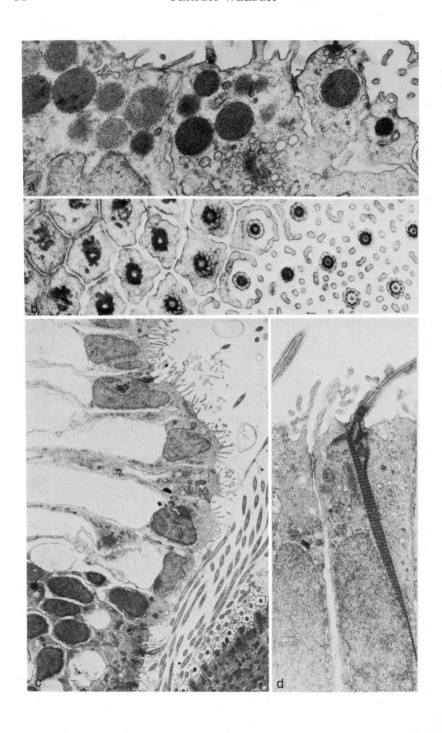

Epibranchial groove

The epibranchial groove is flanked by two epithelial ridges. Two types of cells occur in this area: ciliated cells corresponding to those of zones 1, 3 and 6 of the endostyle, and cells containing a large membrane-bound vacuole in their basal half and a euchromatin-rich nucleus and secretion granules in their apical cytoplasm (Fig. 9a and 9c); their apical surface bears slender microvilli and a small cilium in a loose arrangement. The distribution of these vacuolated cells is apparently not absolutely fixed; generally, however, the central part of the groove is lined by the typical ciliated cells, whereas the vacuolated cells occur preferentially in the lateral parts.

Cyrtopodocytes

Near the epibranchial groove the nephritic canals open into the peribranchial cavity. These canals are perforated by extensions of the cyrtopodocytes (formerly called solenocytes), which have been shown to represent special podocytes (Brandenburg & Kümmel, 1961). This concept has been confirmed by the present investigation using improved methods of fixation. The cyrtopodocytes are specialized coelomic epithelial cells forming numerous extensions, between which an extracellular "slit membrane" has often been observed and which cover the glomerular blood vessels (Fig. 10). The cyrtopodocytes and the blood vessel, which again lacks an endothelium, are separated by a basement lamina and a few filaments. Another set of cellular processes extends from the perikaryon of the cyrtopodocytes into the lumen of the coelomic canal and penetrates the epithelium of the nephridial canal. These processes consist of a long cilium and 10 slender microvillus-like filaments containing extensions surrounding it (Fig. 11a). The microvillus-like processes are not interconnected by any extracellular material. No special structural connexions between podocyte processes and nephritic epithelial cells have been detected passing through the epithelium of the nephridial canal (Fig. 11b). The perikarya of the cyrtopodocytes are characterized by large

FIG. 9a–d. (a) Apical cytoplasm with secretory granules of vacuolated cells in the epibranchial groove (× 20 000). (b) Tangential section through cell apices of the lateral part of the pharyngeal bars. Note that the base of the cilium is at first surrounded by a cytoplasmic collar which transforms into individual microvilli (× 18 000). (c) Vacuolated cells in the epibranchial groove (× 6000). (d) Typical cilium with basal body and rootlet in epithelial cell of pharyngeal bar. Note centriole at the base of basal body inside a loop of the rootlet (× 18 000).

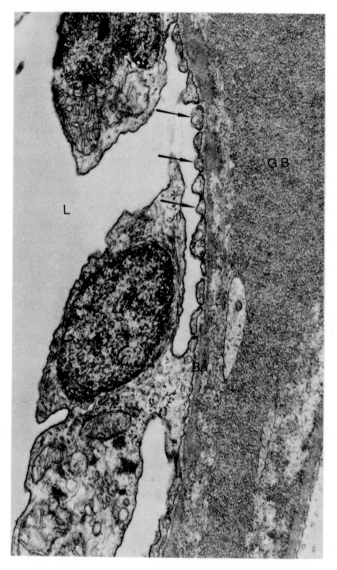

FIG. 10. Cyrtopodocytes on the glomerular blood vessel (B). Ba: Basement lamina; arrows point to the pedicels of the cyrtopodocytes. L: Lumen of subchoral coelomic space (× 18 000).

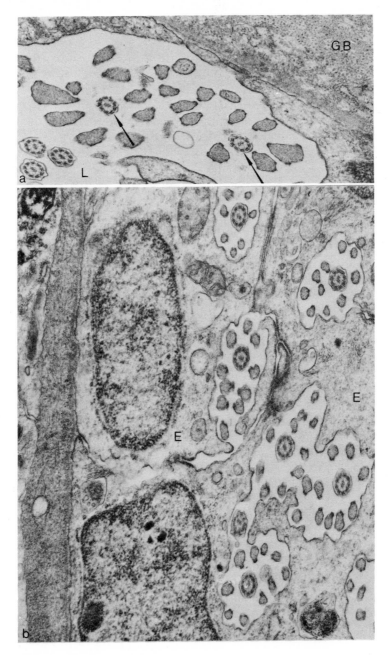

Fig. 11a, b. Cyrtopodocytes. (a) Cross-section through 2 cilia (arrows) of cyrtopodocytes surrounded by 10 microvillus-like long processes which are not interconnected by any extra-cellular material. GB: Glomerular blood vessel, L: Lumen of subchordal coelom (× 18 000). (b) Cilia and accompanying microvillus-like processes penetrating the epithelium of the nephridial canal (E) (× 18 000).

mitochondria, membrane-bound granular inclusions, glycogen particles, a few rough ER cisterns, microtubules, microfilaments, and a Golgi apparatus which is located near the basal body of the cilium. Myofilaments—typical for ordinary coelomic epithelial cells—have not been convincingly observed in the perikarya of the cyrtopodocytes, but possibly they do also occur here.

The main feature of the epithelial cells of the nephridial canals are long cilia and numerous dense bodies with heterogeneous contents.

Digestive caecum

The highly prismatic epithelium of this caecum consists of uniform cells resting on a basement lamina. The nuclei are located in a basal position and are strikingly euchromatin-rich, with a prominent nucleolus. Electron-dense lipid inclusions also occur (Fig. 12b), mainly in a basal position. In a supranuclear position, membrane-bound granules with fine particulate contents occur (Fig. 13c); the electron density of these varies, possibly indicating a process of maturation (Fig. 13b). Occasionally light granules appear to fuse. Further constituents of the supranuclear cytoplasm are lysosomes, which are concentrated in a zone below the apical surface and in the middle of the cells, as the distribution of acid phosphatase has shown (Welsch & Storch, 1969). Also present are various smaller granular and vesicular elements, glycogen particles (α and β ones), the Golgi apparatus, occasional lipid droplets which here are usually electron-transparent, and mitochondria which are often concentrated below the apical surface. A dominant feature of the whole cytoplasm are frequent profiles of rough ER cisterns (Figs 12b and 13a).

Near the opening into the lumen of the gut, these cells are often particularly rich in mitochondria (often in a basal position, Fig. 14b), and usually contain light mucus-like granules and light lipid droplets. The apex often contains a few organelles, above all vesicular elements, and bulges into the lumen (artifact?). The cells regularly bear a cilium with a long rootlet and slender and long microvilli (Fig. 12a).

Profiles of cells or cellular processes have occasionally been observed containing numerous electron-dense granules (Fig. 14a and 14c). It could not always be decided with certainty whether they belong to nerve cells or to endocrine cells. The larger profiles (Fig. 14c), containing also mitochondria and ribosomes certainly are parts of endocrine cells which contain an ovoid nucleus, and a

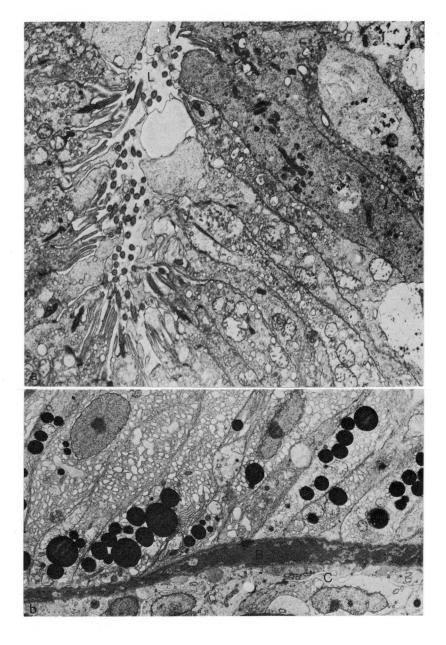

FIG. 12a, b. Digestive caecum. (a) Apices of epithelial cells surrounding the lumen (L). (b) Base of typical epithelial cells containing euchromatin-rich nuclei, electron-dense lipid droplets and rough ER cisterns, B: blood vessel, C: coelomic canal. (a, b) (× 6000).

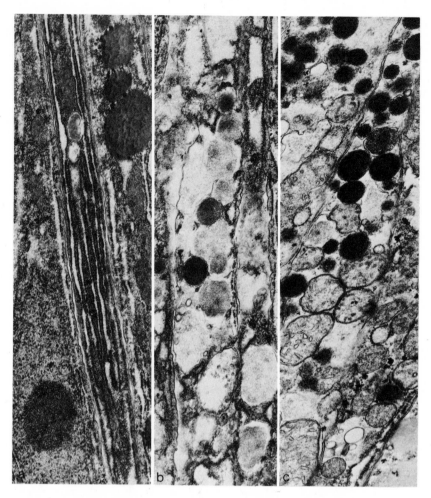

FIG. 13a–c. Digestive caecum, showing details of the epithelial cells. (a) Rough ER cisterns and euchromatin-rich nucleus. (b) Cell with relatively light granular inclusions. (c) Electron-dense granules and mitochondria. (a–c) (× 18 000).

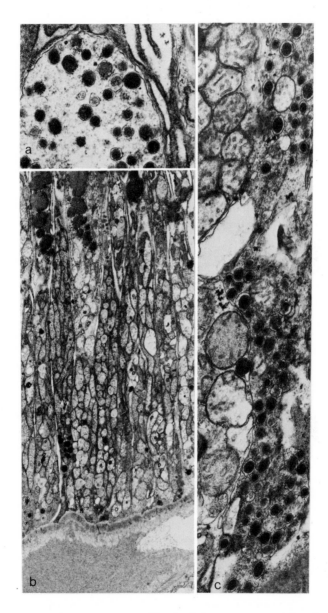

FIG. 14a–c. Digestive caecum. (a) Profile of cellular process containing electron-dense granules (? nerve fibre) (× 20 000). (b) Cells containing numerous mitochondria at their base, B: blood vessel (× 6000). (c) Profile of part of a cell with electron-dense granules (part of an endocrine cell) (× 18 000).

well developed Golgi apparatus which gives rise to the secretory granules.

The coelomic canal accompanying the caecum is again lined by myoepithelial cells, whereas the blood vessel (portal vessel) lacks an endothelium.

DISCUSSION

The fine structural observations on the endostyle are generally in harmony with the light-microscopical and histochemical findings (Krause, 1923; Olsson, 1963; Welsch & Storch, 1969). Olsson (1963) found in zones 2, 4, and 5 indications of protein synthesis and the occurrence of mucous substances. A zone 1b, which according to Olsson is narrow and resembles zone 2, has not been clearly and consistently found in the electron microscope. Zones 2 and 4 are rich in nonspecific esterases (Welsch & Storch, 1969). This corresponds to the demonstration of well developed rough ER cisterns and of secretion granules which, particularly in zone 2, often resemble mucous droplets; these, however, can exhibit great diversity of ultrastructure in invertebrates (Storch & Welsch, 1972). The typical electron-dense secretion granules of zone 4 resemble the granules of polypeptide or protein secreting endocrine cells in vertebrates. However, they also show similarities to the dense apical granules of the dorsal zone 1 in the endostyle of ammocoetes (Fujita & Honma, 1968), and are clearly extruded into the pharyngeal lumen. The numerous rough ER cisterns in the above mentioned zones strongly suggest that their secretory products are predominantly of protein nature. The electron microscopical demonstration of glycogen in most of the endostylar epithelial cells confirms the histochemical findings of Olsson (1963) and gives further support to the idea that in amphioxus many metabolic pathways are anaerobic (Pette, 1970; Meves, 1973). The apical localization of mitochondria and ATP-ase is to be seen in connexion with the activity of the long cilia in zones 1, 3, and 6 of the endostyle and in other areas of the pharynx.

Zone 5 corresponds to the iodination centre of amphioxus (Barrington, 1958, 1964). Unfortunately the ultrastructural study does not add much to our knowledge of this centre. Fine structural autoradiographic studies may produce more information. Such studies on the ammocoetes larva have shown that the iodine-concentrating cells are ciliated and rich in rough ER; mitochondria are concentrated at the cellular apex (Fujita & Honma, 1969). Another characteristic feature are lysosomes (Egeberg, 1965;

Fujita & Honma, 1968). Thus these cells already show points of resemblance to the follicular cells of the vertebrate thyroid gland (Wissig, 1960). The cells in the light microscopical zone 5 of amphioxus correspond in our opinion to the lateral cells of zone 4 which are rich in rough ER and secretion granules, but which—at least in our material—do not differ significantly from the more medially located cells of zone 4. They only contain less lipid and their nuclei are relatively tall, being intermediate in appearance between those of zone 6 and the medial cells of zone 4. Their secretory product is concentrated in the apical region, which corresponds to the light microscopical observations of Barrington (1958) and Olsson (1963). Cells resembling mucus cells which are occasionally found in this area do not in our opinion correspond to any specific zone 5 cells.

Thus the histochemical and ultrastructural results on the area of iodine concentration of amphioxus show a number of agreements with the corresponding tissues of the vertebrates.

The cilia of the amphioxus pharynx structurally agree with those recently described by Lyons (1973).

The blood vessels in the endostyle and pharyngeal bars, and in the nephridia and digestive caecum, generally lack an endothelial lining. Their absence cannot be explained as an artifact. They contain granular amoebocytes and blood particles, similar to those described in enteropneusts (Dilly, 1969) and in polychaetes (Dales & Pell, 1970). For further information about the blood vessels, see Moller & Philpott (1973a, b).

The coelomic canals are always lined by myoepithelial cells, suggesting that these canals possess the ability to contract and to propel fluids. Of particular interest are those of coelomic epithelial cells which have transformed into cyrtopodocytes (Brandenburg & Kümmel, 1961; Kümmel, 1967), and which resemble and presumably have the same function as the podocytes of the vertebrate renal corpuscle. In addition to the findings of Brandenburg & Kümmel (1961), we found that the feet of these cells covering the glomerular blood vessels are interconnected by a slit-membrane. Thus the nephridial filter barrier consists (as in the similar situation in vertebrates) of a slit-membrane, which is difficult to preserve, a basement lamina, and a narrow layer of fibres between blood and coelomic spaces. In contrast to the polychaete solenocytes, no extracellular filter-membrane is to be found between the long microvillus-like processes surrounding the cilium. So the coelomic fluid has (presumably) free access to the central part of this tube-like structure, which is—via the nephritic canal— in direct communication with the peribranchial cavity.

The epibranchial groove is characterized by a relatively strong enzyme activity (Welsch & Storch, 1969) and by the occurrence of two cell types: ordinary ciliated cells with a collar of microvilli around the base of the cilium, and vacuolated cells containing apical secretion granules and a big basally located membrane-bound vacuole of unknown function.

The nerve fibres, which have been found regularly in the epithelium of the pharyngeal bars, probably influence the ciliary beat of the epithelium. Typical synaptic contacts have not been observed, but these need not be expected, as they are also rare in the central nervous system (Meves, 1973). No specific indications exist in regard to the function of the nerves underlying the epithelium of the peribranchial cavity.

The ultrastructure of the digestive caecum suggests that two main functions can be ascribed to it: storage of energy-rich compounds (glycogen and lipid), and protein synthesis (abundant rough ER and secretory granules). These functions correspond to those of the vertebrate liver and exocrine pancreas.* The question of the presence both of nerve fibres and endocrine cells needs further studies. However, the demonstration of small-granule-containing endocrine cells indicates a hormonal control of the activity of this caecum. Endocrine cells have been reported to occur in the intestinal tract of a colonial ascidian species by Burighel & Milanesi (1973).

Acknowledgement

The present investigation has been carried out with the financial support of the Deutsche Forschungsgemeinschaft, We/380-5.

References

Barrington, E. J. W. (1958). The localization of organically bound iodine in the endostyle of Amphioxus. *J. mar. biol. Ass. U.K.* **37**: 117–129.

Barrington, E. J. W. (1964). *Hormones and evolution.* London: English Universities Press.

Brandenburg, J. & Kummel, G. (1961). Die Feinstruktur der Solenocyten. *J. Ultrastruct. Res.* **5**: 437–452.

---

* The concentrations of mitochondria, which can be found in apical and basal location, indicate transport into and through the cells of the caecum.

Burighel, P. & Milanesi, C. (1973). Fine structure of the gastric epithelium of the ascidian *Botryllus schlosseri*. Vacuolated and zymogenic cells. *Z. Zellforsch. mikrosk. Anat.* **145**: 541–555.

Dales, R. P. & Pell, J. S. (1970). Cytological aspects of haemoglobin and chlorocruorin synthesis in polychaete annelids. *Z. Zellforsch. mikrosk. Anat.* **109**: 20–32.

Dilly, P. N. (1969). The nerve fibres in the basement membrane and related structures in *Saccoglossus horsti* (Enteropneusta). *Z. Zellforsch. mikrosk. Anat.* **97**: 69–83.

Egeberg, J. (1965). Iodine-concentrating cells in the endostyle of Ammocoetes. *Z. Zellforsch. mikrosk. Anat.* **68**: 102–115.

Fujita, H. & Honma, Y. (1968). Some observations of the fine structure of the endostyle of larval lampreys, ammocoetes of *Lampetra japonica*. *Gen. comp. Endocr.* **11**: 111–131.

Fujita, H. & Honma, Y. (1969). Iodine metabolism of the endostyle of larval lampreys, ammocoetes of *Lampetra japonica*. *Z. Zellforsch. mikrosk. Anat.* **98**: 525–537.

Krause, R. (1923). *Mikroskopische Anatomie der Wirbeltiere in Einzeldarstellungen.* **3:4**. Berlin: W. de Gruyter & Co.

Kummel, G. (1967). Die Podocyten. *Zool. Beitr.* N.F. **13**: 245–263.

Lyons, K. M. (1973). Collar cells in planula and adult tentacle ectoderm of the solitary coral *Balanophyllia regia* (Anthozoa: Dupsammiidae). *Z. Zellforsch. mikrosk. Anat.* **145**: 57–74.

Meves, A. (1973). Elektronenmikroskopische Untersuchungen über die Zytoarchitektur des Gehirns von *Branchiostoma lanceolatum*. *Z. Zellforsch. mikrosk. Anat.* **139**: 511–532.

Moller, P. C. & Philpott, C. W. (1973a). The circulatory system of Amphioxus (*Branchiostoma floridae*). 1. Morphology of the major vessels of the pharyngeal area. *J. Morph.* **139**: 389–406.

Moller, P. C. & Philpott, C. W. (1973b). The circulatory system of Amphioxus (*Branchiostoma floridae*). 11. Uptake of exogenous proteins by endothelial cells. *Z. Zellforsch. mikrosk. Anat.* **143**: 135–141.

Olsson, R. (1963). Endostyles and endostylar secretions: A comparative histochemical study. *Acta zool., Stockh.* **44**: 299–328.

Pette, D. (1970). *Zellphysiologie des Stoffwechsels.* Konstanzer Universitatsreden. Konstanz: G. Hess.

Storch, V. & Welsch, U. (1972). The ultrastructure of epidermal mucous cells in marine invertebrates (Nemertini, Polychaeta, Prosobranchia, Opisthobranchia). *Mar. Biol., Berl.* **13**: 167–175.

Welsch, U. & Storch, V. (1969). Zur Feinstruktur und Histochemie des Kiemendarmes und der "Leber" von *Branchiostoma lanceolatum* (Pallas). *Z. Zellforsch. mikrosk. Anat.* **102**: 432–446.

Wissig, S. L. (1960). The anatomy of secretion in the follicular cells of the thyroid gland. 1. The fine structure of the gland in the normal rat. *J. biophys. biochem. Cytol.* **7**: 419–432.

Note added in proof. In the posterior midgut of *Branchiostoma japonicum* two types of endocrine cells have been described by Kataoka, K. and Fujita, H. (1974). (The occurrence of endocrine cells in the intestine of the lancelet, *Branchiostoma japonicum*. An electron microscopy study. *Arch. histol. jap.* **36**: 401–406.) Type I of these authors corresponds to the endocrine cells as described in the present study in the digestive caecum.

Symp. zool. Soc. Lond. (1975) No. 36, 43–80.

THE PHYSIOLOGY AND STRUCTURE OF THE NERVOUS SYSTEM OF AMPHIOXUS (THE LANCELET), *BRANCHIOSTOMA LANCEOLATUM* PALLAS

D. M. GUTHRIE

Department of Zoology, The University, Aberdeen, Scotland

SYNOPSIS

Studies on the myotomal motor system indicate the presence of a dual system of motor neurones within the spinal cord connected to the two main types of muscle lamellae. Dorsal motor neurones are associated with the superficial lamellae and slow contractions; ventral motor neurones with the deep lamellae and fast movements.

The myotomal motor output is primarily controlled by a system of secondary inter-neurones with axons extending for two or three neural segments within the ventral areas of the spinal cord. Fine processes of these cells extend into the dorsal (afferent tracts). Serial activation of these elements produces a slow potential wave which passes along the nerve cord at between 0·8 and 1·5 m s^{-1}.

The giant (Rohde) cells of the anterior group are activated by remote as well as local inputs, and provide the major afferent-internuncial interface. The relatively high rates of conduction observed in the Rohde axons (over 5 m s^{-1} preclude their continuous involvement in swimming activity, but recordings show that they are functionally connected to the slow response system, and are involved in its activation. The posterior Rohde cells provide a fast ascending pathway.

Studies on the brain show that specific areas of the brain either inhibit or excite cells in the spinal cord. The strongest effects observed are inhibitory, and some involve the suppression of spikes in Rohde somata. The brain is a differentiated structure with effective control of aspects of spinal function.

Some lancelet sense organs are characterized by high thresholds and low recovery rates.

INTRODUCTION

The aim of this communication is to review briefly some of the more recent published work on the nervous and neuromuscular system of the lancelet, with the addition of unpublished research carried out in the Aberdeen laboratory.

The central theme of the work described here, is the way in which the central nervous elements illustrated by anatomists control swimming movements. Although attempts have been made to revise and review the considerable body of neuroanatomical work that exists (Bone, 1960), many doubts and confusions still remain, and it has been found necessary to augment the earlier descriptions with new research.

Previous studies on motor function, such as those of Parker (1908) and Ten Cate (1938), depended on the observation of whole animal responses. More recent studies have greatly extended their findings, but it is unlikely that problems such as the functional significance of the giant fibres will be fully resolved without much further work. This is not surprising when we consider the obscurities and confusions that remain in our understanding of the spinal physiology of lampreys and teleosts, despite the much greater amount of scientific attention they have commanded.

The myotomal system

The main propulsive forces are provided by the contraction of 64 segmental V-shaped myotomes arranged as a staggered series.

Ten Cate (1938) described three main types of muscular movement. (a) A rapid twitch, involving one muscular wave. (b) First order modulations—one to two waves. (c) Second order modulations—three to four waves.

Descriptions of patterns of muscular movement are important in that they define the problem presented for neurophysiological analysis.

Ciné film of free swimming lancelets confirms the existence of two modes of undulatory movement during forward swimming. Fast swimming usually lasts for up to 10 s at 12°C, and involves forward rates of movement of 4–8 body lengths/second. Webb (1973) has observed rates of 13 body lengths/second, and durations up to 50 s. Two shallow contractions (10–25% body length) typically occur in each ciné frame. Recordings made by means of force transducers indicate that 9–12 complete cycles of lateral displacement occur each second, at the start of fast swimming, slowing to seven to eight cycles per second. The rate of transmission of the muscular wave (important for comparison with central nervous transmission rates) is 0.6–0.3 m s^{-1}; the range corresponding to initial and terminal stages of a swimming burst.

Slow swimming often follows a burst of fast swimming and can be continued for between 5 and 20 min. The forward rate of movement was equivalent to one to three body lengths/second. Between two and four contractions may occur simultaneously, but three is most usual. Contraction waves have a much larger amplitude than in fast swimming; between 25–50% body length. Body flexion angles range up to 110°, compared with maximum angles of 45° in fast swimming. Two or three complete cycles of movement occur in a second, at any point on the trunk. The muscular wave moves at $0\cdot12-0\cdot06$ m s^{-1} along the body.

Backward swimming has been described by Webb (1973). Rates of reversed swimming fall into the category of fast swimming modes. Reversal of direction may occur very suddenly during normal forward swimming.

Lancelets can burrow into unconsolidated sand rapidly with either the head or the tail leading (Webb, 1973).

In the normal sequence of events slow swimming often follows a fast swimming burst, without a pause. Quiescent lancelets suspended freely in seawater commence slow swimming spontaneously after one or two minutes. Fast and slow swimming modes can be observed in suspended lancelets, and are similar in major characteristics to the same modes in free swimming animals. Occasionally, fatigued suspended lancelets swim using only the preatrial myotomes. This can be correlated with the low transmission characteristics of descending spinal pathways in the atrial region, described later.

The muscular basis for the two types of swimming mode appears to depend on a differentiation of the muscle fibres, as in *Myxine* (Jansen, Andersen & Loyning, 1963) and in fish (Barets, 1961; Bone, 1966). Flood (1966, 1968b) was able to show that in the lancelet two basic types of muscle lamella occur: (i) superficial lamellae, with large amounts of sarcoplasm, and abundant glycogen granules, and (ii) deep lamellae with little sarcoplasm or glycogen. These form the bulk of the myotome. An immediate type of lamella occurs between the deep and superficial kinds.

Geduldig (1965) obtained a mean value of 57·5 mV for the lamellar resting potential. Guthrie & Banks (1970b) were able to show that the resting potential of deep fibres was significantly higher than that of superficial fibres (a mean of 56·85 mV, against 39·46 mV); a correlation also observed in fish (Barets, 1961). Recordings of myotome tension reveal two types of contraction: (1) a fast twitch with a rise time of 70–80 ms, and (2) a slow twitch with a rise time of 200–250 ms (Fig. 1). Hagiwara, Henkart & Kidokoro (1971) were able to demonstrate much briefer rise times of the order of 30–40 ms from the fast fibres, using a muscle strip preparation.

Stimulation of the fast muscle lamellae via the central nervous system (Guthrie & Banks, 1970b) or by means of an intracellular microelectrode (Hagiwara & Kidokoro, 1971), results in a spike-like action potential with a positive overshoot of 20–40 mV.

In more superficial regions of the myotomes Guthrie & Banks (1970b) observed small potentials (5–10 mV), with a much greater latency (70–90 ms) than is characteristic of the fast potentials (10–15 ms). It was suggested that these might be the post-synaptic responses of the superficial lamellae. Hagiwara *et al.* (1971) showed that some muscle lamellae were electronically coupled, and small potentials, similar to those observed by Guthrie & Banks (1970a), might occur with increased latency in those lamellae secondarily involved. This work also suggests that the second of two spikes observed by Guthrie & Banks may be derived from adjacent lamellae.

Although Flood (1968b) describes zonale occludentes or tight junctions between adjacent lamellae, Hagiwara & Kidokoro (1971) do not seem to feel that this identification is justified, even though they were able to show that electrical coupling existed between some of the muscle lamellae.

Hagiwara & Kidokoro (1971) were able to show that the fast electrical response depended largely on the influx of Na^+, and that there is an increased permeability to Ca^{2+}. The occurrence of caffeine-induced contractures in Ca^{2+}-free solutions suggested the presence of Ca^{2+} storage sites within, or at the surface of, the lamellae.

Electronmicroscope studies by Peachey (1961) and Flood (1968b) indicated the absence of an internal tubular system, associated with Ca^{2+} storage in other muscles. However, Flood clearly identified sarcoplasmic vesicles, which are possible Ca^{2+} binding sites. Hagiwara *et al.* (1971) were also able to identify sarcoplasmic profiles and observe Lanthanum binding sites in their proximity. As in crustacean muscles, Ca^{2+} influx from the external medium plays a major role in twitch generation, but unlike crustacean fibres, lancelet lamellae do not depend on Ca^{2+} from storage sites. The lancelet lamellae therefore occupy an unusual position in relation to excitation-contraction coupling mechanisms in other types of muscle fibre.

One of the characteristics that the fast lamellae share with fast muscle fibres in *Myxine* is a tendency for fatigue to reduce twitch response amplitudes following repetitive stimulation (Jansen, Andersen & Loyning, 1961; Guthrie & Banks, 1970b). The studies on the excitation of the fast lamellae cited above suggest that this is unlikely to be due to any failure of the electrical response of the muscle membrane, but it may reflect on the properties of the presynaptic terminals of motor neurones. At 40 Hz the decline of the fast twitch follows the third or fourth contractual response following intraneural stimulation, but does not appear to have been observed by Hagiwara *et al.* (1971) using direct stimulation methods up to 35 Hz. These response rates, it should be noted, are well above those normally occurring during fast swimming (12 Hz). The fatigue of the fast system which occurs at low frequencies after many repetitions, is much more likely to be due to properties of the muscle fibres themselves, such as the small amount of intracellular glycogen present, compared with the slow fibres, clearly demonstrated by Flood (1968b).

Flood (1966) was able to show by means of electronmicrographs that a dual system of muscle tails extended from the muscle lamellae to the central nervous system, where they were in synaptic relationship with two distinct types of presynaptic endings. The superficial lamellae have thin (0·5 μm) muscle tails running to the nervous system in a dorsal bundle where they connect with small overlapping synaptic profiles containing small diameter vesicles (mean diameter—700 Å). The deep (fast) lamellae connect by means of thick (1·0 μm) muscle tails with the periphery of the nerve cord, where a single layer of larger presynaptic profiles is visible (mean—1170 Å). The dorsal (superficial) muscle tails, and the ventral (deep) muscle tails, terminate within distinct compartments bound by connective tissue elements.

Bone (1960) figures two major types of motor neurone, together forming chevron-shaped arrays opposite the ventral roots. A form with a vacuolated cell body and a thick ventral process (somatic motorneurone i) can be distinguished from another kind with non-vacuolated cell body and more slender axon (s.m.ii). These were identified in silver-on-the-slide preparations; both appear to run to the ventral region of the zone of presynaptic terminals.

Re-examination of similar material indicates neurones of this type, but details of their endings are obscure owing to the complexity of the branching of the terminals near the synaptic layer. Methylene blue preparations show much more clearly than totally impregnated silver preparations the form of an individual neurone of the s.m.i. type. These cells have at least three major ventral processes, including a posterior axon extending for two or three neural segments posteriorly. There appear to be only a few motor neurones connected to the synaptic layer in each segment, but each neurone has between 15 and 25 terminals.

Exploration of the nerve cord with a stimulating electrode indicates clearly the physical separation of presynaptic zones associated with fast or slow muscular contractions (Guthrie, 1967b); this can be seen in the records shown in Fig. 1. Stimulation within the motor neurone areas indicates the activation of units with their own frequency range of output, maintained independently of stimulation frequencies (Guthrie & Banks, 1970b).

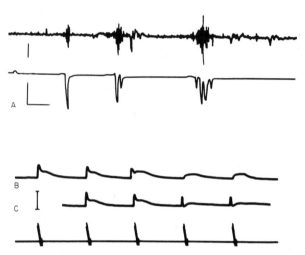

FIG. 1. A. Electrical record (upper trace), and mechanical responses (lower trace) from myotome 35, showing reflex responses to remote mechanical stimulation. All contractions are of the fast twitch type. The electrical record suggests that repeated contractions involve different lamellae. B & C. Mechanical responses recorded from a myotome following electrical stimulation within the adjacent nerve cord. The series in B illustrates the result of altering the position of the electrode tip. In series C the duration of the stimulus pulse was increased from 4 to 9 ms. The stimulus pulse on the lowest trace is repeated at 1 Hz. There is a clear distinction between fast and slow twitch systems with different loci within the central nervous system. Vertical scales: A (upper)—20 μmV, (lower)—3 g; B & C—2 g. Horizontal scale: A—1·0 s.

Recordings from the nerve cord and the lamellae suggest that there is no very actively maintained tonus, but that as in nematodes (de Bell, del Castillo & Morales, 1967), rhythmic ripples of depolarization circulate within the myotome. Nematodes and lancelets have specialized junctions between the muscle tails.

Topical application of acetylcholine in physiological concentrations to the region of the ventral roots produces irregular bursts of activity in the myotome followed by block, but no detailed studies on neurotransmitters have been made.

The spacing of the terminal processes of the motor neurones suggest that adjacent lamellae are not innervated by the same motor neurone. This accords with the idea that an active lamella can excite a few lamellae on either side of it, electronically, and also helps to explain the curious lamellate form of the muscle fibres.

The notochordal system

Early anatomists largely set aside the observations of von Ebner (1895) who described the muscular appearance of the notochordal lamellae. A passive theory of notochordal function has been generally accepted up to the present time.

In 1967 Flood (1968a) was able to show that the fine structure of the lamellae closely resembled that of mollusc adductor muscles. Like these, the notochord is characterized by unusually thick myosin filaments from which paramyosin (tropomyosin A) can be extracted.

Physiological studies (Flood, Guthrie & Banks, 1969) and Guthrie & Banks (1970a) showed clearly that the lamellae contracted. Twitch-like contractions following a single stimulus pulse had rise times of 70–80 ms at 18°C. Decay times for responses to single shocks seldom exceeded 15 s. These muscles are most similar both structurally and physiologically to the striated muscles of the oyster *Crassostrea* described by Hanson & Lowy (1961).

The mechanical responses of notochordal muscle appear to fall into a single type, but there is a bimodal distribution of response amplitudes when these are plotted against stimulus frequency (Guthrie & Banks, 1970a). The lamellae appear to be all of the same structural type (Flood, 1968a).

The notochordal lamellae are produced into dorsally extended muscle tails which terminate below the presynaptic terminations of elements from the nerve cord. Pre- and post-synaptic structures show the typical form observed in other neuromuscular junctions, but the synaptic cleft is unusually wide—0·5 μm. It appears to be partly occupied by the collagenous fibres of the notochordal sheath (Flood, 1968a).

Guthrie & Banks (1970a) found that the notochord could be activated by focal stimulation of the nerve cord, and severe lesions of the nerve cord prevented serial activation of the lamellae. Flood (1968a) illustrates fibres within the ventral region of the nerve cord which run in a well defined longitudinal tract and give off short ventral collaterals to the notochordal synaptic structures. Histochemical methods for demonstrating cholinesterase show a dense staining reaction at the notochordal synapse. Guthrie & Banks (1970a) were unable, after persistent trials, to demonstrate any strong effect of acetylcholine in physiological concentrations when applied to the synaptic areas.

The atrial system

A system of striated lamellar muscles, mainly orientated in the transverse plane, allows the volume of the atrium to be reduced, and this produces an exhalent water current.

These muscles appear to be innervated, unlike the other muscles, by peripheral motor axons running in the dorsal segmental nerves (Bone, 1958). Lele, Palmer & Weddell (1958) demonstrated that the dorsal roots contained three or four thick axons, and their findings were confirmed by electronmicrographs made in this laboratory.

Electrical stimulation of the region of the segmental nerve provokes contractions of the fast twitch type in the atrial muscles with similar time durations to the myotomal contractions. However, the facilitation rate is unusually high. The amplitude of successive contractions increases by 30% at 1 Hz, and overall increases for 10 contractions amount to 230% at this frequency.

Microelectrode penetrations of the lateral atrial muscles demonstrate resting potentials in the low ranges (20–40 mV). Active responses resembled those obscured in myotomal lamellae, and similar spontaneous oscillations of membrane potential occurred.

An important finding was that miniature and plate potentials could be recorded at certain sites in the muscle, correlated with the presence of peripheral neuromuscular junctions.

Graded stimulation of the atrial musculature through the central nervous system reveals at least four components of tension. Central stimulation also reveals that the minimal stimulus interval for the maximal contractual response is approximately 50 s. This corresponds with the periodicity of large spontaneous contractions of the atrium, and must reflect inherent central rhythms of excitability. Similar rhythms of spontaneous exhalation occur in ascidians.

Studies on the central nervous activation of the atrial response indicate that the central pathways conduct at rates of $1 \, \text{m s}^{-1}$, while the much slower conduction times of the peripheral pathway correspond to $0 \cdot 15 \, \text{m s}^{-1}$. The rates of central conduction are equivalent to those characteristic of the descending slow response, and recordings demonstrated that the maximal atrial response and the DSR appeared at the same stimulus intensity. On the other hand a fast small amplitude contraction of the ventral muscles of the atrium coincides with the appearance of the second spike of the compound fast response, in the absence of any slow response.

THE SPINAL CORD

Generalized structure

An anterior brain region lying anterior to the cell body of Rohde cell A can be distinguished from the spinal cord (Edinger, 1906). The boundary lies between the 5th and 6th nerve roots (myotomes 4 and 5).

Serial sections through the spinal cord indicate that it tapers little from its thickest point near the posterior border of the pharynx (myotome 28). At this level the cord has a cross-sectional area of 35 000^2 μm in a 50 mm lancelet. The height and breadth of the cord are both about 270 μm at this point, but the cord tapers to less than 100 μm at the level of myotomes 6 and 55. Estimates made from electron micrographs, and from thin sections reviewed with the light microscope, show that there are over 20 000 membranous profiles present at this level; the great majority represent longitudinally running fibres or fibrous elements.

Counts of fibres falling in different diameter categories reveal that only 10% of the profiles exceed 2 μm. Most profiles are between 0·4 and 0·2 μm across.

Measurements from single whole neurones impregnated with methylene blue and examined under the light microscope show main axons to be 2–3·5 μm thick, with numerous fine side branches of 0·3 μm or less. This accords with the electron microscope picture, and with the distribution of fibre sizes in the enteropneust nerve cord (Dilly et al., 1970).

The small number of fibres over 2 μm in diameter mostly belong to the Rohde cell system.

No sign of blood vessels or blood spaces can be detected in the nerve cord, and probably because of this a majority of cell bodies of neurones and glial cells lie in, or abut on, the central canal. Cell bodies of neurones often have few retroaxonal dentritic processes, while the axonal processes usually subdivide, so that the geometry of the cell bears a certain similarity to a molluscan or an arthropod neurone.

The ependymal glial cells form a major and easily recognizable element in the anatomy of the nerve cord. The major bundles are illustrated by Bone (1960). Under the electron microscope the fibrous extensions of the cells can be seen to contain fine (70 Å) electron-dense fibrils with indistinct striations. The cell bodies stain strongly with toluidine blue, but contain few organelles, and a generally electron-lucent cytoplasm.

Some of the larger nerve fibres are surrounded by flattened profiles forming incomplete sheaths of an irregular nature. The contents of these profiles do not differentiate them clearly from axons. In other areas beaded profiles with fibrous contents occur, and it appears that the radiating bundles of ependymal glia send irregularly branching membrane folds laterally and centrally.

Other types of cell with much denser contents were observed in the lateral and dorsal areas of the cord. These may form spiral folds of glia with a superficial resemblance to myelin sheaths. They appear to correspond to the type 3 glia mentioned by Bone (1960) and originally described by Müller (1900). In addition there are a number of spherical cell types with small cytoplasmic vacuoles, which occur along the borders of the central canal; these were identified as amoebocytes. Lysosomes are often abundant in the cytoplasm, and finger-like processes are visible at the periphery of the cell.

The subcellular organelles associated with the cell bodies and axons of the neurones present a wide range of structures. The large axons contain few inclusions—sparse arrays of neurofilaments, mitochondria, and occasional empty vesicles. The small fibres near the dorsal roots, on the other hand, contain a bewildering variety of vesicular structures and granular particles. Within the neurone cell bodies, it is possible to observe that there are no regular arrays of cisternae forming an endoplasmic reticulum. At the same time rough surfaced e.r. can be observed in irregular groups around the nucleus, and free ribosomes are abundant. As in the nerve cells of enteropneusts (Dilly, Welsch & Storch, 1970), large dense vesicles occur.

Synaptic structures were only occasionally observed in either the fibrous or cellular regions of the spinal cord (Fig. 2). Mostly they occurred in the ventral fibrous regions of the cord, and appeared as axo-axonic junctions. The presynaptic profile contained hyaline or dense-cored vesicles or a mixture of both, ranging in diameter from 500–1500 Å. A single presynaptic profile has a junction with one, or several postsynaptic profiles. The synaptic cleft might be within the usual range of 100–250 Å or much wider (1000 Å). No axo-somatic synapses of the kind

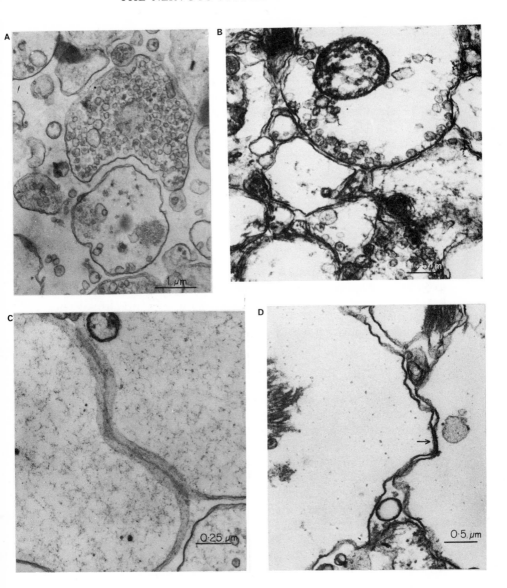

FIG. 2. Cellular junctions in the spinal cord. A & B. Examples of membrane profiles with vesicular contents generally similar in form to chemical synapses. The presynaptic profile in A resembles a calyciform type. The cleft is broad, but much narrower than in the notochordal synapse. B illustrates a more usual type of synaptic structure. C & D. Examples of points of very close apposition between axons of the anterior group of Rohde neurones. Similar junctions to the one figured in D exist between the muscle lamellae which have been shown to be electrically coupled.

described by Dilly *et al.* (1970) were observed. Many of the larger
axons were closely apposed, and zones of especially close contiguity
with the appearance of denser membrane could be observed.
These were similar to the zonale occludentes described by Flood
(1968b) between adjacent muscle lamellae. Hagiwara & Kidokoro
(1971) provided physiological proof of electrotonic coupling between lamellae.

The structure of the Rohde cells

Rohde's description of the giant cells of the lancelet spinal cord
is well known (Rohde, 1891). One group (labelled A–L) have
anterior dorsal cell bodies and posterior dorsal axons. The other
group (M–Z) have posterior dorsal cell bodies and anterior ventral
axons. Cell A is unusual in being a good deal larger than the other
cells and having a median rather than a laterally running axon. In
Rohde's illustrated scheme the anteriorly and posteriorly directed
groups or axons are both shown extending throughout most of the
length of the nerve cord—to ·dorsal root 58, and dorsal root 6,
respectively.

Careful examination of the spinal cord with the light and the
electronmicroscope failed to reveal posterior extensions of the
anterior group of cells beyond myotome 37—just posterior to the
atriopore.

In addition, an indirect method of assessing the point of
termination of the larger axons was used. Measurements of axon
diameter were made from 200 serial sections, for cells A and B.
The rate of taper per mm was found to be regular and the point of
extinction of the axon could be estimated. In the case of cell A this
was at the level of myotome 37; in the case of cell B, at the level of
myotome 35.

Other discrepancies with Rohde's scheme were as follows. (1)
The latero-dorsal axon bundles of cells B–L should consist of five
elements on one side and six on the other, but only three or four
were found to be present. (2) Careful reconstruction of the
anterior axon tracts showed that the constituent fibres ran a spiral
pathway within it. In the case of axons B and C a complete
revolution occurred once in each neural segment. (3) The posterior group of cells M–Z do not necessarily have ventral axons as
shown in Rohde's scheme, but some, including one pair of large
diameter (c. 14 μm), run in the latero-dorsal areas of the cord. (4)
The numbers of large axons in the ventral areas of the cord (over
12 pairs) exceed the numbers that would be expected on the basis

of Rohde's scheme. Some of these may be large collaterals of the type figured by Retzius (1891).

One of the noticeable characteristics of the Rohde cell, shared with some of the smaller types of interneurones, is the comparative scarcity of axonal processes and arborizations. Those that were observed occurred relatively close to the cell body. Electron micrographs, and light microscope preparations of the long posterior median axon, seldom reveal any branches arising from it, or processes terminating on neighbouring elements. Neither were junctions between presynaptic elements and the Rohde axons identifiable.

The anterior and posterior bilateral pairs of Rohde cells may be both analogous and homologous to the anterior and posterior groups of giant cells described by Rovainen (1967) in the lamprey. As in the Rohde cells, there is an initial decussation of anterior Mauthner cells and posterior giant cells of the lamprey. The posterior giant cells in the lamprey also form an alternating or staggered series like the posterior Rohde cells, and 14 cells are believed to constitute both series. The posterior lamprey giant cells have axons that run ventrally at first, before ascending to a more dorsal position. Our studies suggest that a number of anterior and posterior Rohde axons have a similar distribution.

The structure of secondary interneurones

As pointed out by Bone (1960), the classification of the smaller types of interneurone is a difficult task, but the three categories of elements proposed by him—(i) Dorsal commissural cells, (ii) Mid-level commissural cells, and (iii) Edinger cells—provide a useful starting point.

While silver-on-the-slide preparations show up some of the processes of these cells, single elements impregnated with methylene blue allowed the finest processes (less than 0·25 μm in diameter) to be reconstructed from camera lucida drawings.

Most cells observed conformed to the E.C. type, with processes confined to one side of the nerve cord. The axon leaves the cell body laterally, and descends to the ventral region of the nerve cord in a gradual arc. The anterior cells have axons extending for one to three neuromeres posterior to the cell body. The major region of axon branching is from the first quarter of the main axon and includes fine long branches to the dorsal region (somatic sensory area). More posterior branches are few and short, extending dorsally or ventrally. Compared to the motor neurones these are anatomically simple, and the branches of the axon have few subdivisions. Reconstructions of the major axon ramus shows that it lies centrally of the somatic motor neurone collaterals. The Edinger cell type resembles many of the annelid or arthropod neurones figured by various authors (Bullock & Horridge, 1965). No synaptic structures were observed at the surface of the perikaryal membrane, and the cell body is likely to have a largely nutritive function.

A few examples of dorsal commissural cells were observed in detail. In these the retroaxonal part of the cell body has a number of dendritic processes penetrating into the dorsal and lateral regions of the cord. The main axonal process runs as a single ramus to the ventral areas, following much the same course as in the Edinger cell, but lying lateral to it. Again, the major axon branches consist of a few fine processes running up into the dorsal areas of the cord. They leave the axon before it has entered the zone of longitudinal tracts.

From a study of the distribution of the processes of these cells, there is no doubt that they provide a physical pathway between the dorsal sensory tracts and the ventro-lateral tracts of motor neurone fibres. Whether this is a functional pathway will be difficult to show, as the fibrous parts of these cells do not exceed 3 μm in diameter. The extent of the major axonal process was less than three neural segments in both Edinger's and dorsal commissural cells, so that longitudinal transmission in this system would be expected to be relatively slow, assuming that it would depend on the junctions between overlapping series of axons.

Processes between secondary interneurones and the cell body or axon roots of motor neurones were never observed, but close proximity with the more peripheral parts of motor neurones was seen. This situation is reminiscent of the findings of Tauc & Hughes (1963) in *Aplysia* neurones, where input and output may involve only a part of a region of axonal branching.

The electrophysiology of the spinal cord

Spontaneous electrical activity

Bursts of spontaneous electrical activity can be recorded from the nerve cord of quiescent lancelets. Five-second bursts of 15–20 small potentials occur once or twice a minute. Occasionally large potentials occur, accompanied by single muscular twitches of the myotomal musculature, but this does not lead to locomotory activity while the animal remains in contact with the substrate. Part of this activity may be associated with the maintenance of tonus in the trunk muscles, and the potential oscillations recorded across the lamellar membrane that were observed (Guthrie & Banks, 1970b).

Evoked responses

Previous work (Guthrie, 1967b) showed that stimulation of the nerve cord or of the whole animal resulted in muscular responses preceded in the nerve cord by potentials of constant amplitude. These externally recorded potentials are comparatively large $(0.5-1.0 \text{ mV})$, relatively constant in duration (4 m s^{-1}), and form response series directly proportional to the strength of the applied stimulus (light or electric shock). Rates of transmission range from $0.8-1.5 \text{ m s}^{-1}$ according to the size of the animal and the ambient temperature.

Both the duration and the conduction velocity of these potentials (allowing for differences in axon diameter) agree very closely with the values estimated from lamprey giant axons (duration—Rovainen, 1967; conduction rate—Berkowitz, 1953). In the light of these observations, and the close involvement of Rohde fibres in movements of the whole trunk musculature predicted by Parker's experiments (Parker, 1908), it appeared that these potentials were derived from the axons of the Rohde cells. No faster pathways were observed.

Subsequently, however, single unit recordings made with microelectrodes revealed impulse durations of $0.5-1.0 \text{ m s}^{-1}$ at 20°C, and showed that a faster pathway did exist, with conduction velocities of $2-4 \text{ m s}^{-1}$ and sometimes as high as 5.3 m s^{-1} at 20°C. It was also shown that the faster responses were all-or-none phenomena under conditions of recruitment, while graded steps of amplitude preceded maximum amplitude in the slower responses.

Both the fast and the slow responses (Fig. 3) pass anteriorly or posteriorly along the spinal cord so that it is possible to distinguish a descending fast response (DFR), descending slow response (DSR), and ascending counterparts—(AFR) and (ASR).

Fast evoked responses and the Rohde cells

The form of the DFR, recorded by means of glass electrodes within the nerve cord sheath, varies according to the anterior-posterior level of the recording leads. At the level of myotome 15, a

58 D. M. GUTHRIE

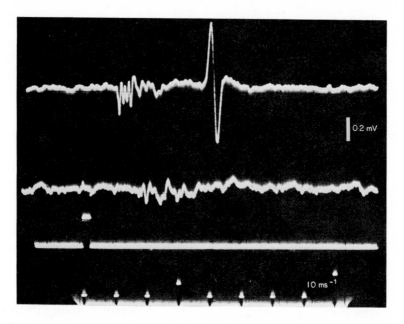

FIG. 3. Fast and slow descending responses from the nerve cord following a stimulus pulse applied to the anterior nerve cord. The two recording points were at the level of myotome 23 (mid-pharyngeal) and myotome 40 (abdominal), and correspond to the upper and second traces shown. The third trace shows the stimulus pulse, and the lowest trace provides a 10 ms time marker. At the pharyngeal level six spikes constitute the DFR followed by a maximally recruited slow response. At the abdominal level the fast response is spread out and reduced in amplitude, and the slow response is absent.

group of five or six closely packed spike-like potentials are observed following stimulation with a single shock of the region occupied by the anterior Rohde cell bodies. Further posteriorly, at the level of myotome 25, the amplitude of the DFR has been reduced by 80%, the individual potentials are much more widely spaced (intervals of over 5 ms), and in some recordings it may be difficult to distinguish more than three of the components. If the stimulating electrodes are placed across one or two neural segments in the region of the anterior Rohde cell bodies, instead of the six neural segments occupied by the complete series, only one or two spikes are observed at anterior levels.

These morphological aspects of the DFR correlate very well with the topography of the anterior Rohde axons, and there can be little doubt that the DFR originates from these elements.

The unitary nature of the DFR components is also suggested by studies using a Biomac averaging computer. Event correlograms provide a 90% recovery of DFR responses within a single 5 ms address, while slow responses show much more scatter. This is shown in Fig. 4.

FIG. 4. Event correlogram obtained by feeding 16 consecutive responses from the pharyngeal nerve cord to an averaging computer, following electrical stimulation of the anterior nerve cord. The first large column represents the fast response, and most fast responses were collected by this address. By contrast the slow responses were more widely scattered—an indication that they are activated indirectly.

The low threshold of DFR (less than 25% DSR threshold) points to the involvement of a large axon system, rather than of fine fibre tracts. This is borne out by penetrations into the zones containing the anterior Rohde axons with microelectrodes, when it is some-times possible to "home in" on the axon. A single large axon was penetrated in this way in the position of Rohde axon A. The external potential from this unit had the characteristic latency and relation to the slow response of a DFR component.

The degree of coupling between the responses comprising the DFR is interesting in view of the electrical coupling described by Rovainen (1967) between lamprey giant cells. As described above, anterior giant cells can be excited independently of one another, but they normally do provide a very regularly spaced series of spikes (Fig. 3), which may not be purely a function of the spectrum of fibre diameters at the recording point.

When only the two anteriormost Rohde axons are excited (A and B), an interesting difference appears between them.

Cell A, which has the higher threshold, presumably due to its exceptionally large dendrite system, has a fixed latency over a considerable range of stimulus intensities. The latency of the response in cell B, on the other hand, is reduced by increased stimulus intensity within certain narrow limits. When the interval between the response of cell B and cell A is less than 1·0 ms, the response in cell B is suppressed. This indicates some form of inhibition acting between the cells determining response intervals.

Mutual inhibition between teleost Mauthner cells has been fully described (Furukawa, 1966).

Microelectrode penetration into the dorsal region of the median fissure of the spinal cord between neural segments 6–12 results in unit contact recordings of the type generally associated with neurone somata. These were rarely obtained elsewhere, and the amplitude of the extracellular potentials suggest that they derive from the largest category of neurones. Responses can be evoked from these cells by electrical stimulation of ascending pathways in the caudal region of the spinal cord (opposite myotomes 50–64). As indicated above, this region does not contain axons of the anterior Rohde cells, so antidromic stimulation is avoided.

With fortunate placement of the electrode it is possible to observe not only the development of the soma spike but also the presynaptic volley in the ascending pathway that precedes it. It is clear from a number of sequences that stimulus intensity, presynaptic frequency, and post-synaptic depolarization are directly related. The Rohde cell appears to integrate synaptic inputs in a similar manner to the mammalian motor neurone.

Examples illustrated (Fig. 5) also show the appearance of a polarization plateau, and potential oscillations following the first soma spike. These damped oscillations are interesting in that they suggest a suppression of repetitive firing, a property of Mauthner cell circuitry described by Furukawa (1966); although no definite hyperpolarization is visible, it may be related simply to the rapid decline in excitability expressed in the dropping away of the depolarization plateau.

Occasional penetrations of cell A were made. The effect of caudal stimulation is to produce a complex series of events. The latency of the responses is much greater than for the cells in the B–L series, and together with the variability of results suggests that cell A is much more indirectly connected with the ascending pathways, or that these play a smaller part in determining its activity. Reconstructions of the dendritic arborizations of cell A

FIG. 5. Microelectrode recording from the anterior Rohde cell column. A–C. The effect of increasing the intensity of single shocks applied to the caudal nerve cord is to produce an augmenting polarization plateau from which primary and secondary soma spikes arise. The stimulus artefact appears to the left. D. The position of the microelectrode in this recording allowed the presynaptic volley as well as the postsynaptic response to be observed. Note the large primary soma spike, and the small latency of the presynaptic response indicating a fast ascending pathway.

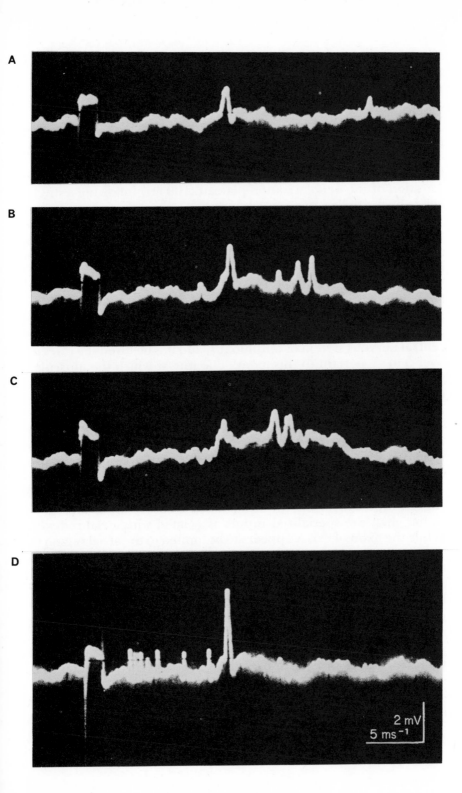

A

B

C

D

2 mV
5 ms⁻¹

agreed broadly with the illustrations of Rohde, and indicate a major connection with brain tracts, rather than posterior tracts.

The sequence of events in cell A comprises a depolarization plateau from which spike-like depolarizations may arise, followed by a spike train in a neighbouring cell or remote part of cell A. The remarkable property of the cell appears to be the very long duration of the depolarization plateau (100 ms) following a single brief shock applied to a remote input.

It has been suggested that cell A is specifically connected with the contraction of the atrial musculature (Bone, 1960). The association of the dendrites of cell A with the dorsal roots containing fibres innervating the velar tentacles, and the juxtaposition of visceral motor cells and the axon of cell A, indicated by this author, lead very reasonably to such a suggestion.

There are, however, a number of other points to consider. The axon of cell A offers the most rapid conduction pathway in the central nervous system of amphioxus, and it is believed the rates of over 5 m s^{-1} pertain to this axon. The rate at which the muscular wave in the atrial musculature passes forward is much slower, less than 1 m s^{-1}, and posteriorward conduction is slower again (0.4 m s^{-1}).

In silver preparations of the spinal cord the larger visceral motor neurones often lie at the point where axons of Rohde cells B–L dip beneath the median axon of cell A, and their processes are in close relation with both types of axon.

The striking anterior dendrites extending into the brain region suggest inputs involving the anterior sense cells of the brain region rather than the specialized inputs associated with atrial reflexes. While the axon of cell A appears to be limited to the atrial region of the nerve cord, so are the other anterior Rohde axons, and no correlation with atrial function may be present.

The secondary spike attributed to Rohde cell B is associated with a small amplitude contraction of the ventral atrial muscles, but major contractions of the atrium appear only when the descending slow response develops.

These points serve to show that it may be unwise to suggest that the giant fibres provide more than the basis for movement responses.

The ascending fast responses (AFR) have been relatively little studied. Anterior cell responses as the result of caudal stimulation indicate the presence of ascending pathways extending most of the length of the spinal cord, with overall conduction rates of

4–4·5 m $^{-1}$. The high rates of conduction observed for the DFR only derive from estimations made from the spinal cord anterior to the atriopore. As shown by Rohde, the posterior group of giant cells are topographically distributed so as to form a chain of cells rather than an overlapping array as is the case with the anterior group. The result of this is that while the posterior group has a greater longitudinal extension, junctions between each successive element cannot be so easily bypassed.

Electrical recordings from the different regions of the spinal cord bear out this anatomical difference. The AFR consists of only one or two spike-like potentials at all levels. Furthermore, the amplitude of the fast potentials are largest near myotome 40, where large-diameter dorsal axons are visible in sections.

It is clear that some form of electrical coupling between posterior giant cells is likely to be associated with the high conduction rates observed. Shocks applied to the terminal region of the nerve cord only, containing cells Y and Z, were as effective in producing maximal responses at the pharyngeal level as were shocks applied across several posterior neuromeres. No recruitment process is involved in activating the more anterior regions of the pathway, which functions like a single pair of greatly extended elements. On the other hand, histological studies indicate that many of the more posterior cells in the series M–Z have axons that do not extend further than five neural segments anteriorly of the cell body. There is no doubt that some form of short delay coupling exists, as occurs between posterior giant cells in the lamprey (Rovainen, 1967).

The important question of the relation between Rohde cell action and the initiation of muscular contractions needs to be considered with reference to the slow responses.

Slow evoked responses

Stimulation of the isolated nerve cord, or of the whole animal, results in the appearance of a potential wave which passes down the spinal cord at between 0·8 and 1·5 m s^{-1}. The maximum amplitude of the slow response recorded with wire electrodes is relatively large—nearly 1 millivolt, so that it can easily be studied in some detail.

Recordings from swimming lancelets indicate quite clearly that the descending slow responses (DSR) always precede contractions in the myotomes, but that they are not derived from the myotomes themselves.

The low rates of conduction characterizing the DSR, and the four ms duration of the major DSR component compared with single unit potentials (0·5–1·0 ms), strongly suggest that the DSR is the result of synchronized activity in many fine fibres.

Penetrations into the nerve cord with fine tipped microelectrodes reveal single unit responses, usually brief axon spikes, in certain areas, and then only rarely. At the same time the massed response is recorded at every point in a complete traverse through the cord.

The first question to be answered concerns the identity of the cells generating the DSR, and a number of pieces of evidence bear upon this point. Careful measurement of the maximum amplitude of the DSR at equal intervals of distance along a traverse through the spinal cord indicates response maxima (25% above the mean) in the central and ventral regions. These areas correspond to the position of the axons and their branches of secondary interneurone types (see the anatomical section), rather than the area enclosing branches of the somatic motor neurones which lie more laterally.

Where single unit responses were observed to be synchronized with the external DSR, they were most often found in the ventral and median regions of the cord where axons of commissural and other secondary interneurone types had been clearly observed as isolated, stained cells. The fact that the massed DSR can be recorded outside the main ventral axon tracts is consistent with the histological demonstration of fine branches of these cells in more dorsal areas.

The form of the DSR varies with stimulus intensity, and the level of the spinal cord from which it is recorded.

The standard fully recruited response consists of a single potential wave of maximum amplitude. This is often most clearly visible in the mid-pharyngeal part of the spinal cord at the level of myotome 25. Raising the stimulus intensity results in repetitive DSR potentials. Graded intensities of electrical stimulation below the threshold for full recruitment reveal a definite recruitment pattern (Fig. 6). At the lowest intensities necessary for a response a single small potential occurs with a long latency. As the stimulus intensity is raised, potential waves with larger amplitudes and shorter latencies appear. Finally, a single large potential results, with a latency slightly less than that of the fastest component observed. If the recruitment pattern is described in terms of the largest component occurring at each stage, it could be said that

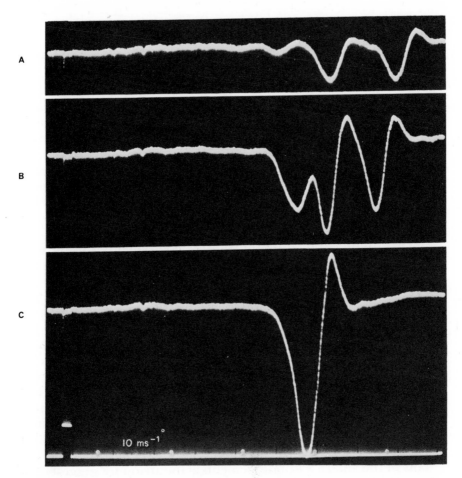

FIG. 6. The recruitment pattern of the descending slow response following the application of single shocks of increasing intensity to the anterior nerve cord. Killed-end preparation. Note the increase in amplitude, and synchronization of the response, involving the disappearance of the slower components. The maximum amplitude was about 500 μmV in this recording.

latency is reduced by 50%, amplitude increases by 300%, and duration by 20%.

At just below full recruitment the major component appears double peaked. One of the most striking aspects of the recruitment pattern is that the initial responses with a long latency disappear completely at higher stimulus intensities.

An explanation of this recruitment pattern is based on the thickness of longitudinal axons, and their distance from the stimulating electrodes on the outside of the nerve cord is as follows. The slowest component (SR 1) derives from fine axons at the periphery of the slow pathway, the fastest component (SR 2) originates in larger or longer axons lying nearer the mid-line, and the largest amplitude component (SR 3) depends on many axons of intermediate thickness lying nearest to the central canal. Morphological correlation can be found.

The appearance of a double peak in SR 3 indicates a tendency for the right and left tracts mediating this response to be slightly out of phase (the difference is less than 1 ms), and may provide the basis for bilateral inhibition of motor neurones as described in teleost spinal cord by Diamond & Yarsagil (1968).

The disappearance of the slowest components (SR 1) with increasing recruitment has been observed by Pickens (1970) in the hemichordates, and he has suggested that some ephaptic type of effect may be involved. The acceleration of impulses in slow crab axons by those travelling in fast ones, described by Katz & Schmitt (1940), is an obvious analogy, given a greater probability by the apparent absence of glial material between the finer axons in electronmicrographs of this region of the lancelet spinal cord.

One of the most striking aspects of the DSR is its tendency to failure at the level of the atriopore, adjacent to myotome 35. Fully recruited responses anterior to this level fail to appear posterior to it in 70% of preparations. This is illustrated in Fig. 3. Repetitive stimulation or increased stimulus intensities are necessary to effect invasion of the post-atrial cord. The termination of the axons of many of the anterior Rohde elements at the atriopore suggests that they may be involved in lowering the threshold for DSR propagation at more anterior levels.

Artificially increasing the afferent activity in the myotomes adjacent to the atriopore has no effect in lowering the threshold at this critical zone. Again Pickens (1970) observed a similar phenomenon in the hemichordate nerve cord, but it involved the ascending pathway. Experiments involving movement of the recording electrodes along the nerve cord make it clear that the passage of the DSR along it involves the activation of serially arranged groups of elements rather than simply a longitudinal transmission through a system of tracts. This is confirmed by the ineffectiveness of axon collision experiments in producing a reduction in DSR amplitude. During swimming, a characteristic sequence of large and small amplitude components of the DSR occurs; this reflects a specific order in which secondary interneurones become active.

The rates at which the DSR can be repeated are considerably below the repetition rates for individual axons. The maximum DSR frequency is between 100–120 times/second. Nevertheless repetition rates required for fast swimming do not exceed 25 Hz.

The ability of the low transmission zone near the atriopore to act as a filter increases with a rise in repetition frequency. The ratio between DSR duration recorded anterior and posterior to the atriopore rises from 2 : 1 to 4 : 1 when the stimulus frequency is raised from 100 to 200 Hz. These results indicate a multisynaptic structure underlying DSR transmission.

Double pulse stimulation indicates strong facilitation where pulse intervals fall between 10–200 ms. At longer intervals there is a decline, reaching a maximum at 10 s, followed by a recovery at five to 10 min. These observations on the decline and recovery of excitability agree with Sergeev's findings on whole animal responses (Sergeev, 1963). Facilitation at small interval is associated with the repetition rates that occur during swimming (40–250 ms). Facilitation at long intervals corresponds to the commencement of spontaneous swimming, which normally occurs at intervals between five and 10 minutes in freely suspended animals.

An ascending slow response (ASR) can be initiated by stimulation of the nerve cord in much the same way as the DSR. Evidence that the ASR involves independent pathways to the DSR came from spike collision experiments, and from the facts that there is no low transmission zone near the atriopore and that repetition frequencies extend to nearly 200 Hz. In other respects the ASR is similar to the DSR.

What is the relation between the fast response pathway mediated by the Rohde cells and the slow response system? The slow response is always accompanied by the fast response due to the much lower threshold of the latter, but this does not imply interdependence. Indeed the considerable difference in their conduction rates means that the interval between the two types of response increases as they pass posteriorly.

Some important evidence of a functional connection comes from microelectrode recordings from the region of the median axon of Rohde cell A. An example is shown in Fig. 7. These show that when both DFR and DSR are recorded a polarization plateau

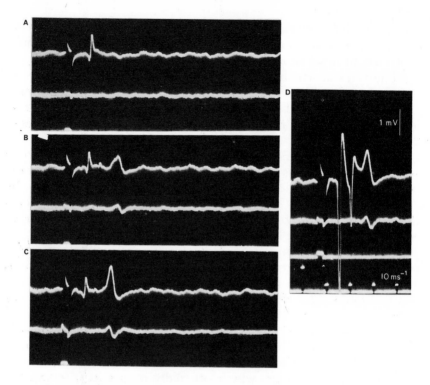

FIG. 7. Evidence for connections between the fast and slow response systems. Upper trace—microelectrode recording from the ventro-medial areas. Lower trace—external wire recording. A–C. Increased intensities of single shocks applied to the anterior nerve cord, result in the appearance in fibre A of the DFR, and then the descending slow response. A polarization plateau exists between the two potentials, and a small intermediate potential arises from this. D. When the microelectrode was advanced, contact was made with fast fibre A, and a closer approach made to the neurone generating the intermediate potential, suggesting the presence of a junction at this site.

links them, and a small intermediate potential arises from this. Recordings of this type strongly suggest the existence of local pathways between the two systems. At the same time it is clear that the Rohde cell cannot by itself activate the slow response system.

THE BRAIN

Structure and regional organization

The region of the neural axis anterior to the cell body of Rohde cell A is considered to be the brain (Edinger, 1906). Edinger has provided a brief description of some of the major tracts, and of the general form of the giant dorsal cells which he shows with long anterior processes. As shown in Fig. 8, the brain can be divided into four major regions, as follows. (1) An anterior region lying in front

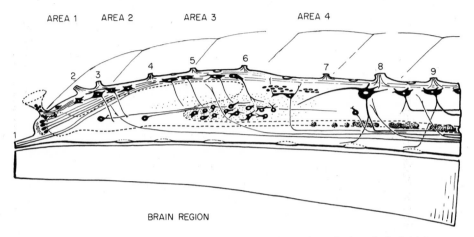

FIG. 8. Diagram of the structure of the brain and anterior spinal cord, derived from sectioned material and whole mounts. For further details see text.

of the second pair of dorsal nerve roots. This region contains the anterior brain ventricle, and the connections with the eye spot (macula), Kölliker's pit, the rostral receptors, and, according to Edinger, the nerves from the frontal organ. (2) A central region lying between the second and the fourth dorsal nerve roots. This contains large dorsal cells, and a very high density of small neurones with subspherical cell bodies. In the more central regions cells with long posterior processes are visible. The infundibular organ lies in the ventral region. (3) The posterior region lies between the fourth and the sixth dorsal nerve roots. There are many large dorsal cells, some with descending processes. The posterior brain ventricle has a conspicuous dorsal extension in this brain region. (4). Linking the brain and the spinal cord there is a zone of longitudinal tracts. It extends between the sixth and seventh dorsal nerve roots. As indicated by Edinger, some of the components of the lateral and dorsal longitudinal tracts originate in the anterior brain region.

The most striking cellular components of the brain are the large dorsal cells of region 2 and 3. The cell bodies are 15–25 µm in diameter and stain weakly with silver nitrate. The peripheral position of these cells, their large size, and their flask-shaped cell bodies, are similarities with the large neurosecretory neurones of the cortex of the ascidian ganglion (Guthrie & Banks, unpublished). However, careful examination with the fluorescence methods for biogenic amines, the paraldehyde-fuchsin staining technique for neurosecretory material, and electron microscopy, failed to show any evidence of neurosecretory function in the lancelet brain cells. Silver preparations of the dorsal cells indicated the presence of ventral processes connecting some of the cells with a median zone of small cells in the posterior part of the brain.

Electronmicrographs of the giant cells indicate that the cytoplasm contains a fragmented and irregular type of endoplasmic reticulum. A few large lysosomal bodies occur. The cell body is surrounded by a sheath formed of several layers of membranous profiles. No evidence of neurosecretory material was visible.

The cells of the infundibular organ, so clearly illustrated by Olsson (1962), are easily identified in the anterior region of the brain. The paraldehyde fuchsin method for neurosecretory material stains these cells intensely while leaving the other brain cells virtually unaffected. This supports Olsson's hypothesis that the cells may be homologous with the vertebrate hypophysial primordium.

Various regions of the brain were examined with the electronmicroscope. The main differences from the spinal cord, of a general nature, were as follows. (1) Synaptic structures are relatively abundant in the brain. (2) Desmosome-like structures were also observed at cell boundaries. These were not found in the spinal cord. (3) No region of the spinal cord is so densely packed as are certain parts of the anterior and central brain regions. Nuclear membranes of adjacent cells are separated by less than $0 \cdot 1$ μm. (4) The whorl-like arrays of membrane profiles are commoner in the brain. These resemble the open type of axonal investment found in arthropods. (5) Axons with diameters exceeding 4 μm were not observed in transverse sections. (6) The fibrous processes of ependymal glia are very abundant; more so than in the spinal cord. (7) The majority of the axons contained electron-dense material.

Functional properties of the brain

Some information concerning the role of the brain can be inferred from the results of disconnecting it from the spinal cord in the otherwise intact animal. The effect on suspended lancelets, which swim spontaneously, is to increase the duration of swimming periods. Decerebrate lancelets lying on the bottom of a dish are more easily aroused by weak mechanical stimulation than intact animals, and then swim vigorously for several seconds, whereas normal animals often swim for less than 2 s.

The threshold for movement responses to stimulation with a beam of light is altered by decerebration. Normal dark-adapted animals would react to light intensities in the range 1·0–0·1 lumens/square foot with a brief period of swimming, but response waned rapidly if the stimulation was repeated at intervals of 10 s. Decerebrate animals reacted more strongly, and repeated stimulation continued to be effective. In dark-adapted normal lancelets, responses to brief illumination at intensities below 0·1 lm/sq.ft. were unreliable, and responses failed completely below 0·08 lm/sq.ft. In decerebrate animals, brief muscular responses could be elicited down to intensities of 0·03 lm/sq.ft. The responses of decerebrate lancelets to photic stimulation in general were more stereotyped and reliable compared to normal animals.

Localized recordings from the central and the posterior brain region in quiescent lancelets indicate regular fluctuations of electrical potential involving many units. The frequency is usually in the range 10–15 Hz. The spontaneous activity appears as a continuous rhythm, rather than isolated bursts as in the spinal cord.

During slow swimming, the spontaneous activity is replaced by a potential wave form repeated at the same frequency as the swimming rhythm, that is 2–3 Hz. At the commencement of swimming, simultaneous recordings from the brain and the spinal cord appeared to show that the swimming rhythm started in the brain 200 ms before it began in the middle region of the cord.

At the end of a period of slow swimming there is a burst of large amplitude potentials in the brain and spinal cord. Their appearance at the two recording sites is nearly synchronous, and their site of origin is not clear, though the fact that the burst is of longer duration in the mid-pharyngeal cord than in the brain supports the idea of a descending response.

Stimulation of the ascending fast pathways in the spinal cord result in weak delayed responses in the brain which can be revealed by signal averaging techniques. As described earlier, the fast pathways produce strong effects in the region of the anterior column of Rohde cell bodies, and experiments were made to show the result of brain stimulation on these effects.

Three kinds of interaction between the brain and the anterior cells of the spinal cord were observed. Localized stimulation of selected brain regions with metal microelectrodes was coupled with glass microelectrode recordings from the spinal cord and remote stimulation of the caudal nerve cord.

(1) A shock applied to the dorsal area of the central region of the brain (area 2) suppresses the large soma spike normally recorded near Rohde cell body B. The presynaptic volley that results from caudal stimulation remains unaltered, and is clearly visible (Fig. 9). The effect is strongest when shocks to the brain and nerve cord are exactly synchronized.

(2) Stimulation of the ascending pathway is associated with the activation of smaller neurones lying slightly ventral to the Rohde cell bodies B–F (myotome levels 6–12). Normally these only respond with one or two spikes. When coupled with stimulation of the medial zone of the central region of the brain, where the small globular cells are most numerous, the responses of the spinal neurones are prolonged to a burst of five or more spikes. The effect is strongest when the stimulus to the brain is delayed by 20 ms relative to the stimulation of the spinal cord.

(3) A single shock applied to the ventral region of the zone of tracts connecting the brain and the spinal cord area 4 results in the activation of cells in the ventral part of the spinal cord not normally affected by ascending pathway stimulation. These cells appear to lie ventral to those described in the previous paragraph. Their characteristic feature is that they fire in bursts containing similar numbers of spikes. The burst repetition frequency is about 15 Hz, or close to the frequency of the DSR during fast swimming. These cells will be referred to as repeater units.

In addition to the effects detailed above, depending on the activation of specific brain regions, more general stimulation could produce striking effects at relatively remote sites within the spinal cord. Recordings made from the lateral areas of the spinal cord between myotomes 15 and 20 revealed unit activity which was totally inhibited by shocks applied across the whole of the central and posterior regions of the brain (Fig. 9). That these are local units that are inhibited, rather than the axons of cells descending from other levels, is suggested by the spike duration and shape. That the electrode is close to the synaptic sites is shown by the precise registration of the inhibitory hyper-polarizing phase, post-inhibitory rebound, etc. This is shown in Fig. 9a. It is to be noted that all these effects resulting from brain stimulation are obtained with single weak shocks. From the observations described above it is clear that the brain of amphioxus, although small and relatively

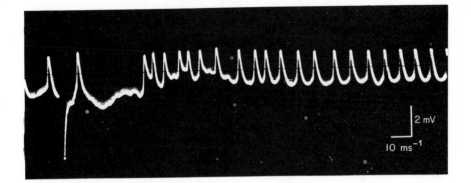

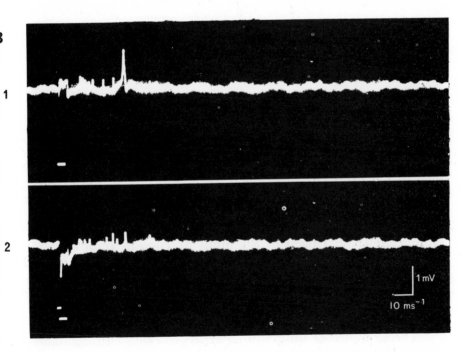

FIG. 9. Inhibitory functions of the brain. A. Microelectrode recording from a spontaneously active unit in the mid-pharyngeal region of the nerve cord. A shock applied across the brain region produces hyperpolarizing i.p.s.ps. followed by a post-inhibitory rebound. B. Microelectrode recording from a junction between an ascending spinal tract and Rohde cell body. (1) Caudal stimulation produces a presynaptic volley, and a postsynaptic response in the Rohde cell. (2) When in addition a 1 ms shock is applied to the dorsal region of brain area 2, the spike in the Rohde cell is suppressed.

unspecialized, is capable of producing both inhibitory and excitatory effects on cells of the spinal cord. A model scheme describing these results is shown in Fig. 10.

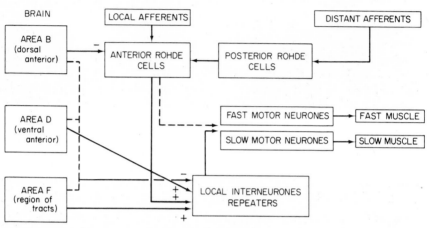

FIG. 10. Functional scheme of pathways suggested by observations described in the text.

THE INPUT ELEMENTS

Although the lancelet lacks major sense organs of the type found in most vertebrates, a considerable range of simple sense organs has been identified microscopically (Dogiel, 1903; Franz, 1924; Van Wijhe, 1913; Bone, 1960). Few functional studies have been attempted, but enough information exists about the photoreceptors and the receptors of the atrial epithelium for us to appreciate some aspects of the input function in amphioxus. In general, lancelets appear very insensitive to most stimuli, as judged by whole animal responses (Parker, 1908; Lele *et al.*, 1958).

The photoreceptors

The presence of longitudinal arrays of cells, with pigmented cups adjacent to the central canal of the spinal cord, was noted by the older anatomists. The general form of these simple sense organs is reminiscent of eye-cup sense organs found in certain of the lower invertebrates. Franz (1924) described the distribution of these cells, and noted that the orientation of the pigmented cups varied between the cells of a group, but his suggestion that this was correlated with rotational swimming has not been generally accepted.

The terminations of the receptor cells have not been described, partly because of their low affinity for the silver stains, but the fine structure of the sense cell body has been illustrated by Eakin & Westfall (1962). An irregular array of tubules near the inner border of the cell appears to constitute the photoreceptive membrane. Salerno (1965) has described the fine structure of the apical macula, or eyespot, in the brain. This structure resembles an aggregation of the spinal eye cups, and its subcellular organization provided evidence of photoreceptive function. However, Parker (1908) found that the brain region was generally insensitive to light.

Parker's study on the regional sensitivity of the spinal cord in whole animals showed that the lowest threshold for movement responses, evoked by shining a narrow beam of light on the cord, were associated with the greatest concentrations of the pigmented cells (Parker, 1908). Sergeev (1963) is in general agreement with these findings, and has described long-term changes in the strength of the light-movement response.

Lele *et al.* (1960) produced evidence to support the view that a part of the light response is a response to a rise in the temperature of the skin, probably mediated by cutaneous nerve fibres.

Illumination of isolated spinal cords demonstrates unequivocally the presence of photoreceptive cells within the structure. Recordings made with wire electrodes placed on or within the nerve cords, reveal the appearance of small potentials with a saw-tooth profile, following exposure to light at $0·1$ lm/sq.ft. If a beam of light is moved transversely across the isolated spinal cord an augmenting train of potentials occurs. This is most easily explained on the basis of the successive stimulation of photoreceptors with pigment cups with different orientations. Recovery times for the photoreceptors are several seconds.

Separate studies with microelectrodes were made. The pigmented cells are clearly visible and it is possible to guide dye-filled microelectrodes into close proximity to them. Stimulation with very brief (less than 5 ms), high intensity (over 1×10^7 lm), flashes of light, evokes a local slow potential wave. This slow wave indicates the presence of a depolarizing potential in the receptor, reaching a peak in 50 ms with a flash intensity of 15×10^7 lm. Decay times were about 100 ms. Reduced light intensities delayed peak depolarization, with a small reduction in the amplitude of the electrical response. The threshold for a response lay between 1 and 5×10^7 lm. Recovery rates were studied with repetitive light flashes. Flash intervals must exceed 10 s for the generator potential to maintain amplitude. Intervals of one second or less lead to the rapid extinction of the response. These experiments show that the high latency and poor recovery, which is characteristic of the response of the whole animal to light, originates at the receptor level.

Attempts to trace the axons of receptor cells were occasionally successful using the intravitam methylene blue methods. Axons of lateral photoreceptors pass horizontally to near the margin of the spinal cord. Then they descend to a position close to the collateral fibres of somatic motor neurones before breaking up into longitudinal processes. This accords with a view supported by other evidence, that the light responses are mediated through the slow rather than the fast pathways.

The atrial receptors

A variety of sensory structures have been described from the inner layers of the atrium, and the surface of the gut, by van Wijhe (1913), Holmes (1953) and Bone (1958). The most striking of these are undoubtedly the large multipolar cells, which bear a strong morphological similarity to the multipolar stretch receptors found in arthropods (Wiersma, Furshpan & Florey, 1953; Finlayson & Lowenstein, 1958; Guthrie, 1967a). However, these cells, as shown by Bone, are mostly thickly clustered in regions, such as the dorsal wall of the gut, which contain little muscle, and he suggests a chemoreceptive function for them.

Careful examination of the gut wall and the inner surface of the atrium with electrical recording methods reveals the presence of high threshold, phasic mechanoreceptors in the lateral regions of the atrial wall. Microelectrode contacts with single receptor units, coupled with tensions applied to the receptor area directly, or indirectly by making the atrial muscles contract, reveal aspects of their function. The receptors only respond to rapid increases in tension (above 20 mg/ms), and they do not provide a spike train that reflects accurately variations in tension above the threshold level. This appears to be due to very rapid adaptation. Normal weak contractions of the atrial musculature produce one or two spikes only, but a strong contraction applied against a maintained tension of 400–500 mg may produce 10–15 spikes that continue into the period of muscular relaxation. The response is essentially of the "on" type, and is unlikely to be involved in any graded functions.

It has not so far been possible to identify the cells involved, but the monopolar cells described by Bone (1958) appear to be the most abundant in the area where most units were encountered.

CONCLUSION

The work described in this communication provides a basis for neural models of systems controlling swimming in the lancelet. A flow diagram illustrating the major connections that would form the basis of such models is shown in Fig. 10. Details of timing would be required to provide a more advanced model.

There are one or two further points. The lancelet resembles insects, gastropods and fish in that locomotory rhythms are not maintained by phasic afferent activity, but depend on the intrinsic properties of central neurones. In the lancelet, preparations lacking 90% of the musculature and all peripheral nervous structures make spontaneous swimming movements.

The giant fibres serve as major sites for the input channels. Their connections with the motor neurones are indirect. As in the cockroach, the largest fibres are not directly connected to the motor neurones, which are activated by a fine-fibred tract of secondary interneurones (Dagan & Parnas, 1970). In the lamprey, this may help to explain the fact that only high frequency repetitive stimulation of the giant fibres elicits muscular movements (Rovainen, 1967).

ACKNOWLEDGEMENTS

My colleague, Mr J. R. Banks, has collaborated with me on much of the original work described in this communication. My thanks are also due to Mr A. Morison, and Mrs Mary Panko for assistance with the electron microscopy of the lancelet nervous system.

I would also like to express my indebtedness to the Science Research Council for grants in aid of this research provided over a four-year period.

REFERENCES

Barets, A. (1961). Contribution à l'étude des systèmes moteur lent et rapide du muscle lateral des teleostéens. *Archs Anat. microsc. Morph. exp.* **50**: 91–187.

Berkowitz, E. C. (1953). Conduction pathways in the nerve cord of the lamprey *Entosphenus. Anat. Rec.* **121**: 264.

Bone, Q. (1958). Synaptic relations in the atrial nervous system of Amphioxus. *Phil Trans. R. Soc.* (B) **99**: 243–263.

Bone, Q. (1960). The central nervous system in amphioxus. *J. comp. Neurol.* **115**: 27–64.

Bone, Q. (1966). On the two types of muscle fibre in elasmobranchs. *J. mar. biol. Ass. U.K.* **46**: 321–342.

Bullock, T. H. & Horridge, G. A. (1965). *Structure and function in the nervous system of invertebrates.* San Francisco & London: W. H. Freeman.

Dagan, D. & Parnas, I. (1970). Giant fibre and small fibre pathways involved in the evasive responses of the cockroach *Periplaneta. J. exp. Biol.* **52**: 313–325.

De Bell, J. T., Del Castillo, J., & Morales, T. (1967). The initiation of action potentials in the somatic musculature of *Ascaris lumbricoides. J. exp. Biol.* **46**: 263–280.

Diamond, J. & Yarsagil, G. M. (1968). The startle response in teleost fish. *Nature, Lond.* **220**: 241–245.

Dilly, P. N., Welsch, V. & Storch, V. (1970). The structure of the nerve fibre layer and the neurocord in the enteropneusts. *Z. Zellforsch. mikrosk. Anat.* **103**: 129–148.

Dogiel, A. S. (1903). Das periphere Nervensystem des Amphioxus. *Anat. Hefte* I **21**: 147–163.

Eakin, R. M. & Westfall, A. (1962). The structure and origin of the photoreceptors of *Amphioxus. J. Ultrastruct. Res.* **6**: 186–191.

Edinger, L. (1906). Einiges von Gehirn des Amphioxus. *Anat. Anz.* **28**: 147–428.

Finlayson, L. H. & Lowenstein, O. (1958). The structure and function of abdominal stretch receptors in insects. *Proc. R. Soc.* (B) **148**: 433–449.

Flood, P. R. (1966). A peculiar mode of muscular innervation in amphioxus. *J. comp. Neurol.* **126**: 181–218.

Flood, P. R. (1968a). The extraction of a paramyosin-like protein from the notochord of amphioxus. In *Electron microscopy 1968* **2**: 291–292. Bocciarelli, D. S. (ed.). Rome: Tipografia Poliglotta Vaticana.

Flood, P. R. (1968b). The structure of the segmental trunk muscle in amphioxus. With notes on the course and the endings of the so-called ventral root fibres. *Z. Zellforsch. mikrosk. Anat.* **84**: 389–416.

Flood, P. R., Guthrie, D. M. & Banks, J. R. (1969). Paramyosin muscle in the notochord of amphioxus. *Nature, Lond.* **222**: 87–88.

Franz, V. (1924). Lichtsinnversuche am Lanzett-fisch zur Ermittelung der Sinnes-funktion des Stirn-oder Gehirnblaschens. *Wiss. Meeresuntersuch. (Abt. Helgo-land.)* **15**: 1–19.

Furukawa, T. (1966). Synaptic interaction at the Mauthner cell of the goldfish. *Progr. Brain Res.* **21A**: 46–70.

Geduldig, D. S. (1965). Resting potential dependency on external potassium in Amphioxus muscle. *Am. J. Physiol.* **208**: 852–854.

Guthrie, D. M. (1967a). Multipolar stretch receptors and the insect leg reflex. *J. Insect Physiol.* **13**: 1637–1644.

Guthrie, D. M. (1967b). Control of muscular contractions by spinal neurones in Amphioxus (*Branchiostoma lanceolatum*). *Nature, Lond.* **216**: 1224–1225.

Guthrie, D. M. & Banks, J. R. (1970a). Observations on the function and physiological properties of a fast paramyosin muscle—the notochord of amphioxus (*Branchiostoma lanceolatum*). *J. exp. Biol.* **52**: 125–138.

Guthrie, D. M. & Banks, J. R. (1970b). Observations on the electrical and mechanical properties of the myotomes of the lancelet (*Branchiostoma lanceolatum*) *J. exp. Biol.* **52**: 401–417.

Hagiwara, S., Henkart, M. P. & Kidokoro, Y. (1971). Excitation-contraction coupling in Amphioxus muscle cells. *J. Physiol.* **219**: 233–251.

Hagiwara, S. & Kidokoro, V. (1971). Na & Ca components of action potential in Amphioxus muscle cells. *J. Physiol.* **219**: 217–232.

Hanson, J. & Lowy, J. (1961). The structure of the muscle fibres in the translucent part of the adductor of the oyster *Crassostrea angulata*. *Proc. R. Soc.* (B) **154**: 173–196.

Holmes, W. (1953). The atrial nervous system of Amphioxus (Branchiostoma). *Q. Jl. microsc. Sci.* **94**: 523–535.

Jansen, P., Andersen, J. K. S. & Loyning, Y. (1963). Slow and fast muscles in the atlantic hagfish. *Acta physiol. Scand.* **57**: 167–179.

Katz, B. & Schmitt, O. H. (1940). Electrical interaction between two adjacent nerve fibres. *J. Physiol.* **97**: 471–488.

Lele, P. P., Palmer, E. & Weddell, G. (1958). Observations on the innervation of the integument of Amphioxus, *Branchiostoma lanceolatum*. *Q. Jl microsc. Sci.* **99**: 421–440.

Müller, E. (1900). Studien über Neuroglia. *Arch. mikrosk. Anat.* **60**: 11–62.

Olsson, R. (1962). The infundibular cells of *Amphioxus* and the question of fibre-forming secretions. *Ark. Zool.* (2) **15**: 347–356.

Parker, G. H. (1908). The sensory reactions of Amphioxus. *Proc. Am. Acad. Arts Sci.* **43**: 416–455.

Peachey, L. D. (1961). Structure of the longitudinal body muscles of amphioxus. *J. biophys. biochem. Cytol.* **10** (suppl.): 159–176.

Pickens, P. (1970). Conduction along the ventral nerve cord of a hemichordate worm. *J. exp. Biol.* **53**: 515–528.

Retzius, G. (1891). Zur kenntnis des centralen Nervensystems von *Amphioxus lanceolatus*. *Biol. Unters.* N.S. **2**: 29–46.

Rohde, E. ('691). Histologische Untersuchungen über das Nervensystem von *Amphioxus lanceolatus*. *Zool. Beitr.* **2**: 169–211.

Rovainen, C. M. (1967). Physiological and anatomical studies on large neurons of the central nervous system of the sea lamprey (*Petromyzon marinus*) II Dorsal cells and giant interneurons. *J. Neurophysiol.* **30**: 1024–1042.

Salerno, V. (1965). [On the apical macula of *Amphioxus*] *Atti Soc. pelorit. Sci. fis. mat. nat.* **11**: 353–359. [In Italian].

Sergeev, B. F. (1963). [The sensory reactions of *Amphioxus* and the effects of anesthesia.] *Fiziol. Zh. SSSR* **49**: 60–65. [In Russian].

Tauc, L. & Hughes, G. M. (1963). Modes of initiation and propagation of spikes in the branching axons of molluscan central neurons. *J. gen. Physiol.* **46**: 533–549.

Ten Cate, J. (1938). Zur Physiologie des Zentralnervensystems des Amphioxus (*Branchiostoma lanceolatum*). *Archs. néerl. Physiol.* **23**: 409–423.

van Wijhe, J. W. (1913). On the metamorphosis of *Amphioxus lanceolatus*. *Proc. K. ned. Akad. Wet.* **16**: 574–583.

von Ebner, H. (1895). Über den Bau der Chorda dorsalis des Amphioxus (*Branchiostoma lanceolatus*). *Sb. Akad. wiss. Wien* (Mat. Naturv. Kl.) **104**: 199–288.

Webb, J. E. (1973). The role of the notochord in forward and reverse swimming and burrowing in the amphioxus *Branchiostoma lanceolatum*. *J. Zool., Lond.* **170**: 325–338.

Wiersma, C. A. G., Furshpan, E. & Florey, E. (1953). Physiological and pharmacological observations on muscle receptor organs of the crayfish *Cambarus clarkii*. *J. exp. Biol.* **30**: 136–150.

Symp. zool. Soc. Lond. (1975) No. 36, 81–104.

FINE STRUCTURE OF THE NOTOCHORD OF AMPHIOXUS

PER R. FLOOD

Institute of Anatomy, University of Bergen, Bergen, Norway

SYNOPSIS

The notochord of amphioxus is a rod-like cellular organ enclosed in a thick collagenous sheath. It extends throughout the entire length of the animal. The main constituent of this organ is the flattened *lamellae* that are stacked together in the antero-posterior direction like a column of coins. Recent structural, physico-chemical and functional analyses have shown that these lamellae are cross-striated paramyosin muscle cells that are innervated through the notochordal processes, where sarcoplasmic extensions of the lamellae penetrate the sheath and approach presynaptic nerve endings at the surface of the spinal cord.

Fluid-filled spaces occur as *intracellular vacuoles* within the lamellae. These are probably remnants of the central vacuole seen in most notochordal cells during the larval development. Highly organized membrane arrays are present in these vacuoles. *Extracellular fluid-filled spaces* occur between the lamellae and as a longitudinal canal at the dorsal border of the lamellae and possibly as a ventral canal at the ventral border of the lamellae.

Between the dorsal canal and the notochordal sheath, and between the two rows of notochordal processes, is a layer of *dorsal Müller cells*. These are very long and closely interconnected fibrous cells oriented in the antero-posterior direction. At intervals they send thin cytoplasmic arborizations through the dorsal canal towards the dorsal border of the lamellae. The ultrastructural appearance of these cells indicates that they may secrete the polysaccharide-containing extracellular fluid of the notochord, but also that they may serve mechanical and conductive functions.

Between the ventral border of the lamellae and the notochordal sheath a layer of *ventral Müller cells* is found. These are less specialized than the dorsal Müller cells, and seem to be related to the production of the extracellular fluid in the ventral canal.

In between the lamellae, at their insertion in the dorso-lateral part of the sheath, some small wedge-shaped *lateral Müller cells* are found. Some of these cells have processes towards the spinal cord like the lamellae. These cells are supposed to be a source of new lamellae added to the old ones during the growth of the animal.

The activation of the notochordal muscle cells will reduce the transverse diameter of the notochord and cause a rise in hydrostatic pressure inside the stretch-resistant collagenous sheath. The increased pressure will result in a stiffer notochord. The functional significance of this adjustable hydroskeleton is poorly understood, but may be related to forward and backward swimming or to swimming and burrowing movements.

When the notochord of amphioxus is compared to that of other chordates it is quite obvious that it has no equivalent. On the other hand, both the ontogeny and some ultrastructural details of the adult notochord indicate that it is homologous to notochords of other chordates. It is perhaps the most highly specialized hydroskeleton ever built and should not be considered as a degenerate organ.

INTRODUCTION

Because of the central position of amphioxus (*Branchiostoma lanceolatum*, Pallas) in chordate phylogeny, and the importance of the chorda dorsalis as an evolutionary landmark, the notochord of

amphioxus has been studied with much interest by many comparative anatomists. Its main structural features were revealed as early
as 1871 by W. Müller, who discerned at least five structural
components: (1) The notochordal lamellae, which make up the
bulk of the organ. These are stacked together like a column of coins
and each lamella covers almost the entire cross-sectional area of the
notochord. The lamellae contain horizontally-arranged birefringent filaments. (2) Less prominent cellular elements (later called
Müller-cells), which are concentrated along the dorsal and ventral
border of the notochord. The cells at the dorsal border have many
processes that constitute a meshwork above the dorsal edge of the
lamellae. (3) Fluid-filled spaces, found as a longitudinal dorsal
canal in the above mentioned meshwork of cellular processes and
as vacuoles between the lamallae. (4) A prominent connective tissue
sheath which surrounds all the above mentioned structures and
encloses the organ entirely. (5) Two rows of regularly spaced
cavities (later called notochordal horns), which are present in the
dorsal part of this sheath throughout the entire length of the
animal.

W. Müller's work (1871) and other early reports (a.o. by J.
Müller, 1842; Ebner, 1895; Joseph, 1895, 1902) thus established a
pronounced structural difference between the notochord of amphioxus and that of higher chordates. However, as is evident from
every textbook of comparative anatomy or zoology, this difference
has not led people to believe that the function of the notochord of
amphioxus should differ from that supposed for other notochords
by serving purposes additional to the mere provision of passive
support.

Since 1967, however, important observations have been made
with modern techniques such as electron microscopy, cytochemis-
try, electrophysiology and cinephotography. As will be reviewed
on the following pages, it may now be regarded as proved that the
notochord of amphioxus is a muscular organ capable of altering its
mechanical properties upon nervous stimulation. However, our
fragmentary knowledge, especially about the Müller-cells and the
compartmentalization of the fluid-filled spaces in the notochord,
still gives room for speculations and controversial hypotheses on
the motor innervation of the lamellae and on the effect of the
muscular contractions. The present communication accordingly
has a threefold purpose. It presents some new ultrastructural
observations, it reviews old ones, and it points out where further
studies are necessary.

THE MUSCULAR NATURE OF THE NOTOCHORDAL LAMELLAE

When examined in polarized light the individual lamella of the cross-sectioned notochord is seen to contain anisotropic and iso-tropic bands running in the dorso-ventral direction, i.e. perpen-dicular to the microscopically visible notochordal filaments (Ebner, 1895; Joseph, 1902; Tenbaum, 1955). The anisotropic and iso-tropic bands resemble those of ordinary cross-striated muscles, except for a more irregular course and a much larger repeat period (Fig. 1). The sarcomere length is about 33 μm against 1–3 μm of most striated muscles. However, cross-striations up to 15 and 25 μm in period are quite common in some invertebrate muscles (Emery, 1887; Jasper & Pezard, 1934; Hoyle, 1967). Haswell (1890) reported the existence of a 33 μm sarcomere length in the proventricular muscle of certain polychaetes. The muscular nature of these large cross-striation periods has been supported by polarized light studies of Schmidt (1936) and by ultrastructural studies of Wissocq (1970). The anisotropic and isotropic bands of the lamellae in the notochord of amphioxus are accordingly not incompatible with muscle sarcomere striations.

The notochordal lamellae have been studied in the transmis-sion electron microscope by Eakin & Westfall (1962), Flood (1967a, 1968a, b), Welsch (1968) and Flood, Guthrie & Banks (1969). In ultrathin sections of material fixed in osmium tetroxide or in glutaraldehyde and osmium tetroxide prior to dehydration, plastic embedding and sectioning, it appeared that the lamellae contain two kinds of filament. One varies in diameter from 100 to 1500 Å and reveals a periodic cross-striation of 140 Å to 145 Å. The other is about 50 Å in diameter and without any cross-striation (Figs 2 and 3). When compared with the polarized light image, the thick filaments are found only in the anisotropic band, and in a highly ordered rectangular pattern corresponding to the middle line of highest birefringence in this band. In this region the thick fila-ments also reach their maximal diameter. They taper gradually off in both directions against the isotropic band. The thin filaments are found between the thick ones in the lateral part of the anisotropic band and in the isotropic band. In the middle part of this band, irregular patches of a moderately electron-dense material are present. This is comparable to the anisotropic material seen here by Tenbaum (1955: Fig. 19). Against the connective tissue sheath, where the plasma membrane of the notochordal lamellae is highly folded, only thin filaments are present. This is quite analogous to

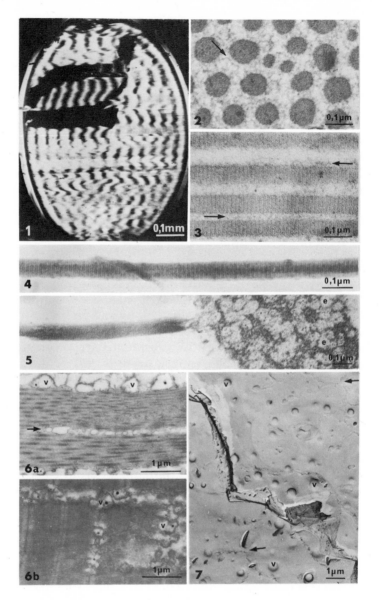

FIG. 1. Polarized light micrograph of the cross-sectioned adult notochord. In most regions several lamellae are superposed on each other, giving a blurred impression of the cross-striation (× 79).

FIG. 2. Transmission electron micrograph of cross-sectioned notochordal filaments. The thick filaments are surrounded by thin ones (*arrow*) (× 76 500).

FIG. 3. Transmission electron micrograph of longitudinally sectioned notochordal filaments. The thick filaments show a regular 140 Å cross-striation. Thin filaments are indicated by *arrows* (× 76 500).

the situation in ordinary myotendinous junctions (Schwarzacher, 1960).

From the available data it was supposed (Flood, 1967a) that the thin filaments were homologous to actin filaments of striated muscle and that the thick filaments, with the 145 Å periodicity, were paramyosin filaments like those found in invertebrate catch muscle (Hanson, Lowy, Huxley *et al.*, 1957; Hanson & Lowy, 1960). In an attempt to find chemical proof of this hypothesis, I have subjected homogenized and ethanol-dried notochordal tissue of amphioxus to extraction in 1 M potassium chloride (in 0·03 M phosphate buffer of pH 6.5). The extract was centrifuged and the supernatant dialysed against 0·15 M potassium chloride (in the same buffer). Highly birefringent, needle-formed crystals precipitated from the solution in the same way as paramyosin precipitates from the solution when homogenates of molluscan catch muscle are subjected to the same treatment (Bayley, 1957). In the electron miscroscope the crystals of notochordal protein showed a 145 Å periodicity (Fig. 4), similar to that found in paramyosin crystals (Hanson, Lowy *et al.*, 1957, and Tsao, Kung, Peng & Tsou, 1965). Extraction at low ionic strength (0·1 M instead of 1 M potassium chloride) gave no crystals. Intact notochordal tissue was subjected to a modified treatment and thereafter prepared for electron microscopy by Flood (1968a). In this material the tissues surrounding the lamellae seemed to act like a dialysing bag. Empty

---

FIG. 4. Transmission electron micrograph of needle-form crystals precipitated from 1M KCl extracts of homogenized notochordal tissue upon dialysation against 0·1 M KCl. The crystals show the same cross-striation as native notochordal filaments (× 82 000).

FIG. 5. Transmission electron micrograph of cross-sectioned notochordal lamellae. The tissue was treated with 1 M KCl and 0·1 M KCl prior to fixation and plastic embedding. Empty spaces (*e*) are present where thick filaments normally occur. Needle-form crystals with 140 Å cross-striation are formed in the extracellular spaces (× 53 000).

FIG. 6a. Transmission electron micrograph of longitudinally sectioned notochordal lamellae with medial row of vesicles (*arrow*) and granulated subsarcolemmal vesicles (*v*) (× 11 000).

FIG. 6b. Transmission electron micrograph of longitudinally sectioned trunk muscle fibres with subsarcolemmal vesicles (*v*) similar to those seen in the notochordal lamellae (× 11 000).

FIG. 7. Transmission electron micrograph of gold-palladium shadowed replica of freeze-fractured and etched notochordal lamellae. Upper right corner shows the internal surface and lower left corner the external surface. Bulges (*v*), corresponding to the subsarcolemmal vesicles, are present at irregular intervals on both sides. The granulated lines (*arrows*) probably correspond to fractured plasma membrane folds (× 5600).

spaces were found where the thick filaments should be, and needle-formed crystals with the typical 145 Å periodicity were found in the extra-cellular space between the lamellae (Fig. 5). As these solubility characteristics (Bayley, 1957) and the cross-striation (Johnson, 1963) are quite special for paramyosin (or tropomyosin A), it seems justified to believe that the thick notochordal filaments are made up of a paramyosin-like protein and, accordingly, that the lamellae are real muscles. The latter conclusion has received experimental support from cinematographic and electron-physiological recordings reported by Flood *et al.* (1969) and Guthrie & Banks (1970).

As the lamellae have thus been proved to be muscles, one should look at other ultrastructural details than the filament array. Above all, special organelles for excitation-contraction coupling and relaxation should be present. In most muscles these functions are performed by a sarcoplasmic reticulum. This consists of membrane-bound vesicles and tubules coupled to the plasma membrane of the muscle cell either at the surface (peripheral couplings) or at a transverse tubule (internal couplings). The vesicles are capable of sequestrating Ca^{2+}-ions and are thus able to regulate the concentration of calcium in the sarcoplasm. The calcium level will govern the degree of actin-myosin interaction and accordingly the degree of contraction or relaxation. (For a general introduction to this subject the reader is referred to the reviews of Sommer & Johnson (1969) and Huxley (1971).

In ultrathin sections of notochord fixed in glutaraldehyde and osmium tetroxide, numerous vesicles are found on both sides of most lamellae. These vesicles contain a granular matrix and are intimately bound to the plasma membrane (Fig. 6a). In both sectioned and freeze-etched material they seem to lack any particular relation to special zones of cross-striation pattern (Fig. 7).

FIG. 8. Transmission electron micrograph of sagittal section through the vacuolated notochord of a 1 mm long larva. Parts of the dorsal nerve cord (*DNC*) and the intestinal epithelium (*IE*) are also seen. N: notochordal cell nucleus (× 4100).

FIG. 9. Transmission electron micrograph of frontally sectioned adult notochord. A cell nucleus (N) makes a cytoplasmic connexion between two lamellae. Note the presence of both extracellular (*ev*) and intracellular vacuoles (*iv*) between the lamellae (× 6300).

FIG. 10. Transmission electron micrograph of cross-sectioned notochordal lamellae of the metamorphosing 6 mm long larva. Intracellular vacuoles (*iv*) are still present in most of the lamellae. Note the fine fibrillar material (*f*) between the myofilament-containing layers of the lamellae (× 6600).

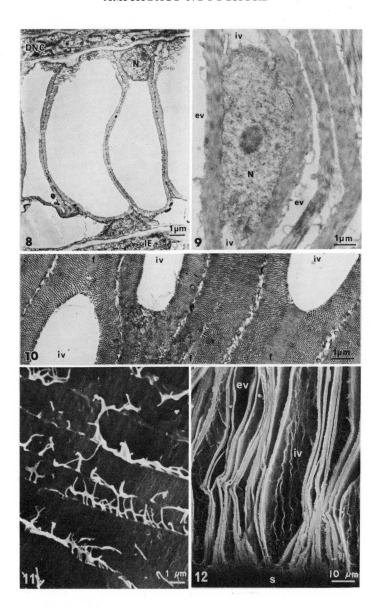

FIG. 11. Scanning electron micrograph of the surface of a lamella. Anastomosing microvilli in parallel rows project towards the extracellular space (× 4000).

FIG. 12. Scanning electron micrograph of longitudinally sectioned notochord and sheath (S). Intracellular vacuoles (iv) contain elaborate membrane arrays whereas extracellular vacuoles (ev) appear empty (× 660).

The vesicles thus show great resemblance to the subsarcolemmal sarcoplasmic vesicles of many invertebrate muscles
(Heumann, 1969; Rosenbluth, 1969) and to the terminal cisternae
coupled to the transverse tubuli in chordate striated muscles (Page,
1968; Kelly, 1969). In chordate striated muscles of quite limited
diameter, however, transverse tubules may be absent. This is the
situation, for example, in the 1 μm thick trunk muscle fibres of
amphioxus. Only sub-sarcolemmal vesicles are present in this
muscle, but in contrast to the situation in the notochordal lamellae,
they are lined up in rows at the Z-line level (Flood, 1969a, in
preparation; Hagiwara, Henkart & Kidokoro, 1971).

When encountered in the same ultrathin section, the subsarcolemmal vesicles of the notochordal lamellae look exactly like the
subsarcolemmal vesicles of the trunk muscle of amphioxus (Fig. 6a,
b). Various osmolarities of the glutaraldehyde fixative give the
same swelling or shrinkage of the two and the same granular
appearance of the vacuolar content. Phosphate-buffered osmium
tetroxide as sole fixative results in disintegrated membranes in
both kind of vesicle. A similar effect of calcium-free fixatives on the
mitochondrial membrane of the muscle fibres has been reported
earlier (Flood, 1967b).

As far as the subsarcolemmal vesicles of the notochord have
been examined, it may accordingly be said that they behave like
typical calcium-sequestrating sarcoplasmic vesicles.

In a few instances narrow tubular structures oriented in the
dorso-ventral direction next to the plasma membrane of the
lamellae have been observed (Fig. 19).

They do not seem to be related to the subsarcolemmal vesicles
and their significance is unknown.

THE CELLULAR UNITS OF NOTOCHORDAL LAMELLAE

As will be evident from a later section, three different modes of
motor innervation of the notochordal lamellae have recently been
proposed. For the evaluation of these hypotheses it seems necessary to know whether each lamella is a separate cell or only part of a
larger cellular unit. I have tried to solve this problem by a
transmission electron miscroscope study of the larval and the adult
notochord.

Larvae up to 1 mm length or about one or two weeks of age
were reared from spontaneously fertilized eggs obtained from
adults kept in aquaria at Laboratoire Arago, Banyuls sur mer,

France, during June 1971 and 1973. Larvae of about 5 mm length were isolated from surface plankton taken at night near Biologische Anstalt, Helgoland, in September 1973. Adults were obtained from the same laboratories. (Methodological details will be given in an extensive report now in preparation).

As described by Hatschek (1882, 1893), the notochord separates from the endoderm at about 12 h after the first cleavage of the egg. At first the cells are filled with yolk granules, but within the next 12 h they arrange themselves in a single row of flattened cells that cover the entire cross-section of the cord. During the rapid elongation of the larvae, from about 24 h after the first cleavage, a single central vacuole appears and enlarges within each cell. At a larval age of about 48 h (the larva is about 0·8 mm long and shows sporadic muscular movements) filaments of paramyosin- and actin-like types appear in the narrow cytoplasmic layer around the central vacuole. This agrees with the observation of a birefringent peripheral layer and a central vacuole in the notochordal cells in larvae of nine days of age (Tenbaum, 1955). The nucleus is usually located inside the filament layer and bulges into the vacuole. The vacuole in the early larva seems to lack a bordering membrane, but this is usually present after the appearance of the filaments (Fig. 8).

At the metamorphosis, however, most of the central vacuoles disappear from the notochordal lamellae. In many cells no trace of a central vacuole persists in the adult. Whether this is a true removal process, or a result of cell divisions, is not known. In other cells a medial layer of vesicles or sarcoplasm persists as a reminiscence of the vacuole. In a few cells the central vacuole persists as a space bounded by a unit membrane. These vacuoles vary enormously in size. Some may cover most of the cross-sectional area of the notochord, whereas others cover only a fraction of it. The small vacuoles are most abundant near the lateral insertion in the notochordal sheath (Figs 10 & 12). In other regions, such lamellae will appear like the ones with just a central layer of vesicles and sarcoplasm. A cell nucleus often makes a narrow cytoplasmic bridge between the two lamellae bordering the large vacuoles (Fig. 9) (Cf. the notochordal corpuscles described with the light microscope a.o. by Tenbaum, 1955). Cytoplasmic continuity between the bordering lamellae is also present at their insertion in the notochordal sheath.

Two additional details support the intra-cellular nature of these vacuoles. Sub-sarcolemmal vesicles, of the typical granulated type described earlier, have not been found near these vacuoles, whereas they are quite abundant at the extracellular side of the lamellae. The fine flocculent material found in the extra-cellular space of the dorsal canal and in some of the extra-cellular spaces between the lamellae has never been found in the vacuoles proposed to be intra-cellular. Finger-like and leaf-like membrane projections from the lamellar membrane are, however, present in the intra-cellular vacuoles as well as in the extra-cellular spaces between the lamellae. These are difficult to interpret in sectioned material, but stand up as highly organized and complicated membrane patterns when the notochordal tissue is examined in the scanning electron microscope (Figs 11 & 12). In the dorsal and ventral direction the limit of the lamellae is more difficult to establish. The presence of numerous delicate processes from the Müller cells (as described in the next section) gives a complicated picture, especially along the dorsal border of the lamellae, but so far no convincing continuity between the Müller processes and the lamellae has been shown. It may accordingly be tentatively concluded that one and two notochordal lamellae constitute cellular units distinct from other cells present in the notochord.

THE MÜLLER CELLS

Cells distinct from the notochordal lamellae are present in the notochord and they are generally called Müller cells in honour of W. Müller (1871). Such cells are preferentially located near the dorsal and ventral edge of the notochord, but some are also found between the lamellae at their lateral insertion in the notochordal sheath.

The dorsal Müller cells are the most prominent ones. Each of these cells is made up of three distinct parts. Against the dorsal notochordal sheath is a very elongated filament-containing cylinder. The axis of this cylinder is parallel to the longitudinal axis of the notochord and 5 to 15 such cylinders from separate cells lie close together and cover the entire roof of the dorsal notochordal canal against the notochordal sheath (Fig. 13). The plasma membranes of the adjoining cylinders are closely apposed to one another, and specialized cell contacts resembling the intermediate junctions described by Farquhar & Palade (1963) may be present at short intervals (Fig. 14) (see also Kelly & Luft, 1966). The filaments seen in the cytoplasm of the cylinders resemble smooth muscle filaments or tonofilaments, and seem to be more abundant against the plasma membrane in the junctional areas (Fig. 14). The two

parallel cell membranes follow a highly folded course between
adjacent cylinders and against the dorsal canal of the notochord.
No partial membrane fusion (tight junctions or maculae occludentes (see Farquhar & Palade, 1963)) was ever seen between
apposed membranes in this region of the notochord. Against the
connective tissue sheath, the cell membrane of the cylinders is quite
smooth.

The nucleus of the dorsal Müller cell is found in a cytoplasmic
pocket ventral to the cylindrical part. Two centrioles, a pronounced Golgi apparatus and numerous vesicles of variable diameter are characteristic features of this part of the cell (Figs 13, 15).
From the nucleated cell body and elsewhere along the cylindrical
dorsal part, numerous slender cytoplasmic processes (the third
distinct part of the dorsal Müller cells) project towards the upper
edge of the notochordal lamellae. They divide now and then and
constitute a delicate meshwork throughout the entire width of the
dorsal canal (Figs 13, 16). Most of the processes radiate towards the
dorsal border of the muscular lamellae and seem to establish
intimate contact with the lamellar membrane (Figs 16 & 17).
Usually two or three processes are seen clinging to each lamella or
preferably to two adjoining lamellae, sealing off the potential space
between them.

Compared to the number of processes seen in the dorsal canal,
the number of processes bordering the lamellae is very high,
suggesting that they run for a considerable distance along the top
edge of the lamellae, rather than just ending in contact with the
lamellae. So far, however, no specialized cell junctions or cytoplasmic continuities between the dorsal Müller cell processes and the
notochordal lamellae have been seen. Neither do the illustrations
in the work of Welsch (1968, Figs 7a, b) convince me that such
continuities exist. The dorsal Müller cells seem to be an early
differentiation of the notochord. They seem to be present in the
larva at about one week of age (1 mm length).

In spite of the highly specialized appearance of the dorsal
Müller cells, it is difficult to propose any specific functional role for
them from the observations reported above. The pronounced
filaments in the longitudinal dorsal cylinders suggest a mechanical
function, providing passive or active opposition to stretch parallel
to the longitudinal axis of the notochord. The complicated interdigitation of the plasma membranes and their specialized junctions
also suggests that this cell layer may be a barrier for diffusion of
various substances. The nucleated cell body, with Golgi apparatus

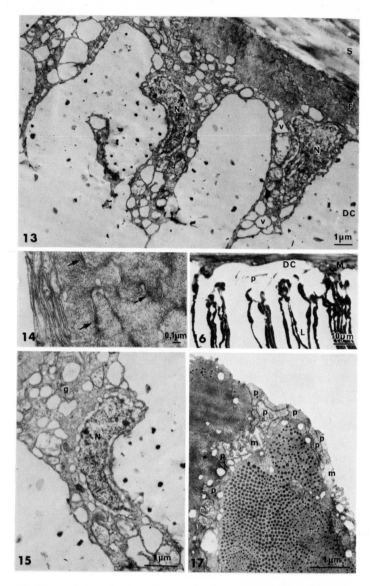

FIG. 13. Transmission electron micrograph of cross-sectioned adult notochord. The sheath (S), the fibrous (F) and vacuolated (v) part of dorsal Müller cells and the extracellular dorsal canal (DC) are seen. N: nuclei of dorsal Müller cells (× 5200).

FIG. 14. Same section as Fig. 13. Detail of the plasma membrane interdigitation between the fibrous part of two dorsal Müller cells. Possible intermediate junctions indicated by *arrows* (× 24 000).

FIG. 15. Same section as Fig. 13. Detail of the vacuolated part of a dorsal Müller cell with the peripheral part of a Golgi apparatus (g) and the nucleus (N). Note the granulated substance in the extracellular space (× 9000).

and numerous vesicles of variable diameter and a vesicular matrix similar to the extra-cellular fluid in the dorsal canal, strongly suggests a secretory function. And finally, the numerous slender processes may indicate some relation to the maintenance of the extracellular fluid of the notochord both in the dorsal canal and between the lamellae. The significance of the extensive contacts between the processes of the dorsal Müller cells and the dorsal edge of the lamellae is at present unknown. They seem to be more extensive than they would need to be if they were just mechanical anchoring sites for the Müller processes. Further examination of these junctions and their development in ontogenetic material may permit further conclusions. At present it can be added that, except for the existence of simple contacts, no ultrastructural observation gives support to the hypothesis that the dorsal Müller cells are a relay station for excitation impulses from the nervous tissue to the muscular lamellae, as proposed by Welsch (1968) and Webb (1973).

The ventral Müller cells differ in appearance from the dorsal ones. Although filaments oriented parallel to the longitudinal axis of the notochord are present in the cytoplasm next to the connective tissue sheath (Fig. 19), no specialized filamentous portions are present as they are in the dorsal Müller cells. Most of the cytoplasm and the nucleus are wedged between the lamellae. Vesicles, Golgi-cisternae and large intracellular vacuoles are frequently encountered, and indicate that these cells may be related to the production of the extracellular fluid of the large spaces in this part of the notochord. These spaces may anastomose and make up what has been called a ventral canal by various authors (a.o. Ebner, 1895; Guthrie & Banks, 1970), or they may occur as separate vacuoles.

The lateral Müller cells are preferentially found in the dorsolateral part of the notochord. Their cytoplasm is highly vesicular and they are wholly wedged between the lamellae without any specialized contacts with these. Some of them have processes that border the notochordal tissue in the notochordal horns (Fig. 18). The degree to which these cells extend in between the lamellae varies,

FIG. 16. Light micrograph of sagittally sectioned adult notochord. The dorsal Müller cells (M) have processes (p) through the dorsal canal (DC) towards the notochordal lamellae (L) (× 625).

FIG. 17. Transmission electron micrograph of cross-section through the dorsal border of the adult notochordal lamellae. Several processes (p) of the dorsal Müller cells are wedged between the lamellae. Tiny microvilli (m) project from the surface of the lamellae (× 11 000).

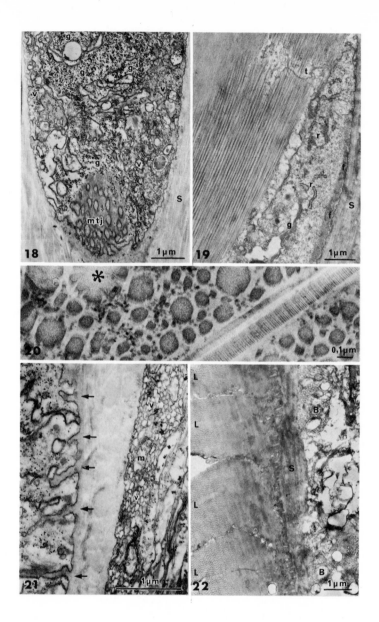

FIG. 18. Transmission electron micrograph of cross-sectioned adult notochordal process as it penetrates the sheath (S). Cytoplasmic profiles of the granular type (g) derived from the lamellae, and vesicular profiles (v) probably belonging to the dorso-lateral Müller cells, are present. mtj: myotendinous junction of notochordal lamellae (× 9500).

FIG. 19. Transmission electron micrograph of cross-sectioned notochord from the 5 mm larva. A ventral Müller cell with granulated endoplasmic reticulum (r), a Golgi apparatus (g) and cross-sectioned filaments (f) against the sheath (S) are seen. An arborizing tubule (t) is present in the notochordal lamellae (× 8700).

and it seems reasonable to suppose that they are recruitment cells for the differentiation of new lamellae. A similar situation seems to exist in the larvae of about 1 mm length, where undifferentiated cells are present at the periphery of the notochord between the lamellae (Fig. 8). Except for their processes in the notochordal horns, no ultrastructural observation speaks in favour of these cells as impulse conducting elements between the nervous tissue and the notochordal lamellae.

As should be evident from the above description, the Müller cells are a heterogeneous gr... p of cells that we ought to know more about before definite conclusions can be drawn.

THE INNERVATION OF THE NOTOCHORD

Two paramedian rows of pits or holes in the connective tissue sheath of the notochord against the spinal cord were described by W. Müller (1871). In adult animals each pit measures about 7 μm in the antero–posterior direction and 25 μm in the transverse direction. Within each row they are interspaced by about 50 μm intact notochordal sheath tissue. Electron microscopists agree that the pits are occupied by cytoplasmic structures derived from the notochord (the notochordal processes) and that a pre-synaptic nerve ending is present at the surface of the spinal cord where this is opposed to the pits (Flood, 1967a, 1968b, 1970; Flood et al., 1969; Welsch, 1968). Because of the ultra-structural appearance of the junction between the notochordal processes and the nerve terminals of the spinal cord (Fig. 21), and the presence of specific acetylcholinesterase at this junction (Flood, 1967a, 1970, in press), it seems uncontroversial to state that the notochordal muscle-lamellae receive an innervation through these pits. Based on the presumed nature of the tissue in the notochordal processes, this innervation has, however, been

---

FIG. 20. Transmission electron micrograph of the adult notochordal sheath. In longitudinal section the fibres show a periodical striation of 640 Å characteristic of collagen. In transverse section some fibres have an unusual irregular outline (*asterisk*) (× 46 000).

FIG. 21. Transmission electron micrograph of section through the junction between a notochordal process and a presynaptic nerve ending with synaptic vesicles (*v*) and mitochondria (*m*) in the adult animal. Note the folded postsynaptic membrane (*arrows*) (× 20 500).

FIG. 22. Transmission electron micrograph of two junctions between the notochordal lamellae (*L*) and presynaptic boutons (*B*) at the surface of the nerve cord of a 5 mm larva. Note the slight difference in thickness of the connective tissue sheath (*S*) in junctional and non-junctional regions (× 7900).

supposed to be indirect via the Müller cells (Welsch, 1968) or direct through cytoplasmic extensions of the lamellae (Flood, 1967a, 1968b, 1970, Flood *et al.*, 1969). Based on the preceding description of the various types of the Müller cell, my earlier published observations on adult material (Flood, 1970), and new observations on larval material (Fig. 22), it seems beyond doubt that at least some of the lamellae have cytoplasmic processes directly opposed to the nerve terminals. Likewise it seems unquestionable that the nerve terminals are usually opposed by cytoplasmic profiles of the same type as those shown to be continuous with the lamellae (Fig. 21 compared to Fig. 18). Processes probably derived from the most lateral dorsal Müller cells, or from the dorsolateral Müller cells, are, however, also present at most junctions. Their significance in relation to the activation of the muscular lamellae seems dubious, as a more direct way of activation is available. As previously described, there are, however, elaborate contacts between the dorsal Müller cell processes and the top of the muscular lamellae. These contacts may be of more than mechanical significance and should be investigated further. The nerve terminals opposed to the notochordal processes seem to originate as short collaterals of axons running longitudinally in the ventrolateral part of the spinal cord (Flood, 1970). It has not been possible to relate these nerve fibres to any of the fibre tracts described by Bone (1960). In the electron microscope the nerve terminals are seen to contain numerous synaptic vesicles between 800 and 1200 Å in diameter. They are circular or slightly oval in outline and in adult material usually have electron-lucent cores (Fig. 21), whereas in larval material ill-defined dense cores are usually present. As vesicles with variable electron density were described by Welsch (1968) in adult material the disappearance of the dense cores in my material may be a fixation artifact. Dense-core vesicles are usually found in monoaminergic neurones and are not supposed to be related to cholinergic transmission (Pellegrino de Iraldi, Gueudet & Suburo, 1971). However, as previously stated, specific acetylcholinesterase is abundant in these junctions. Further studies are accordingly needed to clarify what transmitter substance is most likely to operate in this synapse. Further studies are also needed to clarify the central connections of the nerve terminals and the possibility of axo-axonic synapses as mentioned by Flood (1970).

In a recent paper, Webb (1973: 332) states that in young animals (10–30 mm length) the pits are sparse in the middle part of the trunk and that in these regions nerve fibres pass directly from neurons in the nerve cord through the notochordal sheath to the cells of the Müller tissue, which in turn connect with muscle plates. As these observations were made with the light microscope, one should await their ultrastructural verification before definite conclusions are made. In my larval material the specific junctions between notochordal lamellae and nerve boutons at the surface of the spinal cord have been seen repeatedly without any pronounced pit formation (Fig. 22). This seems to depend on the limited thickness of the sheath in young animals. Nerve fibres traversing the connective tissue sheath between the spinal cord and the notochord have never been observed by me. In any case, detailed knowledge of the growth and multiplication of the notochordal lamellae and other cellular elements of the notochord must be obtained before a detailed understanding of the innervation of the notochord can be established.

THE FLUID-FILLED SPACES OF THE NOTOCHORD

As has been described in connection with the cellular units of notochordal lamellae, the larval notochord consists of a single row of tightly packed and highly vacuolated cells (Fig. 8), whereas in the adult most of the intracellular vacuoles have disappeared and extracellular fluid-filled spaces appear instead. These extracellular spaces constitute, as mentioned in the description of the Müller cells, a continuous dorsal canal, a less developed ventral canal and separate vacuoles between some of the lamellae. No physicochemical data exist on the consistency of the fluids. However, they may be supposed not to move freely within the notochord. This assumption is based on the following observations. All the intracellular vacuoles of the adult lamellae are closed units that are sub-divided into tiny compartments by elaborate membrane systems (Fig. 12). The extracellular vacuoles also contain elaborate membrane folds that originate either from the lamellar membrane (Fig. 11) or from the Müller cells (Figs 13, 16). The extracellular fluid itself may also be gelified, as it seems to contain large amounts of polysaccharides. (It stains bright red with the PAS-reaction, metachromatically with toluidine blue, and appears granular in the electron microscope (Figs 13, 15). These are properties that it shares with, for example, the fluid of the dorsal fin pockets of amphioxus. As will be evident from the functional considerations in a latter section, it seems to be important to know exactly to what degree the fluid can be squeezed forward or backward through the notochord. In addition to the above mentioned factors limiting such movements, Guthrie & Banks (1970) described complete lamellae which form septa across the entire notochord opposite each myocomma.

THE NOTOCHORDAL SHEATH

The notochordal tissue and fluids are surrounded by a very thick connective tissue sheath. This consists of collagenous fibres of unusually large diameter and irregular outline, embedded in a slightly granular matrix (Fig. 20). Towards the notochordal tissue a distinct basement lamina-like layer is found (the elastica interna of Ebner, 1895). Beyond this, no stratification of the sheath is present. The overall thickness of the sheath and the direction of the collagenous fibres in different regions of the sheath seem to have an important relation to the mechanical properties of the notochord as a whole (Guthrie & Banks, 1970).

In brief outline the thickness of the sheath increases gradually up to about 20 μm in adult animals and most of the collagenous fibres follow a spiral course around the notochord (Ebner, 1895; Joseph, 1895). Only at the dorsal and ventral border, where the large fasciae of the spinal cord and the pharyngeal-abdominal cavity join the notochordal sheath, do longitudinally-oriented collagenous fibres seem to occur in significant amounts. The circular and spiral arrangement of collagen allows the notochord to bend without much resistance from the sheath. However, the collagenous fibres will prevent the circumference of the notochord from increasing and, as the sheath is filled with incompressible tissues and fluids, this prevents the notochord from shortening as a result of the trunk muscles activity. Even in the unstimulated state, there seems to be a significant positive hydrostatic pressure inside the notochordal sheath (Guthrie & Banks, 1970). This will maintain a stretched notochordal sheath and give the notochord a certain degree of stiffness. This positive hydrostatic pressure is supposed to be maintained across the cells covering the internal surface of the notochordal sheath i.e. the various types of Müller cells and the notochordal lamellae. The connective tissue sheath itself is not supposed to act as a barrier for slow diffusion of fluids.

FUNCTIONAL ASPECTS

From the previous sections it is evident that the notochord of amphioxus may be briefly described as an organ consisting of transversely arranged muscles and fluid-filled spaces enclosed in a cylindrical tube of dense connective tissue. The orientation of the muscles indicates that the transverse diameter of the notochord will be reduced upon contraction. However, as the fluids inside the sheath cannot escape through it and possibly cannot be squeezed to other parts of the notochord because of its pronounced compartmentalization, a contraction will result in an increased fluid pressure inside the notochord rather than in a flattened notochord. According to the electrophysiological recordings reported by Flood *et al.* (1969) and Guthrie & Banks (1970), both electrical and chemical stimulation of the notochordal muscles result in a stiffer notochord. How and when the animal uses this possibility to alter the mechanical properties of its skeleton, is not known. It may be related to forward and reverse swimming or burrowing movement (Webb, 1973), but direct evidence favouring any particular view is at present scarce. The early onset and the high degree of specialization of the notochord makes me believe that its muscular nature is very important for the success of amphioxus in its everyday life.

To advance any further our understanding of notochordal function in amphioxus it seems necessary to work on several lines. One is to examine under what conditions the animal activates its notochord. Another is to investigate whether the entire notochord is activated simultaneously, or if shorter segments may be activated one after the other. The latter approach seems quite important, as it may tell whether a localized muscular contraction in the notochord will reduce or increase the stiffness of the notochord. If fluid can escape from a region of the notochord where the muscles are activated, this will allow the notochord to reduce its transverse diameter to a degree that might reduce its resistance against bending forces. On the other hand, if the fluid cannot escape, the hydrostatic pressure inside the contracting segment of the notochord will increase, and prevent a pronounced reduction of the transverse diameter. This will result in a stiffer notochord.

PHYLOGENETIC ASPECTS

In the cyclostomes *Petromyzon* (Schwarz, 1961) and *Myxine* (Flood, 1969b, 1973), and in higher chordates (Jurand, 1962; Leeson & Leeson, 1958; Waddington & Perry, 1962; Welsch & Storch, 1971), the classical vacuolated notochordal tissue is found. This consists of tightly packed cells, each with a large central vacuole. Adjacent cells are interconnected by numerous desmo-somes, and abundant tonofilaments are found in a narrow cyto-plasmic zone between the plasma membrane and the central vacuole. Next to the collagenous and elastic sheath surrounding the notochord in these animals, one or two layers of unvacuolated cells are found. Near the central axis of the notochord of at least the cyclostomes, the vacuolated tissue is condensed into the so-called notochordal string.

In *Myxine* highly organized membrane arrays, somewhat simi-lar to those found in intracellular vacuoles of the notochordal lamellae in amphioxus, are present within most of the central vacuoles. In the Tunicata the notochord consists of a single row of either globular cells with multiple large vacuoles (Thaliacea, Flood, unpublished observation) or of tubular or sheet-like cells sur-rounding an extracellular vacuole along the entire length of the notochord (Larvacea and Ascidiacea. Cloney, 1964; Olsson, 1965; Welsch & Storch, 1969). In hemichordates the stomochord consists of a vacuolated epithelium (Welsch & Storch, 1970). From these brief descriptions it may be concluded that the notochord of amphioxus has no equivalent in any other chordate. Except for a convincing homology to other notochords as judged from com-mon ontogenetic principles, it is hard to find any structural similarity between the notochord of amphioxus and that of other chordates. The membrane arrays of the intracellular vacuoles of the lamellae in amphioxus and those of the vacuolated cells of *Myxine* may be the only exception. However, contractile properties have recently been claimed for a variety of intracytoplasmic filaments previously considered as passive supporting elements (Ishikawa, Bischoff & Holtzer, 1969; Thuneberg & Rostgaard, 1969; Cloney, 1972; Rostgaard & Thuneberg, 1972; Rostgaard, Kristensen & Nielsen, 1972). Such a view might reduce the apparent difference between the (tono-)filament-containing vac-uolated cells of higher chordates and the (myo-)filament-containing lamellae of amphioxus. Cytoplasmic filaments seen in notochordal cells of tunicates (Cloney, 1969) may similarly be proposed as being contractile.

Acknowledgements

I wish to express my gratitude to the Norwegian Research Council for Science and the Humanities for financial support (Grant No. C.21.30–8), to Mrs Sigrunn Kjaerstad and Mrs Ingebjørg Steinsbø for the preparation of ultrathin sections, to Mrs Jorunn Spurkeland for typewriting the manuscript, to Mr H. Knutsen and Mr J. Røli for keeping the electron microscopes in excellent working condition, to Mr R. Haakonsen and Mr A. Jensen for photographic assistance, and to the directors and staffs at Biologische Anstalt, Helgoland, Laboratoire Arago, Banyuls s. mer, and The Marine Laboratory, Plymouth, for research facilities and supply of material.

REFERENCES

Bayley, K. (1957). The proteins of adductor muscles. *Pubbl. Staz. zool. Napoli* **29**: 96–108.
Bone, Q. (1960). The central nervous system in amphioxus. *J. comp. Neurol.* **115**: 27–64.
Cloney, R. A. (1964). Development of the ascidian notochord. *Acta Embryol. Morph. exp.* **7**: 111–130.
Cloney, R. A. (1969). Cytoplasmic filaments in morphogenesis. The role of the notochord in ascidian metamorphosis. *Z. Zellforsch. mikrosk. Anat.* **100**: 31–53.
Cloney, R. A. (1972). Cytoplasmic filaments and morphogenesis: Effects of Cytochalasin B on contractile epidermal cells. *Z. Zellforsch. mikrosk. Anat.* **132**: 167–192.
Eakin, R. M. & Westfall, A. (1962). Fine structure of the notochord of amphioxus. *J. Cell Biol.* **12**: 646–651.
Ebner, V. von (1895). Über den Bau der Chorda dorsalis des Amphioxus (*Branchiostoma lanceolatus*). *Anz. k. Akad. Wiss. Wien* (Mat. Naturw. Kl.) **32**: 213–214.
Emery, C. (1887). Intorno a la muscolaturo liscia e striata della *Nephthys scolopendroides*. D.Ch. *Mitt. zool. Stat. Neapel* **7**: 371–380.
Farquhar, M. G. & Palade, G. E. (1963). Junctional complexes in various epithelia. *J. Cell Biol.* **17**: 375–412.
Flood, P. R. (1967a). The notochord of amphioxus. A paramyosin catch muscle? (Abstract). *J. Ultrastruct. Res.* **18**: 236
Flood, P. R. (1967b). The effect of different fixatives on mitochondrial ultrastructure in amphioxus muscle. (Abstract). *J. Ultrastruct. Res.* **18**: 228.
Flood, P. R. (1968a). The extraction of a paramyosin-like protein from the notochord of amphioxus. In *Electron microscopy 1968* **2**: 291–292. Bocciarelli, D. S. (ed.) Rome: Tipografia Poliglotta Vaticana.
Flood, P. R. (1968b). Further observations on the notochord of amphioxus. (Abstract). *J. Ultrastruct. Res.* **25**: 161.
Flood, P. R. (1969a). The sarcoplasmic reticulum of amphioxus trunk muscle. (Abstract). *J. Ultrastruct. Res.* **29**: 569.
Flood, P. R. (1969b). Fine structure of the notochord in *Myxine glutinosa*. (Abstract). *J. Ultrastruct. Res.* **29**: 573–574.
Flood, P. R. (1970). The connection between spinal cord and notochord in amphioxus (*Branchiostoma lanceolatum*). *Z. Zellforsch. mikrosk. Anat.* **103**: 115–128.

Flood, P. R. (1973). The notochord of *Myxine glutinosa* L. related to that of other chordates. *Acta R. Soc. scient. litt. gothoburg.* (Zool.) **8**: 14–16.

Flood, P. R. (In press). Histochemistry of cholinesterase in amphioxus (*Branchiostoma lanceolatum*, Pallas). *J. comp. Neurol.*

Flood, P. R. (In preparation). *The sarcoplasmic reticulum of the trunk muscle fibres in amphioxus* (Branchiostoma lanceolatum, *Pallas*).

Flood, P. R., Guthrie, D. M. & Banks, J. R. (1969). Paramyosin muscle in the notochord of amphioxus. *Nature, Lond.* **222**: 87–88.

Guthrie, D. M. & Banks, J. R. (1970). Observations on the function and physiological properties of a fast paramyosin muscle—the notochord of amphioxus (*Branchiostoma lanceolatum*). *J. exp. Biol.* **52**: 125–138.

Hagiwara, S., Henkart, M. P. & Kidokoro, Y. (1971). Excitation—contraction coupling in Amphioxus muscle cells. *J. Physiol.* **219**: 233–251.

Hanson, J. & Lowy, J. (1960). Structure and function of the contractile apparatus in the muscles of invertebrate animals. In *The Structure and function of muscle.* **1**: 265–335. Bourne, G. H. (ed.) London: Academic Press.

Hanson, J., Lowy, J., Huxley, H. E., Bayley, K., Kay, C. M. & Ruegg, J. C. (1957). Structure of molluscan tropomyosin. *Nature, Lond.* **180**: 1134–1135.

Haswell, W. (1890). A comparative study of striated muscle. *Q. Jl. microsc. Sci.* **30**: 31–50.

Hatschek, B. (1882). Studien über Entwicklung des Amphioxus. *Arb. zool. Inst. Univ. Wien* **4**: 1–88.

Hatschek, B. (1893). *The amphioxus and its development.* (transl. and ed. by Tuckey, J.). London: Swann Sonnenschein.

Heumann, H. G. (1969). Calciumakkumulierende Strukturen in einem glatten Wirbellosenmuskel. *Protoplasma* **67**: 111–115.

Hoyle, G. (1967). Diversity of striated muscle. *Am. Zool.* **7**: 435–450.

Huxley, A. F. (1971). The activation of striated muscle and its mechanical response. *Proc. R. Soc.* **178**: 1–27.

Ishikawa, H., Bischoff, R. & Holtzer, H. (1969). The formation of arrowhead complexes with heavy meromyosin in a variety of cell types. *J. Cell Biol.* **43**: 312–328.

Jasper, H. H. & Pezard, A. (1934). Relation entre la rapidité d'un muscle strié et sa structure histologique. *C. r. hebd. Séanc. Acad. Sci., Paris* **198**: 499–501.

Johnson, W. H. (1963). *Ultrastructure of protein fibres*: 139–176. Borasky, R. (ed.). New York: Academic Press.

Joseph, H. (1895). Über das Achsenskelett des Amphioxus. *Z. wiss. Zool.* **59**: 511–536.

Joseph, H. (1902). Einige anatomische und histologische Notizen über Amphioxus. *Arb. zool. Inst. Univ. Wien* **13**: 125–154.

Jurand, A. (1962). Development of the notochord in chick embryos. *J. Embryol. exp. Morph.* **10**: 602–621.

Kelly, D. E. (1969). The fine structure of skeletal muscle triad junctions. *J. Ultrastruct. Res.* **29**: 37–49.

Kelly, D. E. & Luft, J. H. (1966). Fine structure, development, and classification of desmosomes and related attachment mechanisms. In *Electron microscopy 1966* **2**: 401–402. Uyeda, R. (ed.) Tokyo: Maruzen.

Leeson, T. S. & Leeson, C. R. (1958). Observations on the histochemistry and fine structure of the notochord in rabbit. *J. Anat.* **92**: 278–285.

Muller, J. (1842). Ueber den Bau und die Lebenserscheinungen des *Branchiostoma lubricum* Costa, *Amphioxus lanceolatus*, Yarrell. *Abh. K. Akad. Wiss., Berl.* **1842**: 79–116.

Muller, W. (1871). Beobachtungen des patologischen Instituts zu Jena. 1 Ueber den Bau der Chorda dorsalis. *Jena. Z. Naturw.* **6**: 327–353.

Olsson, R. (1965). Comparative morphology and physiology of the *Oikopleura* notochord. *Israel J. Zool.* **14**: 213–220.

Page, S. (1968). Structure of the sarcoplasmic reticulum in vertebrate muscle. *Br. med. Bull.* **24**: 170–173.

Pellegrino de Iraldi, A., Gueudet, R. & Suburo, A. M. (1971). Differentation between 5-Hydroxytryptamine and Catecholamines in synaptic vesicles. *Progr. Brain Res.* **34**: 161–170.

Rosenbluth, J. (1969). Ultrastructure of dyads in muscle fibers of *Ascaris lumbricoides*. *J. Cell Biol.*: **42**: 817–825.

Rostgaard, J., Kristensen, B. I. & Nielsen, L. E. (1972). Electron microscopy of filaments in the basal part of rat kidney tubule cells, and their *in situ* interaction with heavy meromyosin. *Z. Zellforsch. mikrosk. Anat.* **132**: 497–521.

Rostgaard, J. & Thuneberg, L. (1972). Electron microscopical observations on the brush border of proximal tubule cells of mammalian kidney. *Z. Zellforsch. mikrosk. Anat.* **132**: 473–496.

Schmidt, W. J. (1936). Die Doppelbrechung der quergestreiften Muskelzellen im Proventriculus von *Eusyllis blomstrandi*. *Z. Zellforsch. mikrosk. Anat.* **24**: 526–539.

Schwarz, W. (1961). Elektronenmikroskopische Untersuchungen an den Chordazellen von *Petromyzon*. *Z. Zellforsch. mikrosk. Anat.* **55**: 597–609.

Schwarzacher, H. G. (1960). Untersuchungen über die Skelettmuskel-Sehnenverbindung. 1. Elektronenmikroskopische und Lichmikroskopische Untersuchungen über den Feinbau der Muskelfaser-Sehnenverbindung. *Acta Anat.* **40**: 59–86.

Sommer, J. R. & Johnson, E. A. (1969). Cardiac muscle. A comparative ultrastructural study with special reference to frog and chicken hearts. *Z. Zellforsch. mikrosk. Anat.* **98**: 437–468.

Tenbaum, E. (1955). Polarisationsoptische Beiträge zur Kenntnis der Gewebe von *Branchiostoma lanceolatum*. *Z. Zellforsch. mikrosk. Anat.* **42**: 149–192.

Thuneberg, L. & Rostgaard, J. (1969). Motility of microvilli. A film demonstration. (Abstract). *J. Ultrastruct. Res.* **29**: 578.

Tsao, T. C., Kung, T. H., Peng, C. M. &Tsou, Y. S. (1965). Electron microscopical studies of tropomyosin and paramyosin. 1. Crystals. *Scientia Sin.* **14**: 91–105.

Waddington, C. H. & Perry, M. M. (1962). The ultrastructure of developing urodele notochord. *Proc. R. Soc.* (B) **156**: 459–482.

Webb, J. E. (1973). The role of the notochord in forward and reverse swimming and burrowing in the amphioxus *Branchiostoma lanceolatum*. *J. Zool., Lond.* **170**: 325–338.

Welsch, U. (1968). Über den Feinbau der Chorda dorsalis von *Branchiostoma lanceolatum*. *Z. Zellforsch. mikrosk. Anat.* **87**: 69–81.

Welsch, U. & Storch, V. (1969). Zur Feinstruktur der Chorda dorsalis niederer Chordaten (*Dendrodoa grossularia* v. Beneden und *Oikopleura dioica* Fol). *Z. Zellforsch. mikrosk. Anat.* **93**: 547–559.

Welsch, U. & Storch, V. (1970). The fine structure of the stomochord of the Enteropneusts *Harrimania kupfferi* and *Ptychodera flava*. *Z. Zellforsch. mikrosk. Anat.* **107**: 234–239.

Welsch, U. & Storch, V. (1971). Fine structural and enzymehistochemical observations on the notochord of *Ichthyophis glutinosus* and *Ichthyophis kohtaoensis* (Gymnophiona, Amphibia). *Z. Zellforsch. mikrosk. Anat.* **117**: 443–450.

Wissocq, J. C. (1970). Données ultrastructurales sur le proventricule de *Syllis amica* Quatr. (Annelide, polychète). In *Microscopie électronique 1970* **3**: 801–802. Favard, P. (ed.) Paris: Soc. Française Micr. Electr.

Symp. zool. Soc. Lond. (1975) No. 36, 105–127.

EVOLUTION OF PHOSPHAGEN KINASES IN THE CHORDATE LINE

D. C. WATTS

Department of Biochemistry and Chemistry, Guy's Hospital Medical School, London, England

SYNOPSIS

A phosphagen and its kinases form an ATP regenerating device permitting the sudden utilization of large amounts of energy not possible by other recognized metabolic processes. This is important for the operation of systems characteristic of animal life such as muscular contraction or nervous activity. The most primitive phosphagen kinase is believed to have been a monomeric arginine kinase from which all other di- and tetrameric phosphagen kinases have evolved. This has involved duplication of the kinase gene and selective modification of one product to change the specificity of the guanidine binding site, although not that for ATP. In this way, the adoption of new phosphagens became possible while the original function was retained.

The general distribution of the phosphagens and their kinases in the animal are outlined.

A feature of an enzyme still undergoing evolutionary change may be a relative lack of substrate specificity. Creatine kinase appears to be the major kinase of the Hemichordata but *Saccoglossus* also has traces of arginine and taurocyamine kinase activity. The possible origin of this variability by modification of a single gene or divergent evolution of replicate genes is discussed. The latter appears more probable.

To evolve a new enzyme is only one aspect of evolution; to bring it within the framework of development and metabolic control of the organism may be much more difficult. The distribution of arginine and creatine kinases among the ascidian tunicates only loosely accords with the accepted phylogenetic relationships of the families. Similarly, among the echinoderms, for which much more complete data exist, it is in the advanced forms, rather than the more simple transitional forms, that arginine and creatine kinase occur together in the same individual. An explanation of this apparent paradox is suggested by evidence that creatine kinase first proved of selective advantage in association with a mobile gamete. Control mechanisms permitting its expression in adult forms only slowly evolved, and the way this has occurred, perhaps accompanying morphological reorganization, is examined with reference to *Echinus esculentus*. This evolutionary trend continues through the Cephalochordata into the Chordata, where the selective advantages of creatine as a phosphagen former appear such as to ensure its ultimate dominance.

These data provide biochemical support for Garstang's theory that major evolutionary steps always accompany modification of a primitive biological structure and take effect from an early stage in development.

FUNCTIONAL INVOLVEMENT AND SUBSTRATE SPECIFICITY

The phosphagen kinases, or phosphotransferases as they should more properly be known, catalyse the general reaction

$$\text{guanidine} + \text{MgATP} \rightleftharpoons \text{phosphoguanidine} + \text{MgADP} + \text{H}^+ \quad [1]$$
(phosphagen)

At the present time eight different naturally occurring phosphagens have been found. All but three, arginine, glycocyamine and creatine, are modifications of taurine (Fig. 1) and are confined to the annelid worms (Watts, 1971). It appears that the diversity of products from amino acid metabolism is such that there is no shortage of potential substrates for these enzymes. The concentration in muscle, about 50 mM, is also such that they are a significant component of many carnivorous diets (Van Pilsum, Stephens & Taylor, 1972).

In contrast to the variability of the guanidine substrate, the nucleotide substrate has remained constant throughout the whole of the animal kingdom as ATP, or, functionally, as the Mg-ATP complex. Although other nucleoside triphosphates such as GTP and ITP may act as phosphagen formers *in vitro* the conditions required are usually such as to exclude these having any significance in the living organism. The reason for this lies in the function of these enzymes in acting as an ATP regenerating system, the phosphagen forming what is popularly termed an "energy store". The need for this energy store appears to have arisen with the development of a rapid contractile system which, when triggered off from the resting state, was capable of breaking down ATP to ADP faster than it could immediately be regenerated by the normally available metabolic pathways in a resting tissue. Regeneration of ATP from the phosphagen store bridges the gap until the glycolysis pathway and citric acid cycle can be switched into full activity. The later evolution of complex muscular activity required the concomitant development of a brain and complex nervous system and these, again, are dependent for their rapid integrated function upon a phosphagen store and its associated kinase (Watts, 1971).

The incidental benefits that arise from a device of this sort indicate why the more obvious alternative of simply having a large store of ATP was not adopted. The primary event is the breakdown of ATP with the liberation of a proton

$$ATP^{4-} + H_2O \rightarrow ADP^{3-} + HPO_3^{2-} + H^+ \qquad [2]$$

As an ATP concentration equivalent to the liberation of about 2·5 mM acid may be broken down over the first few seconds of activity this is not an insignificant consideration. However, regeneration of ATP by reaction [1] moving from right to left uses up the proton generated by reaction [2] and thus neutrality inside the cell is maintained after only a transient production of acid. In addition, other problems associated with a high ATP concentration such as a

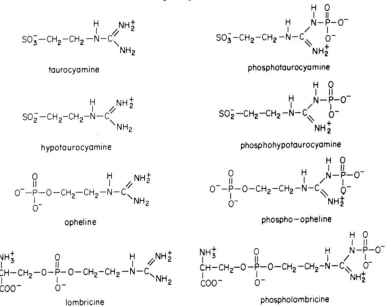

FIG. 1. Structures of the natural guanidine substrates and phosphagens. The new phosphagen, phosphothalassemine (Thoai *et al.*, 1972), has the structure of phospholombricine but with two methyl groups substituting on the free α-amino group.

high ionic strength and the sequestration of free metal ions required for other cellular processes are also avoided.

The importance of the phosphagen regenerating system was shown by the work of Davies (1965) in which it was found that the specific inhibition of creatine kinase in isolated intact frog muscle, by soaking the preparation in a dilute solution of fluorodinitrobenzene, causes almost complete loss of contractile activity even though the contractile machinery is unimpaired.

EVOLUTIONARY ORIGIN AND EVIDENCE FOR HOMOLOGY

In 1930 it was first suggested by Meyerhof that phosphoarginine, as a simple phosphorylated amino acid, was phylogenetically older than phosphocreatine, the only other phosphagen then known. The predominance of phosphoarginine over all others in invertebrates supports this view and hence we conclude that arginine kinase was the first phosphagen kinase from which all others evolved. In time, a fair measure of evidence has emerged in support.

First, the smallest enzymes known are monomer arginine kinases with molecular weights of about 40 000. Some arginine kinases and most of the other guanidine kinases have molecular weights of twice this size and are composed of two similar subunits. Two instances of tetramers have also been recorded.

Second, the amino acid composition is very similar indeed for a variety of kinases from a diversity of animals. This is particularly emphasized when amino acids of similar function are grouped together, as shown in Table I, so that some allowance is made for the conservative replacement of one amino acid by another in evolution.

Third, the site of interaction between subunits has been strongly conserved during evolution. For example, it is possible to take a mixture of a dimer arginine kinase and a dimer creatine kinase and dissociate it into single subunits with urea solution. When the urea is dialysed away the subunits recombine in dimers and a proportion do so to form a hybrid with both arginine and creatine kinase activities. This may be distinguished electrophoretically as shown in Fig. 2. The specificity of this subunit interaction is so great that no other hybrid is formed, even when crude tissue extracts are used as a source of enzyme, indicating that the combining site of a phosphagen kinase subunit is quite distinct from that on any other enzyme subunit in the tissue.

TABLE I

Molar composition of groups of amino acids in some muscle phosphagen kinases (%)

(From Watts, 1971)

	Creatine kinase			Arginine kinase			Glycocyamine kinase	Taurocyamine kinase	Lombricine kinase
	Dogfish	Human	Rabbit	Lobster	American lobster	*Pecten maximus*	*Nephthys caeca*	*Arenicola marina*	Earthworm
Asp, Glu	21·6	20·6	21·8	21·3	23·1	20·9	22·0	21·1	19·1
Lys, Arg, His	18·7	17·5	18·4	15·8	21·9	15·3	17·3	20·6	19·2
Tyr, Phe, Trp	7·3	7·2	7·8	9·1	9·3	7·7	13·6	10·5	11·4
Met, Lys	3·6	3·6	3·8	4·0	3·9	3·5	6·2	4·9	2·6
Ser, Thr	9·0	10·0	10·4	10·9	9·9	10·9	8·1	8·1	7·7
Pro	5·0	6·3	5·1	4·0	3·6	5·3	3·3	3·9	3·9
Gly, Ala, Val, Ile, Leu	34·8	34·8	32·7	34·9	28·3	36·3	27·8	27·1	26·9

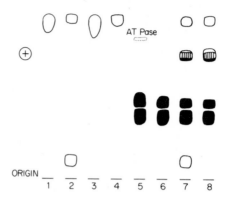

FIG. 2. Electrophoretic patterns of creatine kinases, arginine kinase and their hybrids. The positions of enzyme activity on the gel for creatine kinase, unblocked outlines; arginine kinase, blocked outlines; and hybrid with both enzyme activities, hatched outlines; were measured as described in the text. The samples used were: 1, rabbit brain extract; 2, urea-treated rabbit brain extract; 3, ox stomach creatine kinase; 4, urea-treated ox stomach creatine kinase; 5, *H. forskali* muscle extract; 6, urea-treated *H. forskali* muscle extract; 7, urea-treated mixture of rabbit brain extract and *H. forskali* muscle extract; 8, urea-treated mixture of ox stomach creatine kinase and *H. forskali* extract. (Data from Watts, Focant *et al.* (1972).

Fourth, each subunit has one catalytic site and associated with this is a cysteine side chain that is essential for catalytic activity. The amino acid sequence around this "essential thiol group" is almost identical in four different forms of the enzyme that have been investigated (Fig. 3). The differences may be explained by single base mutations within codons, one codon duplication and one codon deletion, all indicative of close evolutionary affinity and that this feature of the enzyme is of great mechanistic importance to have been so strongly conserved. The thiol itself has the unusual property that it appears to be fixed in a partially ionized state. This is demonstrated in the reaction with an alkylating agent which is independent of pH instead of showing the usual increase in reaction rate as the pH is raised (Fig. 4). In both arginine and creatine kinase the thiol is involved in mediating a conformational change in the enzyme upon binding of the guanidine substrate. This appears to be an essential part of the catalytic process, since if the thiol is reacted with iodoacetamide both the catalytic activity and ability to undergo the conformational change are destroyed.

These and other data, then, provide strong evidence that the phosphagen kinases have a common evolutionary origin. The way in which they have evolved is revealed by studying not only the organisms in which they occur but also, more precisely, the tissues

Creatine kinase (rabbit muscle)   - Asn-His-Leu-Gly-Tyr-Val-Leu-Thr-$[^{14}C]$Cys-Pro — Ser-Asn-Leu-Gly-Thr-Gly-Leu-Arg

Creatine kinase (ox brain)   -Asn-His-Leu-Gly-Tyr-Ile-Leu-Thr-$[^{14}C]$Cys-Pro — Ser-Asn-Leu-Gly-Thr-Gly-Leu-Arg

Arginine kinase (lobster)   -Gln-Thr-$[^{14}C]$Cys-Pro-Thr-Ser-Asn-Leu-Gly-Thr — Val-Arg

Lombricine kinase (earth-worm)   -Leu-Gly-Tyr-Ile — Thr-$[^{14}C]$Cys-Pro-Gly-Ser-Asn-Leu-Gly-Thr — Leu-Arg

FIG. 3. Comparison of the amino acid sequences around the essential cysteine residues of some phosphagen kinases. The gaps in the sequences have been introduced by aligning the amino acids to show maximum homology. Data summarized from Watts (1971) and der Terrossian et al. (1971).

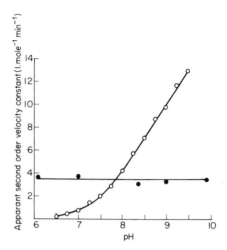

FIG. 4. pH-dependent and pH-independent sulphydryl groups. ○, the reaction of
cysteine with chloroacetamide at 30° (Lindley, 1960); ●, the reaction of creatine phospho-
transferase with iodoacetamide at 25°. Chloroacetamide is used as the alkylating agent for
the reaction with cystein because the reaction of iodoacetamide with a free sulphydryl group
would be too fast to measure. Data from Watts (1965).

in which they function and the stage of development at which they
appear. In this respect the evolutionary line leading to the chordates has provided much valuable information.

DISTRIBUTION

The general picture

The phylogenetic tree shown in Fig. 5 summarizes current
knowledge about the distribution of the phosphagens and their
kinases. The great group of arthropods contains, so far as is
known, only monomer arginine kinase, as do the majority of
molluscs. Of the more primitive phyla, the Platyhelminthes,
Coelenterata, Porifera and Protozoa, very little is known. However,
the arginine kinases of two protozoans, *Stentor coeruleus* and
Tetrahymena pyriformis, show features that are far from primitive.
The *Stentor* enzyme is abnormal in being tightly bound to the
particulate fraction while that of *Tetrahymena* is a dimer. So that
although the Protozoa form the foot of our tree, their modern
representatives may show typically advanced features. On the
other hand the bacterium, *Escherichia coli*, has been found to
contain an arginine kinase of molecular weight 40 000 (Di Jeso,
1967) so that a monomer enzyme may occur in a simple organism.

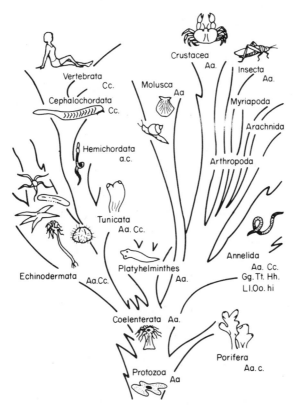

FIG. 5. Phylogenetic tree summarizing the known distribution of the natural phosphagens and their phosphotransferases in the animal kingdom. Abbreviations: capital letters indicate the demonstration of the kinase, the lower case letters that of the phosphagens: A, arginine; C, creatine; G, glycocyamine; T, taurocyamine; H, hypotaurocyamine; L, lombricine; O, opheline; hi, hirudonine. (From Watts, 1971).

As we move up the chordate line we find that a dimer enzyme becomes the rule, the inevitable exception being an even more advanced tetrameric creatine kinase in *Echinus esculentus*. Clearly, the ability to dimerise was an early evolutionary achievement. Both arginine and creatine kinases are found in representatives of the echinoderms, tunicates and hemichordates and it is only with amphioxus and the true chordates that phosphocreatine takes over as the sole phosphagen. It was the discovery of both phosphoarginine and phosphocreatine in echinoderms and hemichordates (Baldwin, 1937) that evoked the inference that here was biochemical evidence for groups transitional between invertebrates and vertebrates. The discovery in the annelids of a wide variety of

TABLE II

Phosphagens and kinases in the Hemichordata

Species	Phosphagen*		Phosphotransferase† (Kinase)			Reference
Balanoglossus clavigerus		PC	—			Baldwin (1953)
B. salmoneus	PA	PC	—			Needham *et al.* (1932)
Saccoglossus horsti		PC	—			Baldwin & Yudkin (1950)
S. kowalevskii		PC	—			Baldwin & Yudkin (1950)
S. kowalevskii	PA	PC	—			Rockstein (1971)
Saccoglossus sp.			CPK	APK(trace)	TPK(trace)	Watts (unpublished)

* PA and PC are the phosphagens of arginine and creatine.
† APK, CPK and TPK are the kinases of arginine, creatine and taurocyamine (see Fig. 1).

phosphagens, including phosphocreatine, undermined this hypothesis but this was because the appreciation of the evidence was essentially naive rather than because it was wrong. As the details were filled in a more complex picture inevitably emerged.

Hemichordata and the evolution of multiple enzyme activities

It was the invitation to address this symposium that brought home the realization that the Hemichordata had been largely neglected since the work of Needham, Needham, Baldwin & Yudkin (1932), using rather primitive identification procedures (Table II). In the intervening period I managed to procure from Plymouth specimens of *Saccoglossus* sp. and found, in support of the early work, that they contained predominantly creatine kinase. The enzyme was very unstable but more interesting was the finding of low activities of arginine kinase and taurocyamine kinase. All this work is on animals from Plymouth (England). Recently, Rockstein (1971) found that *S. kowalevskii* from Woods Hole contained about twice as much phosphoarginine as phosphocreatine. First let us enquire as to how these different enzyme activities arose and then consider the disparity in results (all of which appear to be technically acceptable) from the two sides of the Atlantic. In evolution the complement of potential enzyme activities carried by the genetic material of an organism may change in two ways. The structural gene may become altered so that its enzyme product is able to catalyse a new reaction at the expense of losing the original activity. Part of this process might be expected to be a loosening of the normally tight specificity of the kinase for its natural guanidine substrate to accommodate other natural guanidines. In fact, little evidence has been found to support this for any arginine or creatine kinase, although for some organisms that might be regarded as primitive, or undergoing evolutionary change, a loosening of specificity towards a series of synthetic substrate homologues has been revealed (Table III). This seems not to be found among well established forms such as the Crustacea or advanced forms such as some echinoderms (Table III).

An alternative explanation of the multiple kinase activity in *Saccoglossus* is that the original kinase gene has undergone duplication and that we are seeing the outcome of newly evolved catalytic activities but with the degree of expression being limited by cellular control mechanisms. This second explanation is the more probable one and it would seem that *S. kowalevskii* at Plymouth and Woods Hole is evolving into two sub-species or racially distinct

TABLE III

Specificity of some arginine kinases for a synthetic series of homologues of L-arginine

(Data from Watts, 1965)

Species	Per cent of the reactivity found with L-arginine						
	D-arginine	Glycylglycocyamine	L-α-amino-γ'-guanidino butyrate	L-canavanine sulphate	Guanidino-n-valerate	Nα-benzoyl-L-arginine	L-homoarginine
LAMELLIBRANCHIA							
Pecten maximus (striated muscle)	1·0	0·0	1·7	1·8	0·2	0·2	0·2
Chlamys opercularis (striated muscle)	0·0	0·0	0·0	0·8	2·5	0·7	0·3
ECHINOIDEA							
Echinus esculentus	0·0	0·0	0·0	12·9	0·0	0·0	0·0
HOLOTHUROIDEA							
Holothuria forskali	1·0	0·0	0·0	0·8	0·0	0·0	0·0
CRUSTACEA							
Leander squilla	0·6	0·8	0·0	1·0	0·2	0·2	0·0
L. serratus	0·0	1·6	0·0	1·6	0·0	0·0	0·0
Homarus vulgaris	0·0	0·0	0·0	0·0	0·0	0·0	0·0
Nephrops norvegicus	0·0	0·0	0·0	0·0	0·0	0·0	0·0
Atelecyclus septemdentatus	0·0	0·0	0·0	0·0	0·0	0·0	0·0
Portunus depurator	0·2	0·0	0·0	0·5	0·0	0·0	0·0
Carcinus maenas	0·0	0·0	0·0	0·0	0·0	0·0	0·0
Cancer pagurus	0·0	0·0	0·0	0·0	0·0	0·0	0·0
Maia squinado	2·8	0·3	5·6	7·7	0·0	0·0	0·0
Eupagurus bernhardus	1·2	0·7	2·8	2·9	0·0	0·2	0·2
TURBELLARIA							
Polycelis cornuta	7·1	9·5	2·8	12·0	3·1	0·0	0·0
POLYCHAETA							
Myxicola infundibulum	5·9	0·0	2·1	0·8	0·0	0·0	0·0
Sabella pavonina	27·8	0·0	20·5	52·4	9·0	0·0	26·6

groups distinguishable by the rate at which the more primitive arginine kinase activity is being lost. The American form appears to be more like *Balanoglossus salmoneus* which also has functionally useful concentrations of both phosphocreatine and phosphoarginine and has yet to "learn" how to manage without the latter. A similar situation has been found among the Annelida which has been examined in greater detail (Watts & Watts, 1968).

A third possibility, that an enzyme may acquire an additional catalytic activity without loss of its original catalytic activity, appears not to to occur in practice on the simple principle that no slave can serve two masters adequately. This is because the ancillary problems of control become overwhelmingly complex.

Tunicata

Turning to the tunicates we find (Table IV) that only the Ascidiacea have been investigated, owing to the difficulty of getting fresh material. As with the hemichordates, only arginine and creatine form physiologically important phosphagens.

Historically, the first discovery was of phosphoarginine in *Ascidia mentula*. This was somewhat disconcerting because at that time theory decreed that primitive chordates should have creatine kinase. The discovery of only phosphocreatine and creatine kinase with no evidence for even a trace of arginine kinase in two species of *Pyura* restored the faith but cast doubt on the original observation. When we turned our attention to this problem the first species we examined, *Styela mammiculata*, proved to contain both kinases in almost equal amounts and is the only one so far to do so, although arginine kinase has also been found in other species.

The distribution appears to have no obvious phylogenetic significance except that when arginine kinase does occur it tends to be associated with the less advanced forms. Where enzyme activity was measured no phosphorylation of glycocyamine or taurocyamine was found so that other phosphagen kinases are unlikely to occur in measurable concentrations. At the present time nothing is known about the properties of these enzymes and there is a need to extend the survey to the Larvacea and Thaliacea.

Echinodermata

A paradoxical enzyme distribution

The echinoderms are the best documented group with over 60 species having been investigated. The data, summarized in Table V, show that the sea lilies (Crinoidea) have only arginine kinase. Most sea cucumbers (Holothuroidea) also contain only arginine kinase but the additional presence of phosphocreatine has now been reported for three species, *Cucumaria frondosa* (Verzhbin-

TABLE IV

Distribution of the phosphagens and their kinases among the Tunicata. Only the Class Ascidiacea has been investigated

Family and Species	Phosphagen*	Phosphotransferase*	(Kinase)	Reference
Polyclinidae				
Morchellium argus	—	CPT	—	D. C. & R. L. Watts (unpublished)
Cionidae				
Cioma intestinalis	—	CPT	—	Virden & Watts (1966)
Ascidiidae				
Ascidia mentula	PA	—	—	Baldwin & Yudkin (1950)
Ascidiella aspersa	—	—	APT	D. C. & R. L. Watts (unpublished)
Phallusia mammilata	—	CPT (trace)	APT	D. C. & R. L. Watts (unpublished)
Botryllidae				
Botryllus schlosseri	—	CPT	—	D. C. & R. L. Watts (unpublished)
Styelidae				
Dendrodoa grossularia	—	CPT	—	D. C. & R. L. Watts (unpublished)
Styela mammiculata	—	CPT	APT	Virden & Watts (1966)
Styela rustica	PC	—	—	Borsuk et al. (1933)
Pyuridae				
Pyura stolonifera	PC	CPT	—	Morrison et al. (1956)
Pyura subcalata	PC	CPT	—	Morrison et al. (1956)
Molgulidae				
Molgula sp.	—	CPT	—	Virden & Watts (1966)

* Abbreviations are as in Table II.

TABLE V

Distribution of the phosphagens and their kinases among the Echinodermata

		Phosphagen		Kinase	
Crinoidea	(3 species)	PA(2)	—	APK(2)	—
Holothuroidea	(12 species)	PA(5)	—	APK(9)	—
	(2 species)	PA(2)	PC(2)	—	—
	(1 species)	—	PC	—	—
Asteroidea	(14 species)	PA(1)	—	APK(14)	—
	(4 species)	PA(4)	PC(4)	—	—
	(1 species)	—	—	—	CPK
Ophiuroidea	(9 species)	—	PC(4)	—	CPK(7)
	(1 species)	PA	PC	—	—
Echinoidea					
Aulodonta	(2 species)	PA(1)	—	APK(2)	—
Stirodonta	(1 species)	PA	—	APK	—
Irregularia	(4 species)	—	—	APK(4)	—
Camerodonta	(15 species)	PA(4)	PC(4)	APK(13)	CPK(13)
	(1 species)	—	—	—	CPK

Values in parenthesis indicate the number of species examined. Some species have been investigated for both phosphagen and kinase and some for only one or the other.

skaya, Borsuk & Kreps, 1935), and *Thyone briareus* and *Ludwigthuria floridana* (Rockstein, 1971). The starfish were also thought to contain only arginine kinase until we found that *Henricia sanguinolenta* had only creatine kinase. Four other species are now known to contain some creatine kinase as well as arginine kinase. The phylogenetically late brittle stars (Ophiuroidea), on the other hand, had been found to contain only creatine kinase but, again, an exception has now been reported in *Ophioderma brevispina* which has phosphoarginine as well (Rockstein, 1971). But perhaps the most interesting class is the Echinoidea. Here the more primitive Aulodenta, Stirodonta and Irregularia contain only arginine kinase while the more advanced Camerodonta contain usually both kinases, and in one species, *Psammechinus miliaris*, creatine kinase only (Moreland, Watts & Virden, 1967).

The paradoxical feature of this distribution is that it is only the advanced echinoderms that appear to contain creatine kinase. Surely, if this group lies on the chordate line should not one expect to find creatine kinase in the more primitive forms as well?

A look at the gametes: classical evolutionary theory vindicated

However, evolutionary theory predicts that it is the embryonic form of an animal that is close to the evolutionary stem line rather than the adult into which it develops. When one turns to the phosphagen distribution of the gametes this is fully substantiated, for it turns out that, except for the crinoids which have still to be investigated, in representatives of all the other echinoderm classes the spermatozoa always contain only creatine kinase while the unfertilized egg contains only arginine kinase. This is irrespective of the form found in the adult muscle. For example, *Psammechinus*, already mentioned, contains only creatine kinase in the adult muscle although the eggs contain only arginine kinase. It would thus appear that creatine kinase has a much greater antiquity than we had suspected and that what we see when we study the adult forms is the evolution of the control mechanisms that will switch one kinase on and another off in the course of development. The reason why it is biologically unsound for an enzyme to evolve two different catalytic activities at the same time, assuming this to be mechanically possible, is now all too apparent.

On this interpretation, the diversity and distribution of phosphagens in the echinoderms is readily understood, and the appearance of the occasional species that differs from the rest of its class may well indicate a new successful control gene mutation. The sharp breaks in the major phosphagen types of the classes suggest that a drastic change in the regulatory machinery is most readily evolved when the members of a group are simultaneously undergoing a drastic change in form. Might not the same argument apply to the tunicates and hemichordates? It is interesting that the polychaetes, which also have a flagellate gamete, show the same evolutionary trend as the echinoderms (Table VI); in spite of the diversity of phosphagens that have evolved in this group the spermatozoa only have phosphocreatine.

The emergence of creatine kinase

Let us now pause to consider the selection pressure favouring the emergence of creatine kinase. Since the amoeboid spermatozoa of molluscs and crustaceans (and perhaps of all arthropods) still contain arginine kinase just like a protozoan, it seems reasonable to infer a connection with the development of an active flagellate spermatozoan. An explanation of the selective advantage now emerges, for in the testes, unlike muscle, there would be severe competition for the available arginine between the kinase, trying to

TABLE VI

Phosphagens in the muscle, eggs and spermatozoa of polychaetes.

(Data from Thoai, 1968)

Species	Phosphagen					
	PA	PG	PC	PL	PO	PT
Terebella lapidaria	M, E	—	S	—	—	—
Nephthys hombergii	—	M, E	S	—	—	—
Audouinia tentaculata	—	—	M, S	E	—	—
Glycera convoluta	E	—	M, S	—	—	—
Ophelia bicornis	—	—	S	M, E	—	—
Ophelia neglecta	—	—	S	E?	M	—
Arenicola marina	—	—	S	—	—	M, E

For the naturally occurring phosphagens, see Fig. I.

equip the spermatozoan with an adequate phosphagen store for mobility, and the protein synthesizing system involved in the production of arginine-rich histones. In times of arginine shortages the irreversible synthesis of histones would inevitably proceed at the expense of the reversible phosphagen store. Loss of viability would be in proportion to the resulting decrease in motility. The evolution of creatine kinase simultaneously solves both problems at a stroke. The phosphagen kinase system is now totally isolated from the rest of amino acid metabolism so that it is not possible to drain away the free guanidine by a competing metabolic pathway. At the same time creatine may be synthesized by transamidination to the more commonly occurring glycine, and this requires only catalytic amounts of arginine that may be resynthesized via the urea cycle. Thus the maximum possible amount of arginine is released for protein synthesis. It is not surprising that once creatine had evolved it was destined to become the major phosphagen of the future. Finally, one might ask why creatine rather than glycocyamine, the immediate product of transamidination to glycine (see Fig. 1), is the preferred phosphagen former? The answer to this question almost certainly lies in the greater facility with which the asymmetrical creatine molecule may induce the enzyme to catalyse the transphosphorylation reaction

(Watts, 1973). Absence of the creatine methyl group appears to pose greater mechanistic problems for promoting the necessary conformational changes in the enzyme; problems which have only been solved satisfactorily by the polychaete worms (Thoai, 1968).

Functional association in kinase evolution

If the hypothesis is true, that evolution occurs by changes during development in the expression of features that first appear in the gametes or early embryo, one would expect to find in the transition groups that advanced structural features tended to be associated with arginine kinase while the more primitive ones were associated with creatine kinase. A more detailed study of *Echinus esculentus* has proved to support this conclusion. In chordates generally, the fast, phasic sitrated muscle is considered to be a highly specialized tissue while the slow, more extensible smooth muscle forms earlier in development and is considered more primitive. A parallel evolutionary development has also occurred among the molluscs (Moreland & Watts, 1967; Watts, 1971). The muscles of *Echinus* that have been looked at are those of the masticatory apparatus, called Aristotle's Lantern, and those of the external appendages.

Aristotle's Lantern, which is an advanced and characteristic feature of camerodont echinoids, has four groups of well-defined muscles. Three of these are 5–20 mm long and move the teeth fairly rapidly, searching for food on the surface to which the animal is attached. The remaining group are small, about 2 mm long, and link the teeth together so that they can carry out a slow grinding process on each other. These 'short' interalveolars are also capable of extension in the manner of smooth muscle. Table

TABLE VII

Kinase distribution in the lantern muscles of Echinus esculentus

(Data from Watts, 1971).

Muscle	Function	Creatine kinase	Arginine kinase	Ratio AK/CK
		(μ moles/mg protein/min)		
Interalveolar	Comminution	5·05	1·10	0·2
Adductors	Protractors	0·36	0·4	1·1
Abductors	Retractors	0·28	0·57	2·0
Inter-radials	Lateral movement	0·33	1·0	3·0

VII shows that these small muscles contain mostly creatine kinase while the other larger muscles contain as much or more of arginine kinase. In other words, the creatine kinase is associated with the more tonic function and arginine kinase with the more phasic function, corresponding to the less and more highly evolved muscle types respectively. These muscles have no obviously distinctive histological structures even under the electron microscope (L. H. Bannister, personal communication). However, it would seem that they contain a mixture of fibre types since, as Table VIII shows, the kinase distribution is not homogeneous, with more arginine kinase in the ends of the adductors but not the abductors.

TABLE VIII

Ratio of arginine to creatine phosphotransferases in the lantern muscle of Echinus esculentus

(Data from Watts, 1971).

	A : C	
	Origin and insert of muscle	Middle of muscle
Adductors	3·25	1·74
Abductors	2·25	2·74

TABLE IX

Ratio of arginine to creatine phosphotransferases in tissues of Echinus esculentus

(Data from Watts, 1971).

	A : C
Pedicellariae	
Tridactyle	2·6 : 1
Ophiocephalous	1·4 : 1
Gemmiform	1 : 8
Tube feet	1 : 8
Spines	1 : 20

Similar results are found when one looks at the external appendages of the animal (Table IX). The tridactyles are the most powerful of the pedicellariae and these, to provide the nipping action, contain the only clearly striated muscle found in echinoids. They also contain most arginine kinase. The gemmiform pedicellariae, on the other hand, have poison glands inside the jaws and it is probably their large extensible muscular sacs which account for the high creatine kinase content. The tube feet and spines again have a largely tonic function and again have mostly creatine kinase. It does seem that, in *Echinus,* arginine kinase is used in the faster muscles with a limited range of contractility while creatine kinase is associated with the more extensile muscles of tonic function.

Up to the Chordata

The distribution of phosphagens and kinases in the echinoderms, tunicates and chordates is such that, although the picture is far from complete, it is not improbable that the sort of developmental patterns we see in the echinoderms are common to all three groups. Evolution of the Chordata from a larval stage of a tunicate (Berrill, 1955) would explain the final disappearance of arginine kinase from the adult as the specialized muscles with which it might be associated would be lost in the process. One might expect the evolution of active new adult forms in the Chordata to be associated with a new creatine kinase distinct from that found with the more primitive smooth muscle. This is indeed found. Figure 6 shows the developmental pattern of creatine kinase in striated chicken muscle. At an early stage of development, the myocell stage, the first creatine kinase to be detected is Type I, characteristic of smooth muscle and nervous tissue; as development proceeds through the myotube stage, the Type III enzyme, characteristic of fully developed striated muscle, starts being synthesized. Because, at this stage in development, both kinases are being synthesized together, one finds a new enzyme species, Type II, which is composed of one each of the Type I and Type III subunits. This hybrid dimer formation appears to have no functional significance but simply reflects the close evolutionary affinity of these two molecular species. In the adult the activity of Type I enzyme almost ceases and the striated muscle is characterized by the presence of the Type III enzyme plus a trace of the hybrid form. It is interesting to recall that although the subunit combining sites have been heavily conserved through evolution, so that a hybrid may still be formed between a mammalian creatine kinase

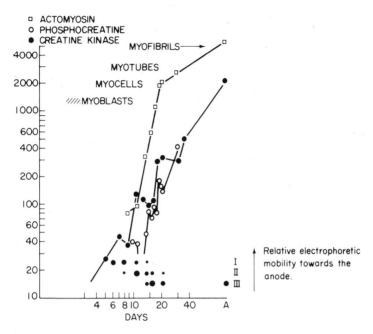

FIG. 6. The pattern of biochemical development in chicken skeletal muscle. The solid circles at the foot of the graph indicate the progressive replacement from day 6 of the Type I creatine kinase by Type III creatine kinase. When both enzymes are present a hybrid is formed consisting of one subunit of each. This is known as Type II creatine kinase. The size of the blobs indicates the amount of each type present. Hatching occurs at about day 10, coincident with the rapid increase in levels of enzyme activity. Data of Eppenberger *et al.* (1964).

and an echinoderm arginine kinase (Fig. 2), in other properties the two chordate creatine kinases have shown considerable divergence. Figure 6 indicates differences in electrophoretic mobility and this reflects marked differences in the contents of aromatic and charged amino acids, such that the Type I or Type III enzymes of different species are more alike than the Types I and III enzymes of a single individual. Thus, within limits, function determines the permitted extent of molecular variation more precisely than does species variation, which is, perhaps, another way of saying that evolution is more likely to proceed from forms of an organism that have not acquired a high level of specialization.

CONCLUSIONS

In this review it has been shown how the phosphagen kinases have evolved from an ancestral monomer arginine kinase. Selection for the nucleotide substrate, ATP, has been rigid and so also, once evolved, the nature of the subunit combining site. The binding site for the guanidine substrate has undergone, and in some groups is still undergoing, marked evolutionary change, thereby creating the potential catalytic activity that may ultimately become incorporated into the organism, forming a new sort of phosphagen store. Studies of the echinoderms suggest that creatine kinase evolved in the chordate line much earlier than was hitherto envisaged, and that it is first expressed in a highly motile gamete. Further evolution, the expression of new enzyme activities, required the development of appropriate control systems and all that this entails for integration into a fully functional adult animal. The limited evidence available is compatible with the idea that similar changes are going on in the tunicates and hemichordates. Perhaps the most striking feature to emerge is biochemical support for Garstang's original theory (Garstang, 1894, 1922), that major evolutionary steps always accompany modification of the most primitive biological structures and take effect from an early stage in development. Thus theories of the origin of chordates, such as that of Berrill (1955), are fully justified in biochemical as well as in biological terms. At present a general survey of the phosphagen kinases is unable to provide a quantitative assessment of the relative affinities of the echinoderms, hemichordates and tunicates to the remaining chordates. A close study of the detailed properties of individual enzymes offers this goal for the future.

REFERENCES

Baldwin, E. (1937). *An introduction to comparative biochemistry.* London: Cambridge University Press.
Baldwin, E. (1953). Biochemistry and evolution. *Symp. Soc. exp. Biol.* **7**: 22–30.
Baldwin, E. & Yudkin, W. H. (1950). The annelid phosphagen: with a note on phosphagen in Echinodermata and Protochordata. *Proc. R. Soc.* (B) **136**: 614–631.
Berrill, N. J. (1955). *The origin of vertebrates.* Oxford: Clarendon Press.
Borsuk, V., Kreps, E. & Verzhbinskaya, N. A. (1933). Chemical changes in muscles of ascidians and annelids. *Sechenov. J. Physiol.* **16**: 773–781.
Davies, R. E. (1965). On the mechanism of muscle contraction. In *Essays in biochemistry* **1**: 29–55. Campbell, P. N. and Greville, G. D. (eds). London: Academic Press.
Der Terrossian, E., Desvages, G., Pradel, L. A., Kassab, R. & Thoai, N. V. (1971). Comparative structural studies of the active site of ATP: guanidine phosphotransferases. *Eur. J. Biochem.* **22**: 585–592.
Di Jeso, F. (1967). Identification of arginine kinase activity in the bacterium *Escherichia coli. C. r. Séanc. Soc. Biol.* **161**: 584.

Eppenberger, H. M., Eppenberger, M., Richterich, R. & Aebi, H. (1964). The ontogeny of creatine kinase isoenzymes. *Devl Biol.* **10**: 1–16.

Garstang, W. (1894). Preliminary note on a new theory of phylogeny of the chordata. *Zool. Anz.* **17**: 122–125.

Garstang, W. (1922). The theory of recapitulation: a critical restatement of the biogenetic law. *J. Linn. Soc.* (Zool.) **35**: 81–101.

Meyerhof, O. (1930). *Die Chemischen Vorgange im muskel.* Berlin: Springer.

Moreland, B. & Watts, D. C. (1967). Molecular weight isoenzymes of arginine kinase in the Mollusca and their association with muscle function. *Nature, Lond.* **215**: 1092–1094.

Moreland, B., Watts, D. C. & Virden, R. (1967). Phosphagen kinases and evolution in the Echinodermata. *Nature, Lond.* **214**: 458–462.

Morrison, J. F., Griffiths, D. E. & Ennor, A. H. (1956). Biochemical evolution: position of the Tunicates. *Nature, Lond.* **178**: 359.

Needham, D. M., Needham, J., Baldwin, E. & Yudkin, W. H. (1932). A comparative study of the phosphagens with some remarks on the origin of vertebrates. *Proc. R. Soc.* (B) **110**: 260–294.

Rockstein, M. (1971). The distribution of phosphoarginine and phosphocreatine in marine invertebrates. *Biol. Bull. mar. biol. Lab. Woods Hole* **141**: 167–175.

Thoai, N. V. (1968). Homologous phosphagen phosphokinases. In *Homologous enzymes and biochemical evolution*: 199–229. Thoai, N. V. and Roche, J. (eds). New York: Gordon and Breach.

Thoai, N. V., Robin, Y. & Guillou, Y. (1972). A new phosphagen, phosphothalassemine. *Biochemistry Easton.* **11**: 3890-3895.

Van Pilsum, J. F., Stephens, G. C. & Taylor, D. (1972). Distribution of creatine, guanidinoacetate and the enzymes for their biosynthesis in the animal kingdom. *Biochem. J.* **126**: 325–345.

Verzhbinskaya, N. A., Borsuk, V. N. & Kreps, E. N. (1935). The biochemistry of muscle contraction in an echinoderm *Cucumaria frondosa. Archs Sci. biol. St Petersb.* (afterwards *Arkh. biol. Nauk*) **38**: 369–382.

Virden, R. & Watts, D. C. (1966). The distribution of guanidine-adenosine triphosphate phosphotransferases and adenosine triphosphatase in animals from several phyla. *Comp. Biochem. Physiol.* **13**: 161–177.

Watts, D. C. (1965). Evolutionary implications of enzyme structure and function. In *Studies in comparative biochemistry*: 162–195. Munday, K. A. (ed.). New York: Pergamon Press.

Watts, D. C. (1971). Evolution of phosphagen kinases. In *Biochemical evolution and the origin of life*: 150–173. Schoffeniels, E. (ed.). Amsterdam: North Holland Publishing Co.

Watts, D. C. (1973). Creatine kinase. In *The enzymes* **8**: 383–455, 3rd edn. Boyer, P. D. (ed.). New York: Academic Press.

Watts, D. C., Focant, B. F., Moreland, B. M. & Watts, R. L. (1972). Formation of a hybrid enzyme between echinoderm arginine kinase and mammalian creatine kinase. *Nature, Lond.* (New Biol.) **237**: 51–53.

Watts, R. L. & Watts, D. C. (1968). Gene duplication and the evolution of enzymes. *Nature, Lond.* **217**: 1125–1130.

Symp. zool. Soc. Lond. (1975) No. 36, 129–158.

PROBLEMS OF IODINE BINDING IN ASCIDIANS

E. J. W. BARRINGTON

Department of Zoology, The University, Nottingham, England

SYNOPSIS

Early work of Cameron (1914, 1915), confirmed and amplified by recent radioiodine studies, shows that ascidians have an impressive capacity for the uptake and organic binding of environmental iodine. The belief that this phenomenon can contribute to our understanding of the origin of thyroidal biosynthesis is probably well-founded, but the argument needs to be developed with much caution.

Some account is given of the chemical composition of the tunic, which is essentially a mesenchymatous tissue with proteins, carbohydrate, a high water content, and a variety of wandering cells. Its structure and mode of secretion are discussed with special reference to *Dendrodoa* and *Styela*.

Radioactive glucose is secreted into the tunic by the mantle epithelium, but there is no evidence that these cells secrete iodinated products. The distribution of iodine in the tunic, where it is associated particularly with the surface cuticle, seems rather to reflect the pattern of distribution of structural protein and of associated oxidase systems.

Iodine binding in the ascidian tunic thus appears as essentially an invertebrate phenomenon, and contrasts markedly with evidence of intracellular iodine binding in the endostyle, which is discussed in more detail in the following chapter. It is thus the endostyle, rather than the tunic, that is more likely to provide evidence directly relevant to the origin of the thyroid gland. In any case, iodine binding in ascidians is at least as much a problem of ascidian biology as of vertebrate endocrinology, while interspecific variation complicates the task of generalizing effectively about such an ancient and highly specialized group.

UPTAKE, DISTRIBUTION AND BINDING OF IODINE

Ascidians have an impressive capacity for the uptake and organic binding of environmental iodine, as was first made clear in the investigations of Cameron (1914, 1915). Working under the direction of the Ductless Glands Committee of the British Association, he found substantial amounts of iodine in sponges, corals, annelid worms and their tubes, ascidian tunics, the opercula of whelks, and the byssus of *Mytilus*. He concluded that iodine was widely distributed in marine invertebrates, and that much of it was in their external secretions. His figures were for total iodine, but he recognized that some of this might be organically bound, and later work has shown that this is certainly so.

He paid particular attention to ascidians, using large numbers of several genera, and found considerable variation of iodine content in these animals, even in different individuals living under identical conditions. The amount of iodine could be very high. For example, the percentage content in the tough and leathery tunic of

Pyura was of the same order as that found in the egg-case of a skate, and in the thyroid of a dogfish, *Squalus sucklii*. Here, and in other ascidian species, almost all of the iodine was in the outer layer of the tunic; an important fact, to which we shall return later. He detected no iodine in the inner tunic, little if any in the body, and none in the endostyle.

More recent work, involving uptake studies with radio-iodide, has confirmed these findings in principle, but improved sensitivity has shown the distribution of iodine in the ascidian body to be less restricted than Cameron believed. Studies of *Ciona* (Roche, Salvatore and Rametta, 1962) shows considerable uptake of iodide from radioiodinated sea water, a plateau being reached after about five days, when the concentration within the animal is 16–18 times that in the surrounding medium. Examination of selected tissues, after removal of inorganic iodine by dialysis, shows that most of the organically-bound iodine is in the protective covering, or cuticle, of the tunic, but some is also detectable in the rest of the tunic, in the endostyle, in the pharynx (branchial sac), and in the body wall and viscera.

The exceptional part played by the tunic, already demonstrated by Cameron, is shown by the fact that 96% of the total bound iodine is present in that tissue. Most, but not all, is localized in the outermost region, the ratio of the concentration of ^{131}I to that in the sea water (tissue/medium, or T/M, ratio) being 50·0 for the cuticle, but only 8·3 for its inner layer. The amounts present in the other tissues of *Ciona* are small, and especially so in the endostyle, which contains only 0·4% of the total iodide content. However, amounts are not necessarily unimportant because they are small. In fact there is certainly some concentration in the endostyle, for its T/M ratio is 6·4, which is not far short of the value for the inner tunic. This figure for the endostyle is the more impressive because autoradiography shows that its bound iodine is restricted to a very limited area of the endostylar epithelium (Fig. 1). It follows, and we shall see the importance of this conclusion later, that this particular area must have considerable concentrating power.

Recent studies of iodine uptake and binding in ascidians have been strongly motivated by the belief that, because these animals are members of the Chordata, the results would contribute to an understanding of the origin of the thyroid gland and of thyroidal biosynthesis. This belief is probably well-founded, especially in view of the undoubted homology of the endostyle of protochordates with the thyroid gland of vertebrates (Barrington, 1974). but

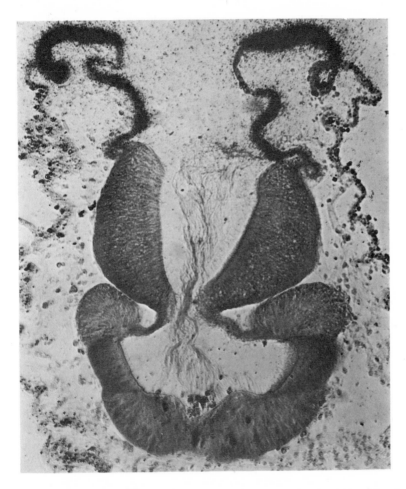

FIG. 1. Autoradiograph (^{131}I) of a transverse section of the endostyle of *Ciona*. Two days immersion in radioiodinated sea water (× 212). (From Barrington & Thorpe, 1965a.)

it presents problems that call for careful handling. As Cameron's results showed, iodine binding is a feature of invertebrate biology in general, rather than of ascidian biology in particular. Moreover, the phylogenetic relationship of ascidians to vertebrates is by no means clear-cut, and, whatever conclusions emerge from this Symposium, it is certain that a long period of independent evolution must separate these two groups from any hypothetical common ancestral stock. In consequence, comparisons between any aspect of the organisation of ascidians and vertebrates need to be formulated with great care, while full regard must also be paid to the inter-specific variability which is to be expected in survivors of so ancient a group.

These considerations apply in full force to the tunic, which, because it is so rich in iodine, has been the focus of much recent discussion, with the implication that its iodination is linked in some way to the establishment of the thyroid gland. Triiodothyronine (T_3) and thyroxine (T_4) have been reported in the tunic of *Ciona*, together with their precursors, monoiodotyrosine (MIT) and diiodotyrosine (DIT) (Roche, Salvatore *et al.*, 1962), in *Clavelina* (Salvatore, Vecchio & Macchia, 1960) and in the thaliacean *Salpa maxima* (Roche, Rametta & Varrone, 1962), but their significance is difficult to interpret, for their points of origin have not been established. Two possibilities (which need not be mutually exclusive) have been proposed (Roche, Rametta & Varrone, 1964): either that they are synthesized *in situ* in the tunic, or that they are transported to the tunic after synthesis elsewhere, most likely in the pharynx.

Our own results stand in some contrast to these, for we have had difficulty in identifying thyronines in ascidians, although we have found small amounts of thyroxine in the endostyle of *Ciona* (Barrington & Thorpe, 1965b). Suzuki & Kondo (1971) have had the same experience with the branchial sac homogenates of *Chelysoma*, in which they readily identify MIT and DIT, but find only traces of thyroxine. They remark that this finding differs from those of Roche, Salvatore *et al.* (1962), but they do not mention that it agrees with our own results from *Ciona*.

More recently, in attempts to find some basis for these disagreements and uncertainties, we have concentrated our attention on two other genera, *Dendrodoa* and *Styela*, both of which have very tough tunics. Initially (Barrington & Thorpe, 1965b) we used Dowex resin columns to purify our homogenates, which were prepared from animals that had been immersed in radioiodinated sea-water. This procedure gives excellent and clear-cut identification of MIT, T_3 and T_4 in the thyroid gland of trout, for example. In the tunic of *Dendrodoa*, however, we found only MIT and DIT, together with a profusion of unknowns and pigments, but no trace of the iodothyronines.

We have therefore resorted to another procedure (Amaral, Morris & Barrington, 1972), involving the partitioning of the tunic homogenate into water-soluble and butanol-soluble fractions (Zappi, 1967), followed by thin-layer chromatography, combining this with the various precautions which are now known to be necessary to exclude loss of iodinated material from mammalian thyroid extracts. The butanol extracts still contain a variety of unknowns, but by elution of areas selected by reference to the positions of known marker compounds, followed by re-chromatographing, developing, and scanning for radioactivity, we have obtained satisfactory evidence for the presence of thyroxine and triiodothyronine in the tunic of *Dendrodoa*. A detailed report of these results will be published later.

The tunic of *Styela*, however, differs somewhat from that of *Dendrodoa*. Using exactly the same procedures, and with amounts of homogenate ten times greater than the *Dendrodoa* material, we have so far failed to find satisfactory evidence of iodothyronines. They may, of course, be present, for we are clearly working close to the limits of sensitivity of our methods, but if they are present, the amounts must be very small indeed, and much less than the amounts present in *Dendrodoa*.

It is against this background of chemical complexity and interspecific variability that we have also been studying, side by side with our chromatography, the structure and mode of development of the tunic in *Dendrodoa* and *Styela*. Our concern, however, is not only with the tunic, which seems too specialized and unusual a tissue to be usefully compared with the thyroid gland. The endostyle, homologous as it is with the thyroid, is altogether a better candidate. This, at least, is the point of view from which Dr Thorpe and I are approaching in these two chapters the problem of iodine binding in ascidians. We aim, in particular, to present some comparison of tunic and endostyle, and to set this in the dual perspective of invertebrate biology and thyroid evolution.

CHEMICAL COMPOSITION OF THE TUNIC

The tunic is a complex tissue (Godeaux, 1964; Stiévenart, 1970, 1971) which can be interpreted in general terms as a form of mesenchyme, into which pass many wandering cells from the blood stream. Some of these cells, as we shall see, probably contribute to its production. It has a very high water content, ranging from 76% in *Halocynthia papillosa* (Stiévenart, 1971) to over 90% in *Phallusia mammillata* (Endean, 1961). Apart from this, its main constituents are protein and carbohydrate, the carbohydrate being present chiefly as cellulose, and as a component of acid mucopolysac-charide (Endean, 1955, 1961; Smith, 1970b). Protein and carbohydrate are present in about equal proportions in *Halocynthia aurantium* (Smith & Dehnel, 1970), but this is one of the features in which there is much variation. Protein is said to be relatively scarce in *Phallusia*, and acid mucopolysaccharide abundant, whereas in *Ascidella* the reverse is true (Stiévenart, 1970). Endean (1955) also finds very little protein in *Pyura*, but this is not easy to reconcile with Cameron's account of the iodine content of this animal, if we can assume that much of the iodine is associated with protein, as it clearly is in other genera. *Pyura* merits further study, using techniques of controlled hydrolysis and amino acid analysis.

Apart from this variability, the protein that is present is of more than one form. Thus in *Halocynthia* (Smith & Dehnel, 1970) the total protein content of the tunic is about 50% of the dry weight, but only 60% of this protein can be hydrolysed by pronase. Acid or base hydrolysis, at high temperatures or for long periods, is required for hydrolysing the remainder.

Results similar in principle were obtained for *Ciona* by Roche, Salvatore *et al.* (1962), in a study of the butanol-soluble iodoamino acids of tunics taken from *Ciona* which had been immersed in radioiodinated sea water. About 20% of the bound iodine was extractable as iodoamino acids without prior hydrolysis. A further 50% of the organic iodine was extractable in butanol after proteolytic digestion, by successive use of pancreatin (for 48 h) and papain (for 24 h) at 38°C, these being longer periods than would now be thought desirable if loss of iodothyronines is to be avoided. The remainder of the bound iodine (25–30% of the total) was extractable in butanol after alkaline hydrolysis of the residue, although with some degradation and production of free iodine. This enzyme-resistant fraction was assumed to comprise some form of structural protein (scleroprotein) localized particularly in the cuticle (Roche, Rametta *et al.*, 1964).

STRUCTURE OF THE TUNIC

These chemical components are distributed amongst three main structural features. One is the hyaline ground substance, in which the water and acid mucopolysaccharides are major constituents (Endean, 1955, 1961; Smith, 1970a, b); a second comprises the fibres which course through this substance; the third is the tough cuticle at the surface of the tunic.

The ground substance contains water and acid mucopolysaccharides, the latter shown by Alcian blue staining and toluidine blue metachromasia. The fibres also respond to these reagents, perhaps because acid mucopolysaccharides act as a binding agent for them (Endean, 1955). However, the best known component of the fibres is the cellulose which makes up much of the carbohydrate content of the tunic. Fibrils in *Phallusia* correspond in diameter to those of the cellulose microfibrils of plant fibres (Endean, 1961). Deck, Hay & Revel (1966) also note their similarity to plant cellulose fibrils, and confirm that they react histochemically like cellulose. However, Smith & Dehnel (1970) find that *Halocynthia* contains a polysaccharide which is resistant to the action of cellulase. They conclude that it is a form of cellulose which has undergone extensive substitution and cross-linkage. Endean (1955), too, finds the fibres of *Pyura stolonifera* to be composed of a very insoluble polysaccharide, which differs from cellulose in certain respects.

Several authors consider that the fibres are not in any case composed solely of carbohydrate, but are a protein-polysaccharide complex. Smith & Dehnel (1970) believe this to be so in *Halocynthia*, as also do Hall & Saxl (1961) in *Ascidella*. These latter authors found evidence for collagen and elastin, but Smith & Dehnel doubt the existence of these in *Halocynthia*, as they found no hydroxy-proline. Barrington & Barron (1960) found histochemical indications of tyrosine and of SH groups in the fibres of *Ciona*, while Barrington & Thorpe (1968) found similar evidence in *Dendrodoa*, where the fibres are much more densely developed, and where histochemical evidence is correspondingly easier to evaluate. By contrast, we have not obtained a positive Millon response from the fibres of *Styela*, although this does not necessarily mean that tyrosine is completely absent from them.

The cuticle, which sometimes bears spines (e.g. in *Halocynthia*, Stiévenart, 1971; *Styela*, Fig. 2), forms a very tough protective covering to the tunic. Often (and this is an important factor in the evaluation of iodine binding) it contains a very high proportion of the total protein of the tunic, amounting to 85% of its protein nitrogen in *Halocynthia* (Stiévenart, 1971). In several genera (including *Ciona, Dendrodoa, Styela*) it gives a strongly positive histochemical response for glycoprotein (PAS reaction), and tyrosine (Millon reaction, Fig. 3). It is, however, negative to Alcian blue and is not metachromatic with toluidine blue. It thus lacks acid mucopolysaccharide, and in this respect it contrasts with the fibres and general matrix of the rest of the tunic. No doubt it consists largely of the scleroprotein-like material already mentioned.

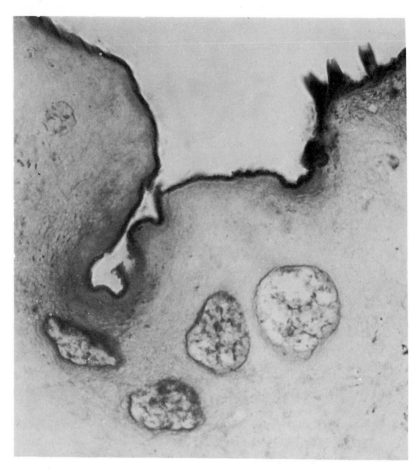

Fig. 2. Transverse section of the outer region of the tunic of *Styela*, to show (above) the cuticle and spines, stained by azocarmine. Formol iodate: Azan (× 570).

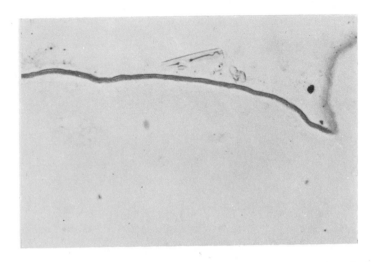

FIG. 3. Transverse section of the outer region of the tunic of *Styela* to show the intense Millon reaction of the cuticle. Formol saline (× 365).

SECRETION OF THE TUNIC

The mode of secretion of the tunic matches the complexity of its structure, but it is agreed that the cells of the mantle epithelium play an important role in this (Smith, 1970b). By light microscopy the ground substance and fibres seem continuous with the tips of these cells, while there is histochemical evidence of acid mucopolysaccharide and RNA in the cytoplasm. There is, however, no evidence of cellulose within the cells. At this level of analysis, the mantle cells of *Ciona, Dendrodoa* and *Styela*, and doubtless of other genera, seem a likely source of both the carbohydrate and the protein moieties of the tunic, with polymerization and fibre formation taking place outside them. *Phallusia*, however, is reported as an exception, for here the epidermal cells lining the tunic vessels produce the mucopolysaccharide, while wandering vanodocytes are thought to produce the cellulose fibres, which appear to arise directly from their surfaces (Endean, 1961).

The importance of the mantle epithelium in *Perophora viridis* and *Dendrodoa* is confirmed by electron microscopy (Deck *et al.*, 1966; and our own observations, Fig. 4), which show conspicuous rough and smooth endoplasmic reticulum, mitochondria, and a Golgi region from which arise presumed secretory vesicles of medium opacity. The vesicles pass towards the apical ends of the cells, where they appear to discharge their contents.

138   E. J. W. BARRINGTON

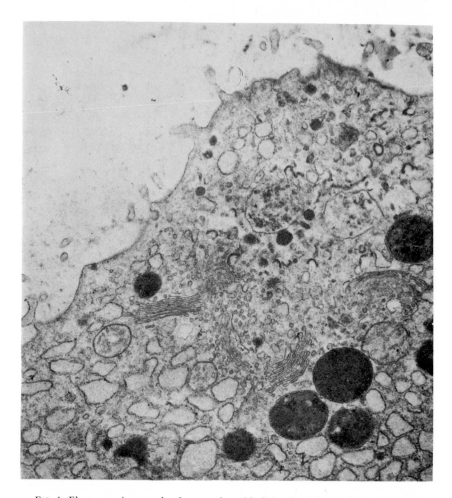

FIG. 4. Electron micrograph of a mantle epithelial cell of *Dendrodoa*. Note the well-developed Golgi region (below centre), with vesicles (presumed secretory) passing from the latter to the cell border. The significance of the larger electron-dense inclusions is not yet clear. 5% glutaraldehyde in 0·1 M cacodylate buffer; postfixation in 1% osmium tetroxide in veronal acetate buffer (× 8575).

    Particularly illuminating in this respect are the preliminary autoradiographical observations of Deck *et al.* (1966) on labelled glucose uptake in *Perophora*, confirmed by our own observations of the same process in *Dendrodoa* and *Styela*. The results are necessarily affected by the age and condition of the specimens, but the supra-nuclear region of the mantle epithelial cells may be well labelled after exposure of the animal for 1–2½ h to tritiated glucose

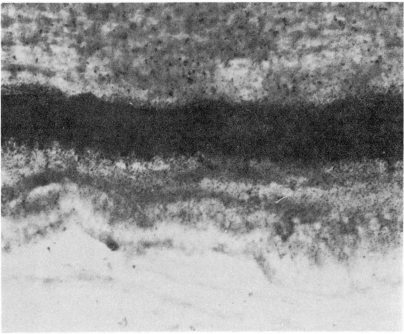

FIG. 5. *A*. Autoradiograph of the mantle epithelium (below) and adjacent tunic of a specimen of *Styela* which had been immersed in sea water containing tritiated glucose (5 μl/ml) for 3½ h, and then left in normal sea water for 2 h. Glucose lies immediately external to the epithelium and partly over the cell surfaces. Bouin's in sea water; Alcian blue and periodic-acid-Schiff (× 365). *B*. Another specimen, similarly treated, but left in normal sea water for 4 days. During the longer interval, glucose has moved away from the mantle epithelium (below) in a dense band (× 365).

in sea water. Then, during the following days, the labelled glucose diffuses out into the tunic (Fig. 5), with some indication of its presence in fibres, although radioactivity is also evident in the general ground substance. Deck *et al.* particularly emphasize that essentially all of the radioactive material that is eventually found in the tunic originates in the mantle epithelial cells, and we can confirm this. Certainly we find some evidence of glucose over the outer surface of the cuticle, but it lies external to this and presumably is simply adsorbed to the surface. The rate of movement of the material through the tunic in young *Styela* indicates considerable secretory activity in these cells. These findings clearly support those of light microscopy, in showing that the mantle cells are the source of the carbohydrate moiety of the tunic. This is an important conclusion, for it contrasts sharply and significantly with the course of iodination.

There is, however, another factor to be considered. This is the contribution made to the tunic by the wandering cells which, as already mentioned, enter it from the bloodstream. Smith (1970b), in his account of *Halocynthia aurantium*, mentions two types which seem of particular significance for the formation of the tunic. One type, called by him the mature morula cell, is particularly concentrated immediately above the mantle epithelium. He suggests that these cells, which break down and lose their characteristic staining properties, are not secretory, for they lack RNA, but are perhaps responsible for the association of the fibres into the regular series of laminae which are a feature of this tunic. The other type, which he calls the dispersed vesicular cell, concentrates at the periphery of the tunic, adjacent to the cuticle. The contents of both types of cell react positively to tests for protein, as also does the cuticle. Smith considers that the positional relationships of these cells (Fig. 6) indicate their functional importance in the production of the tunic, and reinforces this conclusion by showing that the concentration of these cells in the tunic markedly increases after the tunic has been damaged. At first the two types are equally distributed, but by the tenth day after injury there is a significantly higher concentration of dispersed vesicular cells in the outer region of the tunic, while by the 15th day there is a significantly higher concentration of mature morula cells in the inner region (Fig. 7). Unfortunately, there is little histochemical evidence to suggest how their special contributions might be made.

However, our study of *Dendrodoa* (Barrington & Thorpe, 1968) has brought to light another aspect of tunic secretion. In this form,

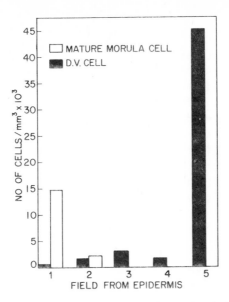

FIG. 6. The concentration of mature morula and dispersed vesicular cells (D.V.) in the tunic of *Halocynthia aurantium*. Field 1 is at the epidermis and field 5 at the external periphery of the tunic. Values are the mean concentrations for a sample of five animals with six replicate counts per animal. (From Smith, 1970a.)

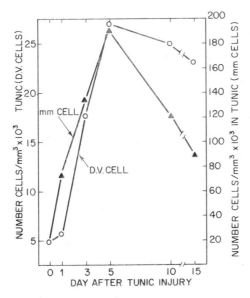

FIG. 7. The concentration of mature morula cells and dispersed vesicular cells ($\times 10^3$) per mm^3 of the tunic of *Halocynthia* as a function of day after injury of the tunic. The ordinate on the right refers to the mature morula cells, that on the left to the dispersed vesicular cells. Each point represents the value for six animals. (From Smith, 1970a.)

two types of granular cell, both of them large and conspicuous, traverse the mantle epithelium and come to lie in a densely compacted layer immediately above the tips of the mantle cells, where they begin to show signs of degranulation. From here they pass through the tunic towards the cuticle, showing disruption and further loss of contents as they do so. They are well seen after fixation in various formalin fixatives, or in Regaud's fluid (Fig. 8), but their granules tend to be destroyed by Bouin's fluid, so that their position close to the mantle cells is then occupied largely by nuclei, together with cell spaces and remnants of granules (Fig. 9).

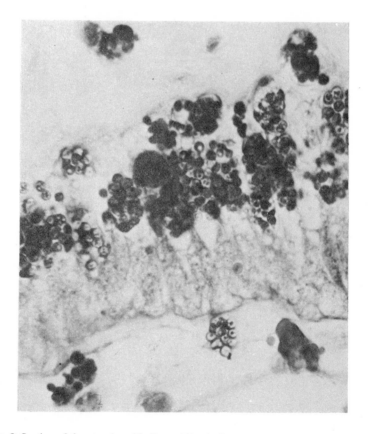

FIG. 8. Section of the mantle epithelium of *Dendrodoa*. Granule-containing cells are seen above and below it. The darker granules are strongly carminophil. The lighter ones belong to the supposed polyphenol cells, and are golden-brown in the original preparation. Formalin iodate, Azan (× 1150). (From Barrington & Thorpe, 1968.)

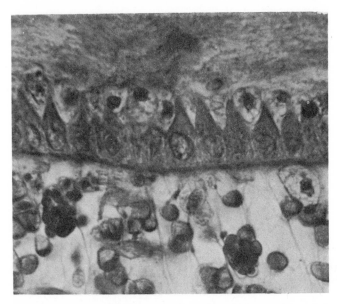

FIG. 9. Section of the mantle epithelium of *Dendrodoa*, which extends across the middle of the field. Between the narrow apices of the epithelial cells are seen the outlines of granule cells, empty except for their nuclei. Cf. Fig. 8. Bouin's in sea water, Azan (× 1160). (From Barrington & Thorpe, 1968.)

One of these cell types has granules that stain strongly with azocarmine. Its function is obscure, although it may perhaps be concerned with the conveying of excretory material, which is said to be deposited in the tunic of various ascidians by wandering cells. The other type has granules that tend to stain with orange G. Neither cell type gives a clear positive response to tests for protein or carbohydrate, but the granules of this second type give strong argentaffin (Fig. 10) and chromaffin responses. These reactions are of particular significance in the context of tunic secretion, because they are suggestive of polyphenols, although not conclusively so. Because of this, and because of the toughness of the tunic of *Dendrodoa*, it seemed possible to us that they might be concerned in the quinone tanning of protein, which depends on the oxidation of polyphenols by a polyphenol oxidase, and which is known to play an important part in the laying down of invertebrate scleroproteins.

The histochemical evidence for this would not be of great force taken by itself, but we have supplemented it with some biochemical evidence (Barrington & Thorpe, 1968). For example, fragments of

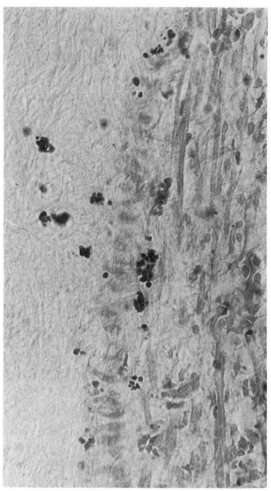

FIG. 10. Transverse section of the mantle epithelium (running centrally and vertically with tunic to the left) and adjacent regions in *Dendrodoa*. The presumed polyphenol cells, darkened by their argentaffin response, are seen below, in, and above the epithelium. Neutral formalin, Masson–Fontana argentaffin technique (× 570). (From a preparation by A. Thorpe.)

tunic rapidly darken to a dark purple-brown after immersion in 0·2% catechol in sea water for 1 h, and this colour is rapidly bleached by subsequent immersion in sodium hypochlorite. These reactions are held to demonstrate polyphenol oxidase (Smyth, 1954). Further, the oxygen uptake of tunic fragments or homogenates is sharply increased on the addition of the polyphenol homocatechol. This effect, which can be eliminated by prior boiling of the homogenate, strongly indicates that a polyphenol oxidase system is present in the tunic.

On this evidence we have concluded that quinone tanning contributes to the toughening of the tunic in *Dendrodoa*, but it is too early to judge whether this process is of general importance in the production of the ascidian tunic. Stiévenart (1970, 1971) finds some histochemical evidence of polyphenols in the tunic of *Ascidiella* and of *Halocynthia papillosa*, but this means little without biochemical support. We ourselves (unpublished) have found marked differences between the tunic of *Dendrodoa* and of *Styela*. In the tunic of *Styela* only one type of cell predominates. It is elongated, with PAS-positive granules, and is perhaps degenerating, for it seems sometimes to lack organelles. It looks like a fibroblast, being often continuous with individual fibres, and in this respect is reminiscent of the vanadocytes of *Phallusia*, referred to earlier (p. 137). This cell type, which extends throughout the tunic, is probably related to somewhat similar cells that cluster immediately above the mantle epithelium of *Styela*, although these seem often to lack contents. Close beneath the cuticle it forms aggregations (Fig. 2), surrounded by fibres and containing what look like disrupted cells and dispersed PAS-positive material. The cuticle might perhaps be formed from such aggregations, although we have no evidence that this actually occurs. In addition to this cell type, there is another, much less abundant, with intensely carminophil granules. It perhaps corresponds to one of the *Dendrodoa* types, but as yet we have no evidence for this (Fig. 11).

Other evidence emphasises the difference between these two genera. We have not obtained any significant argentaffin or chromaffin response from the wandering cells of *Styela*, nor is there any evidence, from colour reaction or from oxygen consumption in the presence of added polyphenol, of a polyphenol oxidase. This need not mean that quinone tanning is entirely absent, but it cannot have the importance that is evident in *Dendrodoa*. Associated with this difference is a difference in the distribution of tyrosine. The cuticle of *Styela*, which is strongly PAS-positive, gives also a strongly positive Millon reaction (Fig. 3), while the rest of the tunic is negative to this test. The cuticle of *Dendrodoa*, by contrast, responds erratically and rather weakly to the Millon tests, while weak responses are also given by the fibres. Tyrosine-rich protein seems, therefore, to be much more localized in *Styela* than it is in *Dendrodoa*, and we shall see that the distribution of bound iodine shows some agreement with this.

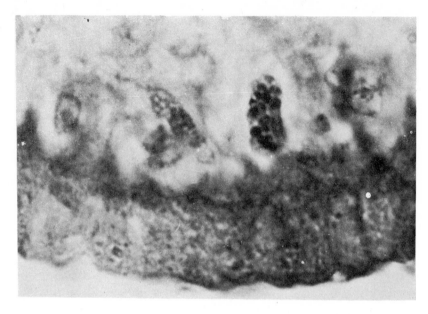

FIG. 11. Transverse section of the mantle epithelium of *Styela*, showing an azocar-minophil granule cell immediately above it. Cf. Fig. 8. Formalin iodate, Azan (× 2280).

Regrettably, these scattered findings mainly expose our consid-erable ignorance regarding tunic organization and development, which urgently need a broadly based comparative study. However, we can discern some common features amongst the diversity of detail. Secretion by the mantle epithelium is clearly one factor, while another is the involvement of wandering cells. The secretion and processing of protein, sometimes associated with the forma-tion of carbohydrate-protein complexes, and with quinone tan-ning, helps to establish the protective and supporting qualities of the tunic. Another factor is the production at the surface of a tough and proteinaceous cuticle, which is sometimes rich in tyrosine, although unfortunately it is not clear how this cuticle is formed. These, then, are the factors which must be taken into account in explaining the presence and distribution of bound iodine in the tunic. At present the explanation can only be tentative, but enough can be said to set this problem of iodine metabolism, and its supposed relationship to thyroidal biosynthesis, in an intelligible evolutionary perspective and against a background of ascidian biology.

IODINE BINDING IN THE TUNIC

Autoradiography of animals that have been immersed in radioiodinated sea water confirms the chemical studies in showing that bound iodine is located mainly in the cuticle. Using specimens of *Dendrodoa* (Barrington & Thorpe, 1968) removed from the iodinated sea water after periods of immersion ranging from 15 min up to at least 24 h, traces of bound iodine are seen over the cuticle within 1 h of immersion, but there are none over the rest of the tunic. By 4 h there is a dense image over the cuticle (Fig. 12), while a scattered and much weaker image is seen over the rest of the tunic, with some concentration around some of the granular cells. There is no trace at all, however, of any image over the mantle epithelial cells. Thus, in contrast to the evidence of carbohydrate secretion obtained with tritiated glucose, there is no evidence whatsoever that iodine is secreted into the tunic from the mantle epithelium.

The same conclusion emerges from our observations on *Styela*. Autoradiographs of the tunic indicate a sharp concentration of bound iodine over the cuticle, and underlying its inner surface (Fig. 13), but there is no trace of radioiodine in the mantle epithelium. In contrast to *Dendrodoa*, however, there is virtually no evidence of iodine in the rest of the tunic, nor does it show any association with the fibroblast-like cells.

The differences between these two genera are as instructive as the resemblances. As regards *Dendrodoa*, the presence in the inner layer of the tunic of tyrosine, and of presumptive polyphenol-containing cells, correlates very well with the autoradiographic evidence of bound iodine in this region, and of its tendency to concentrate around some of the wandering cells, for polyphenols, as we have seen, may well favour iodine binding. In *Styela*, by contrast, bound iodine is restricted to the cuticle, without any of the dispersion over the inner tunic seen in *Dendrodoa*, and with no clumping around cells. This can be correlated with the much higher tyrosine content of the cuticle in *Styela*, and with the lack of any Millon response in the rest of the tunic. It conforms also with the absence of clearly-defined presumptive polyphenol cells, and with the lack of any evidence for quinone tanning. In short, conditions in both genera support the view that the location of bound iodine in the tunic of both of these genera is determined by the pattern of the production and distribution of tyrosine-rich structural protein.

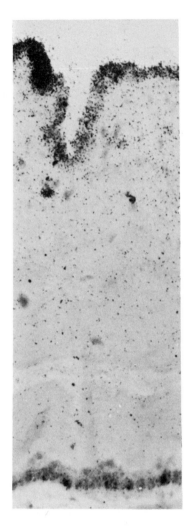

A

Fig. 12 *A*. Autoradiograph (^{125}I) of the tunic of *Dendrodoa* (4 h immersion in 200 μCi/l). The radioactive image of bound iodine is strong over the cuticle, but is very weak in the inner tunic; the small clump at one point in the latter (upper left) is concentrated around a wandering cell. Note the absence of an image over or adjacent to the mantle

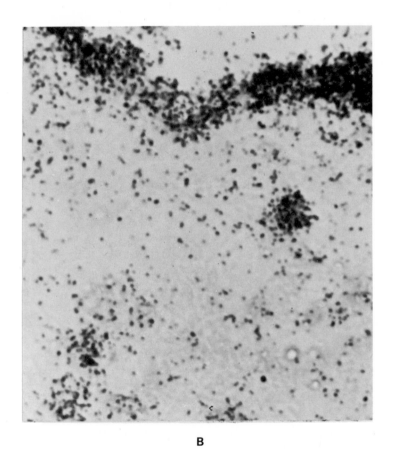

B

epithelium; cf. Fig. 5. Bouin's in sea water (× 365). *B.* A similar preparation (24 h immersion) showing in more detail the concentration of the image over the cuticle, with some concentration also around a wandering cell, at right of centre. Bouin's in sea water (× 1000). (From Barrington & Thorpe, 1968.)

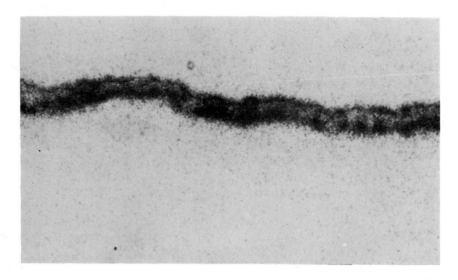

FIG. 13. Autoradiograph of the cuticle of *Styela*. The radioactive image of bound iodine is concentrated over the cuticle, and is virtually absent from the remainder of the tunic, which underlies it. Cf. Fig. 12. Animal immersed for 3 days in $0 \cdot 2 \ \mu Ci/ml$, washed 1 h, fixed Bouin's in sea water ($\times$ 365).

However, even if there is no quinone tanning in *Styela* (and, of course, it may not be entirely lacking), one must assume that some oxidase system is present, presumably associated particularly with the cuticle. Evidence that this must be so is found in autoradiographs prepared from isolated pieces of tunic treated in one or other of three ways: (a) immersion in radioiodinated sea water; (b) immersion in radioiodinated sea water with added thiouracil (which inhibits iodine binding in the thyroid gland through its anti-oxidant action; (c) boiled and then immersed in radioiodinated sea water. Procedure (a) results in a concentration of bound iodine over the cuticle and nowhere else; (b) and (c) result in complete absence of bound iodine. It is difficult to account for these results except on the assumption that enzyme-mediated action is involved in the binding. They certainly speak strongly against the possibility that the autoradiograph images might result solely from adsorption of iodine to the cuticular surface. As already mentioned, such adsorption probably does account for the images of glucose overlying the outer surface of the cuticle of specimens treated with radioactive glucose.

THE ASCIDIANS AS INVERTEBRATES

The evidence outlined certainly suggests a close parallel between ascidians and other invertebrates in respect of iodine binding. Consider, for example, the situation in molluscs, as analysed by Beedham & Trueman (1958), by Roche, André & Covelli (1960), and by Tong & Chaikoff (1961). These several studies were prompted by the fact that tyrosine is an important constituent of the scleroprotein in the periostracum and outer shell layers of lamellibranchs, and by the demonstration by Gorbman, Clements & O'Brien (1954) that protein-bound iodine can be located by autoradiography in the shell and underlying mantle of the lamellibranch *Musculium partumeium*.

In *Mytilus*, for example, uptake at 18°C is at first very rapid with a tendency to plateau after about 15 days (Roche, André *et al.*, 1960). It is inhibited by thiocyanate and, more markedly, by thiourea, but is resumed when these inhibitors are removed. Of the total radioactivity, 55% is in the byssus and byssus gland, 37·7% in the valves, 3·3% in the mantle, 2% in the gills, and 2% in the remaining soft tissues. These values compare quite well with figures, quoted earlier, for the various tissues of *Ciona*. Within the valves, the newly formed marginal region of the periostracum is the most active, while the secretory margin of the mantle is more active than the rest. Roche, André *et al.* (1960) concluded from these data that the binding of iodine in *Mytilus* is part of the process of scleroprotein formation, and is not a form of thyroidal biosynthesis. In accord with this, they found, as also did Tong & Chaikoff (1961), that the binding led only to the production of iodotyrosines, and that no iodothyronines were formed.

The autoradiographical studies of Beedham & Trueman (1958) are also in accord with this, and provide a remarkably close parallel to the situation in ascidians. They showed that ^{131}I was taken up into the alimentary tract and the digestive gland, but found no evidence that it was incorporated into the shell by secretory activity of the mantle. The iodine is mainly located on those surfaces of the shell which are exposed to the isotope in the surrounding sea water, and this is true also of the byssus, which autographs strongly, but only when it is fully exposed to the sea water. No iodine is found in the byssus threads while they are being secreted, nor is there any in the secreting tissues. Beedham & Trueman suggest that the distribution of iodine on these external surfaces may be in part accounted for by adsorption, but this can hardly be the complete explanation, for silver grains overlie the inner substance of the byssus, just as they do parts of the inner layer of the tunic of *Dendrodoa*. Presumably here, as in the endostyle (Thorpe & Thorndyke, 1975), the binding results from the meeting of several requirements: the presence in the tissue of tyrosine-rich protein, the exposure of the tissue to iodide, and the presence of an oxidase to activate the iodide. This is why uptake is inhibited by antioxidant goitrogens. However, one should not exclude the possibility that activated iodine is more readily adsorbed than the unactivated iodide.

But in any case, it appears that iodine binding in the tunic of ascidians has the characteristics of an essentially invertebrate process, closely comparable to the iodine binding found in lamellibranchs, and, like that, unrelated to thyroidal biosynthesis. Further, it is not different in principle from the situation in certain anthozoans and sponges, where the binding of iodine to scleroprotein results in the production of monoiodotyrosine and diiodotyrosine, accompanied by only minimal amounts of iodothyronines (Roche, Rametta *et al.*, 1964). Nor is it different in principle from the situation in the nemertine worm *Lineus*, where iodine becomes bound to the mucus secreted over the body surface. Major, Hanegan & Anoli (1969), who have reported this, assume that it must be due to the presence of a peroxidase. They find that triiodothyronine (but not thyroxine) is formed, in addition to MIT and DIT, which shows that this invertebrate type of iodine binding is not necessarily restricted in its results to the production of iodotyrosines. Indeed, these authors rightly remark that the process that they describe in nemertines may be widespread in invertebrates, and might well have been the origin of true thyroidal biosynthesis.

This, of course, is the possibility with which we began. To what conclusion have we now been led with respect to the thyroidal significance of iodine binding in ascidians? To answer this question we must take account not only of the tunic but also of the endostyle, for it is surely in this organ, the homologue of the thyroid gland, that we are most likely to find the answer.

THE ENDOSTYLE

The endostyle is dealt with in detail in the next chapter; all that I need to do here, therefore, is to emphasize the sharp contrast that we find between this organ and the tunic with respect to iodine binding. It is well shown in our time studies of iodine binding, which, as already mentioned, gave no evidence of any image over the mantle epithelial cells, and no evidence, therefore, of the production of any iodinated secretion by those cells. In the endostyle (Fig. 1), by contrast, there is one clearly defined region (zone 7 in our nomenclature, and to some extent zone 8) where bound iodine is certainly present within the cells.

Autoradiographs of *Ciona* (Barrington & Thorpe, 1965a) show a delicate image over the surface of these zones in specimens that have been immersed in radioiodinated sea water for only 30 min. Nothing similar is seen anywhere else in the endostyle, so that even from this very early stage these two zones emerge as having special relationships with iodine binding. At 1 h, 1½ h, 2 h and 4 h, images continue to be visible over the cell surface, but in addition there are images over the cell bodies of zone 7. After 8 h there is a dense image over zones 7 and 8, and by this time bound iodine appears to be present within the cell bodies of zone 8, although this is always much more evident in zone 7. Conditions in *Styela* (Figs 14 and 15) are similar. Our impression, which will be amplified in the next chapter, is that an iodinated product is present within the cells of zone 7, that it is being discharged (Fig. 15), and that the image over zone 8, which is a strongly ciliated zone, is due in large measure to secreted material adhering to the cilia.

IODINE BINDING AND THE ORIGIN OF THYROIDAL BIOSYNTHESIS

I suggest, in conclusion, that iodine metabolism in ascidians includes two distinct processes. One of these is the binding of iodine to tunic protein, which is not associated with secretory activity of the mantle cells, and is essentially extracellular. The other is intracellular binding to a secretion formed within the cells of a specialized region of the endostyle. (The possibility that it may also occur in parts of the pharyngeal epithelium is not excluded.) This second process could well be related phylogenetically to the thyroidal activity of vertebrates, as will be explained in the next chapter.

Here is a field in which we clearly need much more research, and, especially, comparative studies of a number of genera. But let us suppose that some form of thyroidal biosynthesis is indeed

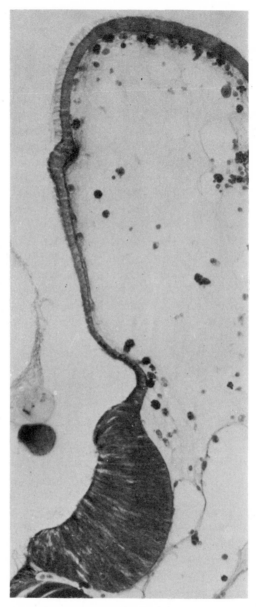

FIG. 14. Transverse section of one side of the upper region of the endostyle of *Styela*. The uppermost glandular zone (zone 6) lies below; above it extends the slender zone 7 and above this the ciliated zone 8, passing into the pharyngeal epithelium. Formol saline, Azan (× 365).

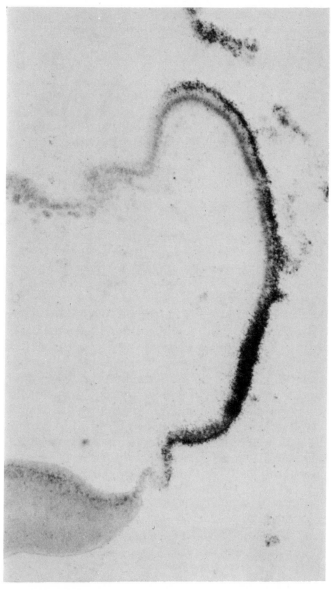

FIG. 15. As in Fig. 14, but an autoradiograph from an animal that had been immersed in 0·2 μCi/ml for 3 days, and placed in normal sea water for 1 h. Note the image of intracellular iodine in zone 7 and an image overlying the surface of zone 8, together with evidence of discharge into the pharyngeal lumen. Bouin's in sea water, haematoxylin (× 365).

established in the ascidian endostyle. Is there, in this case, any
relationship between tunic and endostylar iodination? The problem is complicated by Kennedy's (1966) demonstration that the
blood of *Ciona* is rich in iodine, and is capable of binding iodine *in
vitro*. She does not suggest an explanation for this, and the
information is too limited to permit discussion at this stage.
Kennedy implies that the iodination is fortuitous, but she rightly
emphasises that the blood must at least be capable of transporting
iodine, and that the tunic might, as suggested by Barrington &
Thorpe (1965a) and Roche, Rametta *et al.* (1964), utilize iodine
derived from elsewhere in the body. Perhaps this accounts for the
presence in the tunic of *Ciona* (Roche, Salvatore *et al.*, 1962) of
enzymatically hydrolysable iodoproteins, and of free iodoamino
acids. On the other hand, of course, it may be that they do indeed
originate in the tunic as a result of the binding of iodine to a protein
fraction which is not inert scleroprotein.

However this may be, the uncertainty regarding the origin of
the tunic iodine makes it impossible at present to decide whether or
not there is any functional relationship between iodine binding in
the tunic and iodine binding in the endostyle, or whether their
presence in the same animal is fortuitous. Nevertheless, it makes
for economy of hypothesis to suppose that peroxidase-mediated
pathways of iodine binding exist in ascidians, and that the products
resulting in the various regions of the body are determined by the
nature and reactivity of the proteins that are present. Iodine
binding may have arisen in the endostyle because this organ was
heavily involved in protein metabolism. Its products may then have
proved to be physiologically advantageous, perhaps because conditions in the endostyle favoured the production of iodothyronines,
or perhaps because the products were incorporated into the food
and could therefore be absorbed from the intestine (Barrington,
1969). Such favourable circumstances would then have been
encouraged and developed by the action of natural selection. At
present this is as far as we seem able to go in an analysis of the
evolutionary significance of ascidian iodine binding, and of its
relevance to the origin of the thyroid gland. Let me repeat,
however, that I do not consider this to be the most important aspect
of the matters that I have been reviewing. I prefer to regard the
problems of iodine binding in ascidians as an aspect of ascidian
biology that transcends issues of comparative endocrinology.
Looked at in this way, it presents a range of fascinating problems
that offer much reward to enterprising researchers.

ACKNOWLEDGEMENTS

Dr R. Morris, Mr J. Carey and Mr T. Owen have kindly allowed me to refer to some unpublished results of research in progress in this Department.

REFERENCES

Amaral, A. D. do, Morris, R. & Barrington, E. J. W. (1972). Thin-layer chromatography of iodinated compounds in the ascidian *Dendrodoa grossularia* (van Beneden). *Gen. comp. Endocr.* **19**: 370–372.

Barrington, E. J. W. (1969). Unity and diversity in comparative endocrinology. *Gen. comp. Endocr.* **13**: 482–488.

Barrington, E. J. W. (1974). Biochemistry of primitive Deuterostomians. In *Chemical zoology* **9**. Florkin, M. & Scheer, B. T. (eds.). New York: Academic Press.

Barrington, E. J. W. & Barron, N. (1960). On the organic binding of iodine in the tunic of *Ciona intestinalis* L. *J. mar. biol. Ass. U.K.* **39**: 513–523.

Barrington, E. J. W. & Thorpe, A. (1965a). An autoradiographic study of the binding of ^{125}iodine in the endostyle and pharynx of the ascidian, *Ciona intestinalis* L. *Gen. comp. Endocr.* **5**: 373–385.

Barrington, E. J. W. & Thorpe, A. (1965b). The identification of monoiodotyrosine, diiodotyrosine and thyroxine in extracts of the endostyle of the ascidian, *Ciona intestinalis* L. *Proc. R. Soc. (B)* **163**: 136–149.

Barrington, E. J. W. & Thorpe, A. (1968). Histochemical and biochemical aspects of iodine binding in the tunic of the ascidian, *Dendrodoa grossularia* (van Beneden). *Proc. R. Soc. (B.)* **171**: 91–109.

Beedham, G. E. & Trueman, E. R. (1958). The utilization of ^{131}I by certain lamellibranchs, with particular reference to shell secretion. *Q. Jl microsc. Sci.* **99**: 199–204.

Cameron, A. T. (1914). Contributions to the biochemistry of iodine. I. *J. biol. Chem.* **18**: 335–380.

Cameron, A. T. (1915). Contributions to the biochemistry of iodine. II. *J. biol. Chem.* **23**: 1–39.

Deck, J. D., Hay, E. D. & Revel, J.-P. (1966). Fine structure and origin of the tunic of *Perophora viridis*. *J. Morph.* **120**: 267–280.

Endean, R. (1955). Studies of the blood and tests of some Australian ascidians. II. The test of *Pyura stolonifera* (Heller). *Aust. J. mar. Freshwat. Res.* **6**: 139–156.

Endean, R. (1961). The test of the ascidian, *Phallusia mammillata*. *Q. Jl microsc. Sci.* **102**: 107–117.

Godeaux, J. (1964). Le revêtement cutané des Tuniciers. *Studium Generale* **17**: 176–190.

Gorbman, A., Clements, M. & O'Brien, R. (1954). Utilization of radioiodine by invertebrates, with special study of several annelids and mollusca. *J. exp. Zool.* **127**: 75–93.

Hall, D. A. & Saxl, H. (1961). Studies of human and tunicate cellulose and of their relation to reticulin. *Proc. R. Soc. (B.)* **155**: 202–217.

Kennedy, G. R. (1966). The distribution and nature of iodine compounds in ascidians. *Gen. comp. Endocr.* **7**: 500–511.

Major, C. W., Hanegan, J. L. & Anoli, L. (1969). Organic binding of iodide in nemertean mucus, *in vivo* and *in vitro*. *Comp. Biochem. Physiol.* **28**: 1153–1160.

Roche, J., André, S. & Covelli, I. (1960). Sur la fixation de l'iode par la Moule (*Mytilus galloprovincialis* L.) et la nature des combinaisons iodées élaborées. *C. r. Séanc. Soc. Biol.* **154**: 2201–2206.

Roche, J., Rametta, G. & Varrone, S. (1962). Sur la présence d'hormones thyroïdiennes chez un Tunicier pélagique, *Salpa maxima* Forskål. *C. r. Séanc. Soc. Biol.* **156**: 1964–1968.

Roche, J., Rametta, G. & Varrone, S. (1964). Métabolisme de l'iode et formation d'iodothyronines (T3 and T4) au cours de la régénération de la tunique chez une ascidie, *Ciona intestinalis* L. *Gen. comp. Endocr.* **4**: 277–284.

Roche, J., Salvatore, G. & Rametta, G. (1962). Sur la présence et la biosynthèse d'hormones thyroïdiennes chez un tunicier, *Ciona intestinalis* L. *Biochim. Biophys. Acta* **63**: 154–165.

Salvatore, G., Vecchio, G. & Macchia, V. (1960). Sur la présence d'hormones thyroïdiennes chez un Tunicier, *Clavelina lepadiformis* (M.Edw.) var. *rissoana*. *C. r. Séanc. Soc. Biol.* **154**: 1380–1384.

Smith, M. J. (1970a). The blood cells and tunic of the ascidian *Halocynthia aurantium* (Pallas). I. Haematology, tunic morphology, and partition of cells between blood and tunic. *Biol. Bull. mar. biol. Lab. Woods Hole* **138**: 354–378.

Smith, M. J. (1970b). II. Histochemistry of the blood cells and tunic. *Biol. Bull. mar. biol. Lab. Woods Hole* **138**: 379–388.

Smith, M. J. & Dehnel, P. A. (1970). The chemical and enzymatic analyses of the tunic of the ascidian *Halocynthia aurantium* (Pallas). *Comp. Biochem. Physiol.* **35**: 17–30.

Smyth, J. D. (1954). A technique for the histochemical demonstration of polyphenol oxidase and its application to egg-shell formation in helminths and byssus formation in *Mytilus*. *Q. Jl microsc. Sci.* **95**: 139–152.

Stiévenart, J. (1970). Recherches histologiques sur la tunique de deux ascidiaces phlébobranches: *Ascidiella aspersa* Müll. et *Phallusia mammillata* Cur. *Annls Soc. r. zool. Belg.* **100**: 139–158.

Stiévenart, J. (1971). Recherches sur la structure de la tunique d'une Ascidie stolidobranche: *Halocynthia aurantium* (Pallas). *C. r. hebd. Séanc Acad. Sci., Paris* **272**: 1873–1875.

Suzuki, S. & Kondo, Y. (1971). Demonstration of thyroglobulin-like iodinated proteins in the branchial sac of tunicates. *Gen. comp. Endocr.* **17**: 402–406.

Thorpe, A. & Thorndyke, M. C. (1975). The endostyle in relation to iodine binding. *Symp. zool. Soc. Lond.* No. 36: 159–177.

Tong, W. & Chaikoff, I. C. (1961). [131]I utilization by the aquarium snail and the cockroach. *Biochim. Biophys. Acta* **48**: 347–351.

Zappi, E. (1967). Group separation of an aqueous solution of some iodinated amino acids and derivatives by means of solvent extraction. *J. Chromat.* **30**: 611–613.

Symp. zool. Soc. Lond. (1975) No. 36, 159–177.

THE ENDOSTYLE IN RELATION
TO IODINE BINDING

A. THORPE and M. C. THORNDYKE*

*The Department of Zoology and Comparative Physiology,
Queen Mary College, London University, Mile End Road, London, England*

SYNOPSIS

There are distinct zones of cells in the protochordate endostyle which contribute to the production of a membrane essential to the filter-feeding habit. [125]I autoradiography shows that the cells of zone 7 in tunicates and zone 5 in amphioxus are able to bind iodine, whilst chromatography of endostylar extracts reveals the presence of monoiodotyrosine and diiodotyrosine with limited amounts of the hormones thyroxine and triiodothyronine. There are slight histochemical differences between the iodine-binding cells of tunicates and amphioxus, although both seem capable of producing a glycoprotein suitable for iodination. At the ultrastructural level, tunicate cells usually have small electron-lucent vesicles emanating from an active rough endoplasmic reticulum and Golgi apparatus. They appear either free in the apical cytoplasm or associated in multivesicular bodies. Electron microscopy combined with ^{125}I autoradiography suggests that the iodination process involves (a) the apical plasma membrane (b) the cytoplasmic vesicles (c) the multivesicular bodies (d) the residual bodies (a developmental stage of the multivesicular body) and (e) the intra-lumenal material. Electron microscope histochemistry shows, furthermore, that a peroxidase enzyme, known from vertebrate thyroid studies to be essential for the initial conversion of iodide to its active state and for the subsequent coupling of iodotyrosine to the hormonal molecules, is present in certain of the electron-lucent vesicles, whereas others are thought to contain the iodinating protein.

Evidence points to the iodine-binding process occurring within the cell, and notably within the multivesicular body, possibly due to the aggregation here of a mixed population of reacting vesicles, from whence the iodoprotein is ultimately released to the exterior following lysis. Glycoprotein could also be iodinated immediately following its release into the lumen where, presumably, the presence of the peroxidase enzyme is also essential. This extra-cellular process is likely to occur in the immediate vicinity of the apical plasma membrane.

Although reabsorption of iodinated material into the zone 7 cell is not ruled out, it seems more likely that digestion of the protein, and release of any hormones, occurs when the filtering membrane and enmeshed food particles reach the intestine.

It is seen that there are obvious and interesting parallels with vertebrate thyroid biosynthesis, the evidence favouring the iodine-binding cells of the protochordate endostyle as fore-runners of the thyroid gland cells.

INTRODUCTION

The endostyle is a characteristic pharyngeal organ of protochordates, producing a complex membrane essential to their filter-feeding habit. The discovery that certain of its cells are able selectively to bind iodine (Barrington & Franchi, 1956; Thomas,

* Present address: The Department of Zoology, Bedford College, London University, Regents Park, London.

1956; Barrington, 1957) has considerable interest with regard to the possible evolution of the thyroid gland. Thus, it is well known that the endostyle of the ammocoete larva, a structure homologous to that of the protochordates, also has iodine-binding cells and that part of the organ transforms into the thyroid gland of the lamprey at metamorphosis. Could it be that the cells of the protochordate endostyle are even earlier thyroidal cells?

In a group of animals which have long interested students of chordate phylogeny (Berrill, 1955), it is not surprising that several research programmes have attempted to answer this question. The evidence resulting from some of these studies on ascidians and amphioxus (*Branchiostoma lanceolatum*) is the subject of this account.

The endostyle of *Ciona intestinalis* (Fig. 1a) may be taken as representative of the ascidians in general, for, excepting a few minor variations, the organ is remarkably uniform throughout the group (Sokólska, 1931; Berrill, 1950). Essentially, there are eight clearly-defined zones of cells (Barrington, 1957), each of which presumably plays a rôle in the production of the filtering membrane which is ultimately passed over the perforated walls of the pharynx.

Although there is some duality of function, the major ciliated zones are undoubtedly 3, 5 and 8, whereas zones 2, 4 and 6 have a dominant secretory rôle. The iodine-binding cells are chiefly those of zone 7 although zone 8 cells may have some minor capacity (Barrington & Thorpe, 1965b).

The endostyle of amphioxus (Fig. 1b), an animal believed by Berrill (1955) to be a specialized offshoot of the main chordate stock, differs quite markedly from that of ascidians, the iodine-binding cells being those of zone 5.

Despite the differences in endostylar design, it must be assumed that their rôles are basically the same. In some quite extraordinary way the various zonal secretions, including the iodinated products, interact to form a structured membrane which is moulded by the cilia and finally moved up the pharyngeal wall. The iodination process occurring in the specialized region, and the release of iodinated products into the endostylar lumen, may be considered as a basic part of membrane formation. The subsequent release of any resultant "thyroidal" hormones from parent protein

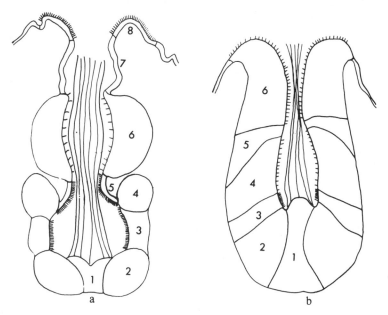

FIG. 1. Schematic diagram of protochordate endostyles to show zonation. (a) *Styela,* (b) *Amphioxus.*

molecules could occur when the membrane and its enmeshed food particles are digested in the intestine. Thus, it is quite conceivable that such biologically-important molecules as thyroxine could find their way into the interior tissues of protochordates along with the products of digestion (Barrington, 1969). The possibility also exists that some "thyroidal" products are liberated within the cells and released through the basal region to the circulatory system, or through the apical region to the endostylar lumen.

MICROSCOPICAL OBSERVATIONS

^{125}I autoradiography

Autoradiographical studies using ^{125}I on a variety of ascidians have established quite clearly the unique properties of the cells of zone 7 (Barrington & Thorpe, 1965b; Kennedy, 1966). The marked iodination occurring initially at or around the apical membrane of these cells in *Ciona* is seen nowhere else in the endostyle. Only after several hours immersion in radio-iodinated sea water can any iodinated material be seen in zone 8, and even then it is difficult to decide whether it has actually been formed there or whether it has been passed forwards from zone 7. At no time is there a significant autoradiographical image over the other endostylar zones. Barrington & Thorpe (1965b) concluded from their light microscope autoradiographical studies on *Ciona* that "the zone 7 cells are specialized in a way that promotes the synthesis of an iodinated product". They also suggested that the cells discharged this product, which later accumulated on the cilia of zone 8.

Autoradiographical images in amphioxus endostyles are found only at the apices of zone 5 cells (Thomas, 1956; Barrington, 1958).

Histochemistry

From the rather limited number of histochemical studies that have been carried out on tunicates, several general observations relevant to the iodine-binding phenomenon can be made. Thus, the secretions of the major glandular tracts, 2, 4 and 6, are rich in protein (Olsson, 1963; Lévi & Porte, 1964; Thorpe & Barrington, 1965; Thorpe, Thorndyke & Barrington, 1972) and it is of interest that in zone 6 in particular the protein is shown to be very rich in tyrosine residues (Thorpe, Thorndyke & Barrington, 1972). In view of the consistently negative results from autoradiography in this zone, it is evident that the iodine-binding process requires more than the mere presence of a tyrosine-rich protein.

Zone 1 primarily, and zone 3 to a limited extent, produce acid mucopolysaccharide and also some other carbohydrate or carbohydrate-protein complex. Zones 5 and 8 are dominant ciliated zones and histochemical evidence suggests only limited secretory function.

The zone 7 cells, which, in comparison with the major glandular tracts appear to have a minor secretory role, nevertheless show, at least in *Ciona* and *Styela*, clear-cut responses to histochemical tests for RNA, protein and a carbohydrate moiety as evidenced by the PAS test (Thorndyke, 1971; Thorpe, Thorndyke & Barrington, 1972). They share with vertebrate thyroid cells, therefore, the ability to elaborate a glycoprotein.

The iodine-binding zone 5 cells of amphioxus also produce material which is PAS-positive (Barrington, 1958; Olsson, 1963), although the dominant product here would seem to be acid mucopolysaccharide rather than glycoprotein.

THE IDENTIFICATION OF IODINATED PRODUCTS FROM THE ENDO-
STYLE

Chromatograms and electrophoretograms from endostylar extracts of *Ciona* (Roche, Salvatore & Rametta, 1962) and amphioxus (Tong, Kerkhof & Chaikoff, 1962; Thorpe, 1964) have shown the presence of the hormones thyroxine and triiodothyronine together with their precursors mono- and diiodotyrosine. Since the only cells to show positive autoradiographical images are those of zone 7 and zone 5 respectively, we must assume that the iodinated molecules are produced as part of the secretion of these cells. The proportions of hormones to their precursors in protochordates are dissimilar to the vertebrate thyroid products, and, indeed, Barrington & Thorpe (1965a) working with *Ciona*, and Suzuki & Kondo (1971) with *Chelysoma*, whilst showing the presence of the iodotyrosines and thyroxine in endostyle and branchial sac extracts, were unable to demonstrate the presence of triiodothyronine in these ascidians.

Such considerations could be regarded as evidence for the non-thyroidal nature of the iodine-binding process at this stage of evolution. Nevertheless, the fact that the iodothyronines are formed at all is in marked contrast to the results of the majority of studies on the iodination of scleroprotein in structural tissues. Roche, André and Covelli, (1960), working on the iodine binding of the tyrosine constituent in the shell scleroprotein of *Mytilus* suggested that, because only iodotyrosines were formed, the process was not indicative of thyroidal biosynthesis. Tong & Chaikoff (1961) reached the same conclusion from their studies of invertebrate scleroprotein. The much more significant point of the iodination of scleroproteins in the ascidian tunic has been admirably discussed by Barrington (1975) in the preceding paper of this Symposium, and, whilst there are certain species whose tunics show very limited amounts of iodothyronines, he concludes that the process is closely comparable with the iodine binding found in lamellibranchs and other invertebrates, and is unrelated to thyroidal biosynthesis.

Whilst, therefore, the exact nature of both endostylar and tunic iodine binding is doubtless open to discussion, there is a basic difference between the two phenomena. The iodination of intracellular protein within the endostyle cells, as opposed to scleroprotein laid down in skeletal tissues, makes them a very likely starting point for thyroid cell evolution and worthy of special examination.

ULTRASTRUCTURAL OBSERVATIONS

The normal appearance of iodine-binding cells

Electron microscope observations on the endostyle of *Ciona* provide additional evidence for the view that zone 7 cells have unique properties (Aros & Virágh, 1969; Thorpe, Thorndyke & Kirkham, 1969; Fujita & Nanba, 1971; Thorpe, Thorndyke & Barrington, 1972). Unlike the glandular tracts which have large, electron-dense granules as an obvious manifestation of their protein-secreting character, zone 7 cells have small electron-lucent vesicles, both free in the apical cytoplasm and aggregated in multivesicular bodies (Fig. 2). The other features of interest in the

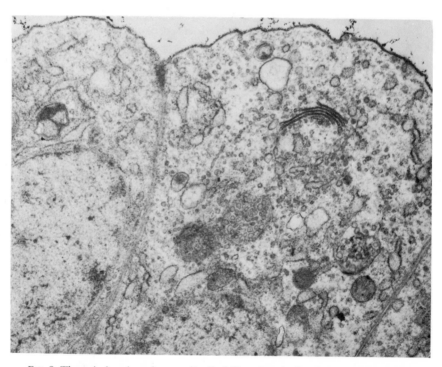

FIG. 2. The apical region of a zone 7 cell of *Ciona intestinalis*. The main features are the scattered rough endoplasmic reticulum, Golgi apparatus, multivesicular bodies and numerous electron-lucent vesicles in the cytoplasm (×10 000).

iodine-binding cells of *Ciona* are an active Golgi apparatus producing the small (60–100 nm) electron-lucent vesicles, occasional residual bodies (which are thought to result from breakdown of the multivesicular bodies (Thorpe, Thorndyke & Barrington, 1972)), and a well developed rough endoplasmic reticulum.

Additional studies on a range of tunicates by Thorndyke (1971), whilst confirming the basic properties of the zone 7 cell, nevertheless suggest that minor inter-specific variations may occur. Thus, whilst *Molgula manhattensis*, *Dendrodoa grossularia* and *Styela clava* all have the small electron-lucent vesicles and multivesicular bodies typical of *Ciona*, *Styela* has, in addition, larger electron-dense droplets (Fig. 3). These quite possibly represent a coales-

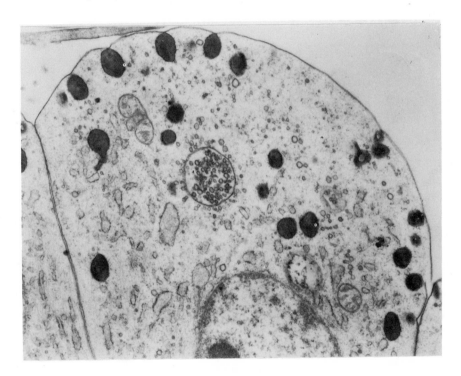

FIG. 3. The apical region of a zone 7 cell of *Styela clava*. The contents are similar to *Ciona* except for the electron-dense droplets (×18 000).

cence of the smaller vesicles, but, equally, may imply a more fundamental difference in cellular metabolism and its products, which must reflect on any general hypothesis relating to iodine binding.

Certainly, the problem of iodination is not straightforward, for the iodine-binding cells of amphioxus have a quite different ultrastructural appearance from their tunicate counterparts. Thus, the apices of these cells contain large secretory droplets typical of mucus-secreting cells (Thorndyke, 1971) (Fig. 4). Welsch & Storch (1969) have also described globular inclusions in zone "4/5" of amphioxus, (Welsch, 1975 finds zone 5 not clearly distinguishable), "which presumably represent mucus droplets". This has further been confirmed by Olsson (1969).

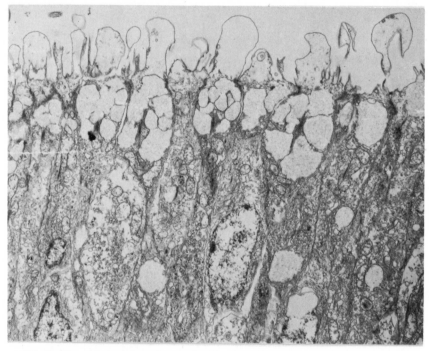

FIG. 4. The apical region of zone 5 cells of amphioxus showing prominent mucus droplets (×5000).

Whilst a detailed comparison with the ammocoete endostyle is considered to be outside the scope of the present paper, it is noteworthy that the ultrastructural and autoradiographical details of the iodine-binding cells of this primitive larval vertebrate (Fujita & Honma, 1968, 1969) appear to resemble more closely the tunicates than amphioxus.

Electron microscope ^{125}I autoradiography

Some of the problems of vertebrate thyroidal biosynthesis (not yet themselves completely resolved) have been studied using ^{125}I or

^{131}I in a series of pulse-labelling and other tracer experiments (Novikoff, 1963; Stein & Gross, 1964; Lupulescu, Andreani & Andreoli, 1967). This technique has been applied to the tunicate problem in studies of *Ciona* (Thorpe, Thorndyke & Barrington, 1972) and *Styela* (Thorndyke, 1971).

Initial experiments were designed to show the effects of immersion in radioiodinated sea water for periods of 30 min to 8 h. In later experiments, radioiodination periods of up to 4 h were followed by varying periods in non-radioactive sea water in order to study the sequential events of iodine incorporation.

As with light microscope autoradiography, images are observed only over the cells of zones 7 and 8. Following short immersion periods of up to 1 h, very little evidence of bound iodine is seen in the zone 7 cells, but where an autoradiographical image is observed, it is present either in close association with the apical plasma membrane or with the apical electron-lucent vesicles. From 1–4 h there is a progressive increase in the number of silver grains within cells and they are frequently observed overlying the multivesicular bodies (Fig. 5). During later periods, the residual bodies are seen to have an autoradiographical image, and there is good evidence from the pulse-labelling experiments that this iodinated material is ultimately discharged from the cells towards the lumen (Fig. 6).

The results of independent ultrastructural and autoradiographical studies by Fujita & Nanba (1971) on the iodine-binding phenomenon in *Ciona* are in general agreement with those outlined above from our own laboratory, although the interpretation placed upon them differs in certain respects. In particular, these authors raise the interesting possiblity that material which has been iodinated in the endostylar lumen may be reabsorbed and concentrated within the multivesicular bodies. This idea is certainly in accord with events in the vertebrate thyroid cells but, whilst we do not ignore this possibility, our own results from pulse-labelling and electron microscope histochemistry, to be reported later, strongly imply that the multivesicular bodies are more directly concerned with the iodination process. The evidence we have obtained from zone 8 also favours the view that iodinated material is largely discharged from 7 cells since it can clearly be seen lying amongst the cilia in this adjoining region (Fig. 9).

We agree completely with Fujita & Nanba (1971), however, that secretory material can become iodinated whilst it is present in the lumen of the endostyle, although the precise moment at which this occurs is in doubt. Evidence for this extracellular iodine binding is

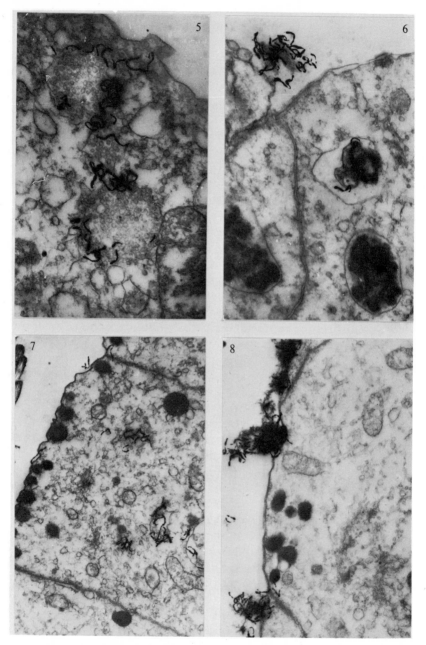

FIG. 5. An electron microscope ^{125}I autoradiograph of a zone 7 cell of *Styela* following 4 h immersion in radioiodinated sea water (500 μCi/litre). The label appears over the multi-vesicular bodies ($\times$ 18 630).

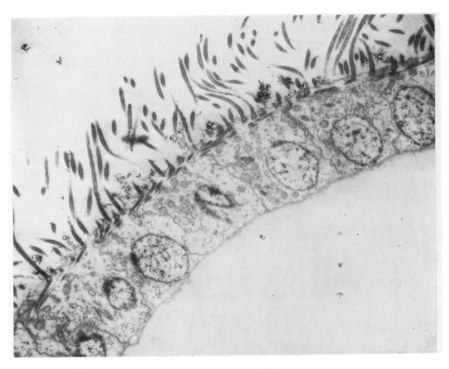

FIG. 9. Zone 8 of *Ciona* after 4 h immersion in ^{125}I. Iodine appears to be limited to material trapped between the cilia (×4500).

most clearly seen in *Styela,* where the electron-dense droplets of the zone 7 cells do not appear to acquire an autoradiographical image until they are released into the lumen (Figs 7 and 8).

THE PROBLEMS OF VERTEBRATE THYROIDAL
BIOSYNTHESIS AND THE RÔLE OF THYROID PEROXIDASE

Before discussing the possible phylogenetic significance of the iodine-binding phenomenon in protochordates, it is necessary to

FIG. 6. An electron microscope ^{125}I autoradiograph of a zone 7 cell of *Ciona* 8 h following a period of immersion in radio-iodinated seawater. Residual bodies can be seen, one of which is labelled. The iodinated material in the lumen is of the same electron density and appears to have been discharged from the cells (×20 700).

FIGS 7 and 8. Electron microscope autoradiographs of zone 7 cells of *Styela* after 4 h immersion in radio-iodinated seawater. Although, in Fig. 7, the label is clearly seen over multi-vesicular bodies and certain electron-lucent vesicles free in the cytoplasm, the electron-dense droplets are not labelled. Only when the material is released from the apical membrane (Fig. 8) does it acquire a label (×12 150 and 16 200).

review briefly the major features of vertebrate thyroidal biosynthesis.

The first stage in the process necessitates an ATPase-sensitive ion-pump (Pitt-Rivers & Cavalieri, 1964), whereby iodide ions are actively pumped from the circulation into the thyroid cells. Before iodide can iodinate the tyrosine residues of thyroglobulin, it must first be oxidized to a "different ionic species" (Strum & Karnovsky, 1970). The exact nature of this activated iodide is not known, although it may be an iodinium ion (I^+) or a hypoiodite ion (IO^-) (Strum, Wicken, Stanbury & Karnovsky, 1971). There is general agreement, however, that the oxidation is carried out by a thyroid peroxidase enzyme in the presence of endogenous hydrogen peroxide (Taurog, 1970; Lamas, Dorris & Taurog, 1972).

Current thinking suggests that such an enzyme is produced by thyroid cells and becomes associated in some way with the apical plasma membrane or localized within small electron-lucent vesicles in the apical cytoplasm (Strum & Karnovsky, 1970; Strum et al., 1971). In view of this, it is of the greatest importance that most of the thyroidal binding is seen to occur either at the cell/colloid interface (Lupulescu et al., 1967; Stein & Gross, 1964; Fujita, 1972) or within the subapical vesicles (Strum et al., 1971), which may also contain thyroglobulin (Nadler, Young, Leblond & Mitmaker, 1964). Thus, the site of oxidation of iodide is also apparently the site of iodination, for, as van Zyl & Edelhoch (1967) point out, the chemical activity of the oxidised iodide intermediate would be too great for its free existence in the cells. Furthermore, Lamas et al. (1972) have provided evidence that a peroxidase enzyme (possibly one of a series present within the cell) is essential for the enzymatic conversions of the iodotyrosines to the thyroid hormones.

The iodinated thyroglobulin is reabsorbed by micropinocytosis into the follicle cells to form the "colloid-droplets" (Seljelid, 1967; Stein & Gross, 1964). Finally, the colloid droplets fuse with lysosomes and, under the influence of the appropriate enzymes, thyroid hormones are released from their parent molecules and pass into the blood through the basal plasma membrane.

ULTRASTRUCTURAL LOCALIZATION OF PEROXIDASE
IN THE TUNICATE ENDOSTYLE

In order to make adequate comparisons of the iodine-binding process in the zone 7 cells and true thyroidal cells it became imperative to know where, if at all, the peroxidase enzyme was to be found in the tunicate endostyle. Recent electron microscope histochemical studies by Thorpe & Thorndyke (unpublished) on the tunicate *Styela* have indeed shown that peroxidase is produced in considerable amounts by the cells of zone 7, and in very limited quantities in zone 8. It is apparently completely absent from all other zones. The enzyme is localized in certain of the small apical vesicles in the cytoplasm and in the multivesicular bodies, and also in association with the apical plasma membrane (Fig. 10).

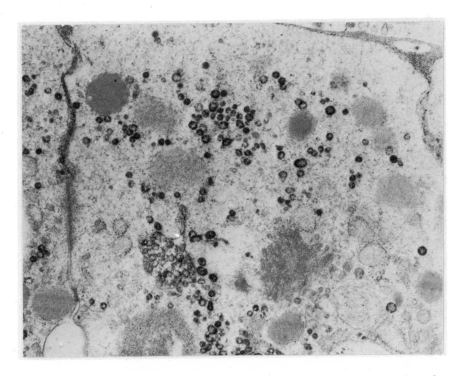

FIG. 10. The apical region of a zone 7 cell of *Styela*. The dark reaction product shows the sites of peroxidase activity and is evident in certain of the small vesicles both free in the cytoplasm and in the multivesicular body. The apical plasma membrane also shows limited activity (×25 000).

CONCLUSIONS AND PHYLOGENETIC CONSIDERATIONS

The present evidence, which has arisen from several different experimental approaches, supports the view that the iodine-binding cells of protochordates represent the fore-runners of the thyroidal cells. Naturally, since the protochordates are such a diverse group of animals, further work on a greater number of species is essential before generalizations can be made. In this regard, the results of electron microscope ^{125}I autoradiography and histochemistry on amphioxus will be of interest. Another extremely important consideration, of course, is that modern protochordates must be regarded as very specialized animals which have become closely adapted to their individual modes of life (Barrington, 1965). To what extent the iodine-binding cells have undergone change from their former primitive state as a part of the process of specialization and adaptation is difficult to estimate.

Whatever the outcome of the phylogenetic argument, the fact remains that many problems of interpretation and experimental technique remain to be solved in protochordate endostyle iodine binding. The finding that the peroxidase enzyme is present in tunicate zone 7 cells is an important parallel with the thyroid cell. Thus, it provides the basis for the initial iodination as well as the later conversion of the iodotyrosines into the thyronine hormones. The great difficulty lies in establishing the sequence of events in iodine-binding and their relation to the ultrastructure of the cells. The studies of Thorpe, Thorndyke & Barrington (1972) on the endostyle of *Ciona* and Thorndyke (1971) on *Styela* show very clearly that the electron-lucent vesicles of the apical cytoplasm and the multivesicular bodies possess an autoradiographical image when treated with ^{125}I. Pulse-labelling experiments suggest further that breakdown of the multivesicular bodies occurs, with the release of iodinated material to the exterior. One could deduce from these facts that iodide enters the cell from the sea water present in the endostyle lumen and that it is converted in the apical cytoplasm into an active form by the peroxidase enzyme present there in the ultrastructural form of electron-lucent vesicles. Also present in the cytoplasm and multivesicular bodies are electron-lucent vesicles which do not contain peroxidase but which, we can assume, contain the glycoprotein ultimately to be iodinated. Limited iodination may occur within vesicles free in the cytoplasm but it is more likely that the multivesicular body acts as the major intracellular location in which the mixed population of vesicles may ultimately react. The resulting iodinated glycoprotein is finally released from the cell following multivesicular body breakdown. It is also likely from all the evidence that peroxidase enzyme vesicles and glycoprotein vesicles (or larger droplets in the case of *Styela*) can be released separately to the exterior, providing for iodination either during passage through the apical plasma membrane, or immediately exterior to it in the lumen. The suggestion by Fujita & Nanba (1971) that iodinated material is actually reabsorbed into zone 7 cells of *Ciona* for subsequent hydrolysis is an attractive one on phylogenetic grounds, in that it more closely resembles events in the thyroid gland, but our own studies do not readily support this view. We feel it more likely that iodinated material produced in one of several possible ways by zone 7 cells is a secretory component of the feeding mechanism and that only later enzymatic action in the intestine would enable any thyroid hormones to be released and absorbed into the circulatory system.

Electron microscope autoradiographical studies on amphioxus will be very useful in deciding this problem. The large mucus droplets of these particular iodine-binding cells makes the possibility of reabsorption of iodinated material extremely difficult to envisage.

The development of the follicle and the provision of a colloid store in the vertebrates makes possible an extremely efficient

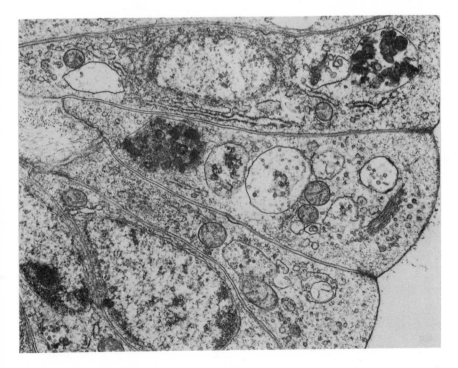

FIG. 11. Zone 7 cell of *Ciona* after 5 days treatment with thiourea (0·01%). The increased numbers of multivesicular bodies and transitional forms are obvious (×22 000).

biochemical process for homone synthesis and release into the blood stream. If, in the protochordates, the process, as outlined above, is inefficient, depending as it does on long range enzymatic hydrolysis of iodinated protein, we must consider whether the iodine-binding process is at all "thyroidal" in the vertebrate sense. Further studies are needed in this direction, but initial experiments (Thorndyke, 1971) suggest that iodine-binding cells do respond to mammalian thyroid stimulating hormone (TSH) and goitrogens (Figs 11, 12). Whether or not these are simply drug-induced

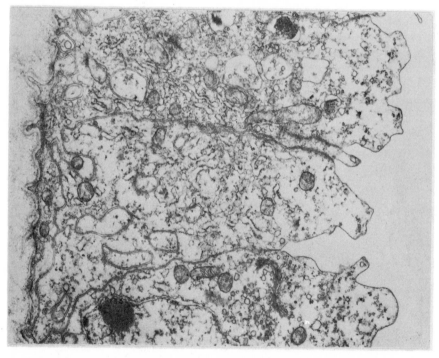

FIG. 12. Zone 7 cell of *Ciona* after 72 h treatment with thyroid stimulating hormone (5 mg/litre seawater-activity 0·5–2·0 IU/mg). Large and extensive "pseudopodia" and an increase in secretory vesicles are apparent (×17 500).

pathological conditions is debatable, but it is interesting that TSH causes pseudopodial development at the apical plasma membrane in much the same way that it does in thyroid cells (Seljelid & Nakken, 1968). Likewise, goitrogen treatment of tunicates induces in the iodine-binding cells a pattern of lysosomal activity similar to that seen in vertebrates (Wissig, 1964; Coleman, Evennett & Dodd, 1968a,b).

Thorndyke (1973), investigating the possibility of a functional rôle for thyroid hormones in tunicates, showed that the enzymes involved in the tanning of the tunic were stimulated by the presence of triiodothyronine in the incubating medium. It would be useful to study other possible parameters of thyroid hormone action in the protochordates as an aid to solving the problems of the evolution of the thyroid gland.

ACKNOWLEDGEMENTS

We are very grateful to Mr J. F. Pacy of the Biological Electron Microscope Unit of Queen Mary College for his assistance in preparing the photographs used in this paper.

We would also like to thank Mr D. Houghton of the Admiralty Exposure Trials Station, Portsmouth, for allowing us to use the facilities of his Department and for assistance in the collection of tunicate specimens.

REFERENCES

Aros, B. & Viragh, S. (1969). Fine structure of the pharynx and endostyle of an ascidian (*Ciona intestinalis*). *Acta biol. hung.* **20**: 281–297.

Barrington, E. J. W. (1957). The distribution and significance of organically bound iodine in the ascidian *Ciona intestinalis* L. *J. mar. biol. Ass. U.K.* **36**: 1–16.

Barrington, E. J. W. (1958). The localization of organically bound iodine in the endostyle of *Amphioxus. J. mar. biol. Ass. U.K.* **37**: 117–126.

Barrington, E. J. W. (1965). *The biology of Hemichordata and Protochordata.* Edinburgh: Oliver and Boyd.

Barrington, E. J. W. (1969). Unity and diversity in comparative endocrinology *Gen. comp. Endocr.* **13**: 482–488.

Barrington, E. J. W. (1975). Problems of iodine binding in ascidians. *Symp. zool. Soc. Lond.* No. 36: 129–158.

Barrington, E. J. W. & Franchi, L. L. (1956). Organic binding of iodine in the endostyle of *Ciona intestinalis. Nature, Lond.* **177**: 432.

Barrington, E. J. W. & Thorpe, A. (1965a). An autoradiographic study of the binding of [125]iodine in the endostyle and pharynx of the ascidian, *Ciona intestinalis* L. *Gen. comp. Endocr.* **5**: 373–385.

Barrington, E. J. W. & Thorpe, A. (1965b). The identification of monoiodotyrosine, diiodotyrosine and thyroxine in extracts of the endostyle of the ascidian, *Ciona intestinalis* L. *Proc. R. Soc.* (B.) **163**: 136–149.

Berrill, N. J. (1950). *The Tunicata.* London: Ray Society.

Berrill, N. J. (1955). *The origin of vertebrates.* London: Oxford University Press.

Coleman, R., Evenett, P. J. & Dodd, J. M. (1968a). Ultrastructural observations on the droplets of experimentally-induced goitres in *Xenopus laevis* with especial reference to the development of Uhlenhuth colloid cells. *Z. Zellforsch. mikrosk. Anat.* **84**: 490–496.

Coleman, R., Evennett, P. J. & Dodd, J. M. (1968b). Ultrastructural observations on some membranous cytoplasmic inclusion bodies in follicular cells of experimentally-induced goitres in tadpoles and toads of *Xenopus laevis* Daudin. *Z. Zellforsch. mikrosk. Anat.* **84**: 497–505.

Fujita, H. (1972). Electron microscopic autoradiography on the iodine metabolism of the thyroid gland in phylogenetic aspect. *Acta histochem. cytochem.* **5**: 207.

Fujita, H. & Honma, Y. (1968). Some observations on the fine structure of the endostyle of larval lampreys, ammocoetes of *Lampetra japonica*. *Gen. comp. Endocr.* **11**: 111–131.

Fujita, H. & Honma, Y. (1969). Iodine metabolism of the endostyle of larval lampreys, ammocoetes of *Lampetra japonica*. *Z. Zellforsch. mikrosk. Anat.* **98**: 525–537.

Fujita, H. & Nanba, H. (1971). Fine structure and its functional properties of the endostyle of the ascidian, *Ciona intestinalis*. *Z. Zellforsch. mikrosk. Anat.* **121**: 455–469.

Kennedy, G. R. (1966). The distribution and nature of iodine compounds in ascidians. *Gen. comp. Endocr.* **7**: 500–511.

Lamas, L., Dorris, M. L. & Taurog, A. (1972). Evidence for a catalytic role for thyroid peroxidase in the conversion of diiodotyrosine to thyroxine. *Endocrinology* **90**: 1417; 1426.

Lévi, C. & Porte, A. (1964). Ultrastructure de l'endostyle de l'Ascidie *Microcosmus claudicans* Savigny. *Z. Zellforsch. mikrosk. Anat.* **62**: 293–309.

Lupulescu, A., Andreani, D. & Andreoli, M. (1967). Thyroglobulin synthesis, iodination and hydrolysis. *Folia endocr.* **20**: 385–405.

Nadler, N. J., Young, B. A., Leblond, C. P. & Mitmaker, B. (1964). Elaboration of thyroglobulin in the thyroid follicle. *Endocrinology* **74**: 333–354.

Novikoff, A. B. (1963). Lysosomes in the thyroid epithelium of untreated, TSH-stimulated, and ^{131}I-irradiated rats. *Biol. Bull. mar. biol. Lab., Woods Hole* **125**: 358.

Olsson, R. (1963). Endostyles and endostylar secretions: a comparative histochemical study. *Acta zool., Stockh.* **44**: 299–328.

Olsson, R. (1969). General review of the endocrinology of the Protochordata and Myxinoidea. *Gen. comp. Endocr.* suppl. **2**: 485–499.

Pitt-Rivers, R. & Cavalieri, R. R. (1964). Thyroid hormone biosynthesis. In *The thyroid gland*. Pitt-Rivers, R. & Trotter, W. R. (eds.). London: Butterworth and Co.

Roche, J., André, S. & Covelli, I. (1960). Sur la fixation de l'iode par la Moule (*Mytilus galloprovincialis* L.) et la nature des combinaisons iodées élaborées. *C. r. Séanc. Soc. Biol.* **154**: 2201–2206.

Roche, J., Salvatore, G. & Rametta, G. (1962). Sur la présence et la biosynthèse d'hormones thyroïdiennes chez un tunicier, *Ciona intestinalis* L. *Biochim. Biophys. Acta* **63**: 154–165.

Seljelid, R. (1967). Endocytosis in thyroid follicle cells. II. A microinjection study of the origin of colloid droplets. *J. Ultrastruct. Res.* **17**: 401–420.

Seljelid, R. & Nakken, K. F. (1968). Endocytosis of thyroglobulin and the release of thyroid hormone. *Scand. J. clin. Lab. Invest.* **22**: suppl. **106**: 125–143.

Sokólska, J. (1931). [Contribution à l'histologie de l'endostyle des Ascidies.] *Folia morph.* **3**: 1–34. [In Polish, with French summary].
Stein, O. & Gross, J. (1964). Metabolism of ^{125}I in the thyroid gland studied with electron microscope autoradiography. *Endocrinology* **75**: 787–798.
Strum, J. M. & Karnovsky, M. J. (1970). Cytochemical localization of endogenous peroxidase in thyroid follicular cells. *J. cell Biol.* **44**: 655–666.
Strum, J. M., Wicken, J., Stanbury, J. R. & Karnovsky, M. J. (1971). Appearance and function of endogenous peroxidase in fetal rat thyroid. *J. cell Biol.* **51**: 162–175.
Suzuki, S. & Kondo, Y. (1971). Demonstration of thyroglobulin-like iodinated proteins in the branchial sac of tunicates. *Gen. Comp. Endocr.* **17**: 402–406.
Taurog, A. (1970). Thyroid peroxidase and thyroxine biosynthesis. *Recent Prog. Horm. Res.* **26**: 189–247.
Thomas, I. M. (1956). The accumulation of radioactive iodine by *Amphioxus. J. mar. biol. Ass. U.K.* **35**: 203–210.
Thorndyke, M. C. (1971). *Ultrastructural studies on protochordate endostyles.* Ph.D. Thesis, University of London.
Thorndyke, M. C. (1973). An *in-vivo* stimulatory effect of 3,3',5-tri-iodo-L-thyronine on polyphenol oxidase activity in an ascidian. *J. Endocr.* **58**: 679–680.
Thorpe, A. (1964). *Studies on iodine binding in the protochordates.* Ph.D. Thesis, University of Nottingham.
Thorpe, A. & Barrington, E. J. W. (1965). The histochemical basis of iodine binding in *Ciona. Gen. comp. Endocr.* **5**: 710.
Thorpe, A., Thorndyke, M. C. & Barrington, E. J. W. (1972). Ultrastructural and histochemical features of the endostyle of the ascidian *Ciona intestinalis* with special reference to the distribution of bound iodine. *Gen. comp. Endocr.* **19**: 559–571.
Thorpe, A., Thorndyke, M. C. & Kirkham, J. B. (1969). Ultrastructural observations on the iodine-binding cells of tunicate endostyles. *Gen. comp. Endocr.* **13**: 534–535.
Tong, W. & Chaikoff, I. L. (1961). ^{131}I utilization by the aquarium snail and the cockroach, *Biochim. Biophys. Acta* **48**: 347–351.
Tong, W., Kerkhof, P. & Chaikoff, I. L. (1962). Identification of labelled thyroxine and triiodothyronine in *Amphioxus* treated with ^{131}I. *Biochim. biophys. Acta* **56**: 326–331.
van Zyl, A. & Edelhoch, H. (1967). The properties of thyroglobulin. XV. The function of the protein in the control of diiodotyrosine synthesis. *J. biol. Chem.* **242**: 2423–2427.
Welsch, U. (1975). The fine structure of the pharynx, cyrtopodocytes and digestive caecum of amphioxus (*Branchiostoma lanceolatum*). *Symp. zool. Soc. Lond.* No. 36: 17–41.
Welsch, U. & Storch, V. (1969). Zur feinstruktur und histochemie des Kiemendarmes und der Leber von *Branchiostoma lanceolatum* (Pallas). *Z. Zellforsch. mikrosk. Anat.* **102**: 432–446.
Wissig, S. L. (1964). Morphology and cytology. In *The thyroid gland*: 32–70. Pitt-Rivers, R. & Trotter, W. R. (eds.). London: Butterworth and Co.

Symp. zool. Soc. Lond. (1975) No. 36, 179–212.

THE DISTRIBUTION OF AMPHIOXUS

J. E. WEBB

Department of Zoology, Westfield College, Hampstead, London, England

SYNOPSIS

Amphioxus is widespread in temperate and tropical seas. The adult lives in immense numbers in some sands, gravels or shell deposits, but is absent from others of apparently similar texture. The larva is pelagic for 75 to over 200 days and may drift across oceans, but is mostly restricted in its distribution by coastal currents. Recruitment of adults at a particular locality depends on a supply of larvae usually from another site within the current system. The local distribution of a population changes with alterations in current pattern or deposit composition and, in temperate seas, according to the season.

Amphioxus prefers stable, well-ventilated, smooth-textured deposits free from organic decay, but in such deposits may be present at one point and absent from another less than a metre away. This is due to the animal's reaction to subtle differences in the rate of flow of water and the volume of the flow pathways through the deposit (specific permeability). In summer adults congregate where there is a high rate of flow in the deposit and a low pathway volume (high specific permeability) and are thus brought into close proximity for spawning. In winter they disperse into deposits with a high pathway volume (high capillary space) and incidentally a low rate of flow.

Since adults congregate at specific localities and recruitment depends on drift of larvae it is possible to use the distribution of amphioxus to map current patterns. Taxonomic recognition of the adult populations is first required and then a knowledge of the characteristics of preferred deposits. This is important because absence of amphioxus may be due to either the physical unsuitability of the deposit or the failure of larvae to reach the ground which can then be assumed to be outside a particular current system.

INTRODUCTION

Amphioxus once thought to be rare is now known to be common though curiously local in its distribution. It occurs in large numbers in specific and often widely separated "amphioxus grounds" but only occasionally or not at all in the intervening regions. The adult lives in a range of sediments from comparatively fine sand to shell gravel in shallow seas, usually not much more than 30 m deep, and in certain circumstances intertidally, but it is not found everywhere even within this range of habitats. This extraordinary distribution poses a number of problems particularly concerning the way in which it is maintained. Meadows & Campbell (1972) made the point that the distribution of animals in the sea is largely determined by their repertoire of behavioural responses to environmental stimuli. This is probably true of all animals at all levels of distribution, but gives no clue to the nature of the significant stimulus or the manner of the response. The work of the past 20 years on the ecology and behaviour of amphioxus permits some assessment of its distribution in these terms so that it is now worthwhile to review the relationship between amphioxus and its environment with this interpretation in mind. The distribution of amphioxus is of intrinsic interest to zoologists, but there is another reason for a survey at the present time. There is growing evidence that the distribution of populations of the adults may give a valuable indication of the pattern of water flow in the seas and oceans. This happens because amphioxus has a long planktonic larval phase of two to six months or more and in this time is carried in circulating waters which determine its area of distribution. It would follow conversely, therefore, that the distribution of a population composed of a sequence of aggregates of adults on suitable grounds linked by a free interchange of larvae should represent a current system and could be used to map the pattern of water movement.

THE PLANKTONIC PHASE

The planktonic phase begins when the larva shown in Fig. 1 swims upward into the open water from the sand or shell deposit in

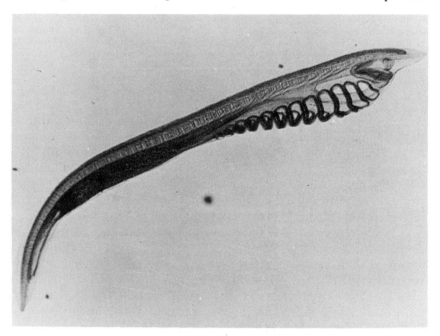

FIG. 1. A late larva of *Branchiostoma lanceolatum* from Helgoland.

which development up to the 1 to 6 gill pouch stage has taken place (Webb, 1958a; Gosselck & Kuehner, 1973). The behaviour of the planktonic larva characteristically consists of periods of upward swimming by means of undulatory body movements alternating with passive sinking in the horizontal position with the mouth on the left side directed downward (Rice, 1880; van Wijhe, 1927; Webb, 1969a). These movements are thought to be associated with the capture of food material consisting either of relatively large organisms such as copepods and other suspended material of similar size (Webb, 1969a) or of phytoplankton and detritus (Wickstead & Bone, 1959; Gosselck & Kuehner, 1973). The alternate upward swimming and sinking tends to maintain the larva in or near the surface waters where it is subject to drift with the currents. The position of the larva relative to the sea surface depends on the rates of movement in and duration of the active and passive phases.

In a special case the larvae of *Branchiostoma nigeriense* from the coastal waters at Lagos in West Africa reacted to salinity and temperature gradients. Here outflow from the rivers reduced the salinity of the surface waters of the sea which was at a temperature of 27–28°C. Larvae sinking passively into deeper and colder water were stimulated to activity at 24°C and swam upward until the salinity fell to 20 p/10³, when swimming ceased and the larva sank back into the deeper water. Although this behaviour was determined in laboratory experiments it was estimated that in these West African coastal waters the larvae would have been restricted by this behaviour to depths between 3·5 and 36·5 m (Webb & Hill, 1958). Now it is clear that water conditions such as these are not the general case and that the larva reacts to variables other than salinity and temperature. For example Chin (1941) found that the larvae of *B. belcheri* at Amoy were present at all depths at night but during the day were absent from the plankton, and Wickstead & Bone (1959) with the same species found similar effects off Singapore. Here in water 82 m deep larvae were found in the daytime from a depth of 27 m to the bottom, but prior to sunset the larvae increased their range to cover all depths from the surface. They maintained this distribution throughout the hours of darkness except that the zone of maximum numbers rose from the bottom to the surface shortly after sunset and then fell to about 27 m during the night. This suggests that these larvae either directly or indirectly were reacting to light. In these cases in relatively shallow seas the bottom provided the lower limit to vertical movement, but in

deeper waters light intensity during daytime might be expected to
determine that limit. Wickstead & Bone (1959) reported larvae of
B. malayana in surface waters where the sea was over 1000 m deep
and where there was no possibility of the larvae traversing that
depth either through continuous upward swimming or passive
sinking within the 12 h periods of dark or light. On the other hand
Hartmann & John (1971) with *B. lanceolatum* in the North Sea and
Gosselck & Kuehner (1973) with *B. senegalense* in the Cap Blanc
area of the Spanish Sahara coast, found no diurnal vertical
movement of larvae probably because of tubulence.

THE DISTRIBUTION OF LARVAE

Behaviour leading to vertical movement evidently ensures that
the larva is planktonic for a substantial proportion of the time, if
not entirely, depending on the depth of the water. Allowing that
surface currents of 1 knot are not unusual, it is possible for a larva
normally planktonic for 75 days and drifting at this speed to be
transported distances approaching 3000 km. This distance would
be considerably extended at higher current speeds and where the
larval period is prolonged. There is evidence to suggest that
journeys on this scale are made with some frequency as shown in
Fig. 2. *B. belcheri* occurs off the coast of China, in the East Indies

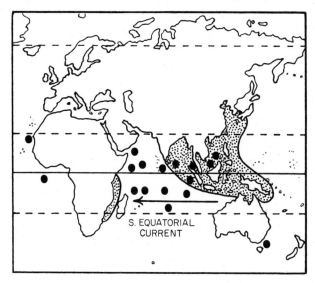

FIG. 2. The distribution of *Branchiostoma belcheri* (stipple) and the sites of capture of
various species of amphioxides larvae (●).

and North Australia. It is also recorded from the east coast of
Africa at Zanzibar (Franz, 1922) and from two localities in Mozambique (Webb, 1957). Specimens from one of these in the Morrumbene Estuary are very similar to those from Amoy on the China
coast and can only be distinguished from them by very minor
differences if at all. It seems probable that larvae from the East
Indies are carried to East Africa by the east-west Equatorial
Current which strikes the coast of Madagascar and passes around
the island both to the north and the south. Transfer of larvae
evidently takes place with sufficient frequency to maintain the
similarity between the populations and involves transport across
some 8000 km of ocean. There is evidence that, where conditions
for settlement are not suitable, metamorphosis may be delayed. It
has been shown that, for reasons which are not understood, in *B.
nigeriense* the larval period can be increased from 75 to 140 days
(Webb, 1958a). In *B. senegalense*, too, young larvae are found off
Cap Blanc at the beginning of June (Gosselck & Hagen, 1973), are
fully grown three months later at the beginning of September, but
do not settle until November or December giving a pelagic phase
that can be in excess of 200 days (Gosselck & Kuehner, 1973). It
would be possible, therefore, if metamorphosis and settlement
were similarly delayed, for the larvae of *B. belcheri* to drift from the
East Indies to East Africa within the span of a single larval phase.
Branchiostomid larvae with delayed metamorphosis and in some
cases developing gonads (Wickstead, 1973) are known as amphioxides larvae and have been collected from a number of sites in the
Indian Ocean far from land (see Fig. 2).

In most cases, however, the distribution of amphioxus larvae is
on a much smaller scale. On the west coast of Africa, for example,
there are six species in sequence from the Spanish Sahara coast in
the north to the Cape, as shown in Table I, and three of these (*B.
senegalense, B. gambiense* and *B. leonense*) have been collected in the
same area off the Gambia (Webb, 1955, 1956a,b, 1957, 1958b). The
major current systems of the West African coast are the
Canaries Current off the Spanish Sahara and Senegal, the warm
Guinea Current flowing east along the Guinea Coast and the
Benguela Current flowing north from the Cape. It is perhaps
surprising, therefore, that these species of *Branchiostoma*, five of
which are relatively closely spaced along a comparatively uniform
coast (see Table I), have remained in isolation except off the
Gambia. A reason for this may lie in the nature of the inshore
components of the Canaries and Guinea Currents. In at least some

TABLE I

The distance and difference factor between samples of a sequence of species of Branchio-stoma *from the coasts of Europe and Africa*

Species	Locality	Distance (km)	Difference factor
B. lanceolatum	Plymouth		
		4000	25·6
B. senegalense	Dakar		
		160	20·6
B. gambiense	Gambia		
		720	13·1
B. leonense	Sierra Leone		
		1440	6·2
B. takoradii	Takoradi		
		560	26·9
B. nigeriense	Lagos		
		5120	31·1
B. capense	Cape Town		
		2100	41·9
B. belcheri	Mozambique		

regions of the coast these are reciprocal and give rise to water masses oscillating parallel to the shore, while in other regions they form seasonally changing gyres to which the distribution of larvae can be related (Gosselck & Hagen, 1973). Gosselck & Kuehner (1973) failed to recover larvae of *B. senegalense* from the bottom itself. In shallow waters, however, with high light intensities and no upwelling turbulence, larvae might be restricted to near the bottom during the daytime and, progressing by saltations, tend to drift with the current only at night. Under these conditions it is possible to see how larvae could become limited to specific coastal regions.

SETTLEMENT OF LARVAE

Although very many larvae are produced in the three month spawning period, they usually disperse during the long pelagic life so that, except under special circumstances, they are seldom found in the plankton in large numbers. In most places where it is possible to collect amphioxus larvae, such as at Helgoland in the North Sea, only a few are usually found in a plankton sample and then only at the end of August and the beginning of September. Exceptions to this are in the harbour at Lagos in the months of October,

November and December, where up to 400 larvae have been taken in a single vertical plankton haul in water 8 m deep (Webb, 1958a) and off the Spanish Sahara coast in September where as many as 37 000 larvae were caught in a plankton sample taken at 0–25 m from the bottom (Gosselck & Kuehner, 1973). It is therefore, perhaps, remarkable that in spite of the low densities of larvae generally present in the plankton high concentrations of adults can occur in suitable sand deposits. For example there can be up to 800 animals per m^2 at Le Racou on the Mediterranean coast of France (Webb, 1969b) and up to 9000 per m^2 in a mussel-shell bank off the coast of the Spanish Sahara (Gosselck & Kuehner, 1973). There is cause to wonder whether the planktonic larvae are specifically attracted to sites already occupied by adults. Evidence that this may be so was given by Webb (1971) for the population at Helgoland.

The amphioxus ground at Helgoland in the German Bight is 5 km to the N.E. of the island and is about 1 km or so in diameter. The distribution of the adults on the ground in September, in places with a concentration of more than 400 per m^2, is given in Fig. 3a. At this time of year settlement takes place and it is clear from Fig. 3c that the distribution of newly metamorphosed adults is not at random. The direction of the current (Kuhl, 1972) and hence the line of approach of the larvae is from the south. The highest concentration of newly metamorphosed individuals, again in excess of 400 per m^2, coincides with the area of deposit already occupied by adults and falls off rapidly at the northern margin of the ground. On the other hand the distribution also agrees, though perhaps less accurately, with that of the sand deposits (Fig. 3d). Larvae will metamorphose in a glass container in the laboratory in the absence of either sand or adults. Moreover the annual colonization of sand deposits in Lagos lagoon certainly takes place in the absence of adults, for these have all been killed seven months earlier by falling salinity (Webb, 1958a). But it seems from the distribution and large numbers of newly metamorphosed individuals in the sand at Helgoland compared with the low concentrations of the larvae in the plankton that larvae about to metamorphose may preferentially settle when they reach a sand or shell gravel deposit, particularly if there are adults already in the sand. However, experimental evidence is needed in confirmation.

THE ESTABLISHMENT OF ADULT POPULATIONS

In view of the prolonged transport by currents it must be comparatively rare for the larvae from a particular spawning to

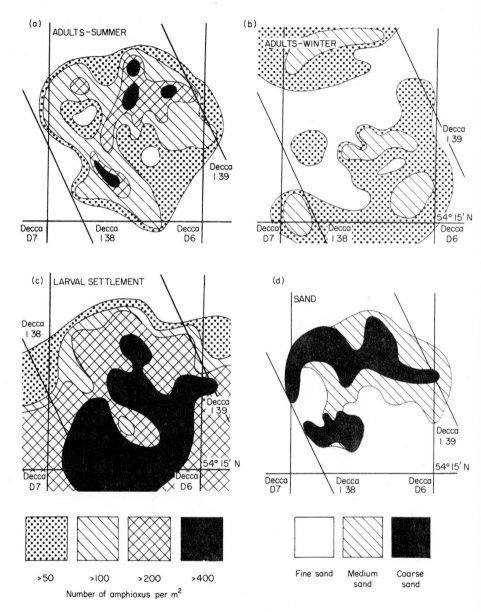

FIG. 3. The distribution of *Branchiostoma lanceolatum* and the grades of deposit at Helgoland, (a) adults in the summer, (b) adults in the winter, (c) larval settlement, (d) grades of sand. (From Webb, 1971).

return to their ground of origin. An historical exception to this in the latter half of the 19th century was the population confined within the land-locked salt water Lago di Faro in the north east of Sicily (Willey, 1894). Here transport by currents was precluded and the life-cycle completed within the lake. The maintenance of adults at a particular open sea locality normally implies either a regular interchange of larvae between sites linked in sequence by a current pattern or, at least, a supply of larvae from another site transported by currents. If this were not so adult populations would eventually die out through lack of recruitment. This effect is demonstrated in Fig. 4a & b. As two factors operate, first the existence of deposits

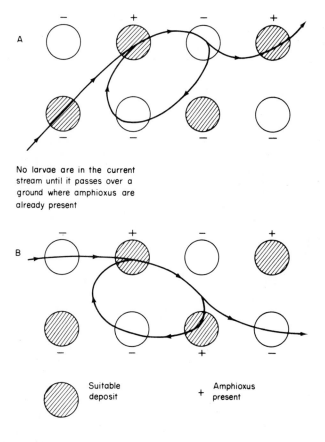

FIG. 4. Diagram showing the effect of a change in the current pattern and hence in larval dispersal on the distribution of amphioxus. Some suitable deposits lying outside the circulating current system are not colonized.

suitable for the adult and secondly the current pattern determining larval distribution, not all suitable deposits are colonized and the incidence of colonization is liable to change with alterations in the current pattern (Fig. 4b). It is suspected that changes of this kind in recent years were responsible for the disappearance of amphioxus from localities in the Bay of Naples which had long been a source of specimens, and their occurrence at new localities, for example off the island of Ischia, where they were not previously known. But there is no certainty of this, as other factors may have been in operation such as changes in composition of the sediment arising from alteration in rates of current flow and heavy pollution from Naples itself.

We may therefore consider a population of amphioxus as the individuals on a sequence of grounds lying within a circulating current system so that the group of animals on each ground is maintained by recruitment of metamorphosing larvae from other grounds within the system. It is probable that such a system exists in the German Bight and includes the Helgoland group of animals, another off the island of Sylt and no doubt others in that area. Other populations recognizably different from the Helgoland form have been described from the North Sea and English Channel (Webb, 1956a). There is probably some interchange of larvae between these different populations either at a reduced level or of an intermittent character according to the extent to which

TABLE II

The distance and difference factor between samples of populations of
Branchiostoma lanceolatum

Locality	Distance (km)	Difference factor
Kattegat		
	750	0·7
Helgoland		
	400	0·8
East Anglia		
	640	0·2
Plymouth		
	3900	2·3
Naples		
	5550	0·9
Kattegat		

there is mixing of water from the various current systems or sporadic changes in current pattern (see Table II).

THE GENERAL CHARACTERISTICS OF "AMPHIOXUS SANDS AND GRAVELS"

In the coastal waters and lagoon system at Lagos in Nigeria a wide range of deposits is available for colonization by the local species of amphioxus, *Branchiostoma nigeriense*. Shell gravel is found offshore and coarse wave-disturbed or highly scoured sands are present on exposed coastal beaches and in channels subject to tidal rip currents. Finer sands and sandy muds of all intermediate grades occur in the sheltered creeks (Hill & Webb, 1958; Webb, 1958c). A broad survey of the distribution of the animal has shown that not all of these deposits provide suitable habitats. In some, amphioxus occurs in large numbers, while in others it is absent, although it is evident that in these cases the animal is not excluded by non-arrival of metamorphosing larvae (Webb, 1958a).

Several factors have been shown to contribute to this pattern of distribution. The first is disturbance by wave action which produces unstable conditions in the surface sands. Amphioxus has an acute tactile sense which, when stimulated by the agitation of coarse sand particles, causes the animal to leave disturbed sands. A second factor is the presence of fine particles in the deposit. It has been shown that amphioxus is not found in sand where 25% by weight of the grains are less than 0·2 mm in diameter or the silt fraction of the deposit exceeds 1·5% (Webb & Hill, 1958). Increasing the fine particle content beyond these levels causes very rapid decrease in permeability of the sand as shown in Fig. 5 (Webb, 1958d). Ciliary currents passing in through the mouth and out of the atriopore provide food, supply oxygen and remove metabolic waste. Amphioxus is dependent, therefore, on a free flow of water between the grains of the sand in which it lives and adopts a different posture in the sand according to its permeability (see Fig. 6). In coarse sands the body is usually entirely beneath the surface, in medium sands the mouth is exposed at the surface while in fine sand only the posterior part of the body is buried and both mouth and atriopore are clear of the sand. Two other general aspects of the deposit are important. Grains in which the surfaces are coated with micro-organisms are preferred and sharp sands such as river sands of comparatively recent origin are avoided. This again is due to the stimulation of the tactile sense by rough surfaces and sharp edges and corners on the grains as the animal moves between them.

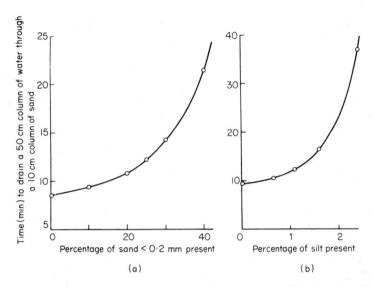

FIG. 5. Graphs showing the relation between the percentage (a) of grains <0·2 mm
diameter and (b) of silt present and the permeability to water of sand from Lagos, Nigeria.
(From Webb, 1958d).

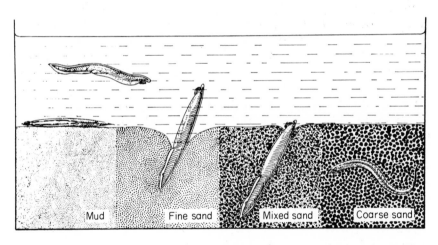

FIG. 6. The behaviour of amphioxus in deposits of different particle size. (From Webb &
Hill, 1958).

Amphioxus will therefore tend to be found in deposits that are
stable and well-ventilated and composed of smooth-textured
grains. In general these conditions provide an environment free
from organic decay to which the animal is also sensitive. It follows

that otherwise suitable sites for the adult may be rendered unfavourable by pollution with decaying organic materials such as untreated sewage. Choice chamber tests have shown clearly the reaction of *B. nigeriense* against deposits with rough or sharp grains or decaying organic matter and its preference for smooth grained deposits free from decay and supporting an active microfauna and flora (see Fig. 7). It seems that amphioxus congregate in favourable

Distribution in the tank of Ikoyi sands modified with regard to living and dead organic content and sharpness of grains		
Ikoyi sand autoclaved and stored in sterile jar.	Ikoyi sand stored in an unsterilized jar.	Ikoyi sand formolized, neutralized with ammonia and stored in a sterile jar.
Artificial mixture of sieved and graded grains equivalent to Ikoyi sand stored under sea water for 4 weeks		Ikoyi sand washed to remove silt, dried and incinerated to remove all organic matter and stored dry.
Artificial mixture of sieved and graded grains equivalent to Ikoyi sand stored dry	Ikoyi sand washed to remove silt and dried (not incinerated)	Ikoyi sand washed and incinerated and stored under sea water for 4 weeks

Numbers of lancelets recovered and the growth of micro-organisms, the sharpness of grains and the organic decomposition in each sand		
2 Organic growth very rich: rounded grains: decay present.	2 Organic growth rich: rounded grains: decay present.	3 Some organic growth: rounded grains: decay present.
3 Organic growth rich: sharp grains: no decay.		4 Very little organic growth: rounded grains: no decay.
2 Very little organic growth: sharp grains: no decay	2 Organic growth very rich: rounded grains: decay present	13 Organic growth rich: rounded grains: no decay

FIG. 7. A choice chamber test showing the reaction of *Branchiostoma nigeriense* to deposits with sharp grains, decaying organic matter, or a rich growth of micro-organisms present. (From Webb & Hill, 1958).

deposits as a result of trial and error and may remain quiescent for long periods until conditions deteriorate or the animals are disturbed. In deposits that do not provide optimal conditions amphioxus is sooner or later stimulated to move elsewhere. The number of animals present in a deposit, therefore, is a measure of the time for which they will tolerate those particular conditions.

THE SPECIFIC PERMEABILITY AND CAPILLARY SPACE OF SAND DEPOSITS

It has long been known by collectors that even in deposits of apparently similar texture satisfying the conditions described amphioxus may be present in large numbers at one point and absent from another less than a metre away. The fishermen from the Stazione Zoologica at Naples dredge for amphioxus with a bucket and before they bother to sift through the gravel in search of specimens test for the presence of the animal by the odour of the sand, which is similar to the iodoform-like aroma of *Balanoglossus*. If none is caught, then another dredging is made a few metres away. It seems, therefore, that either the animals react to some characteristic of the deposit that changes from place to place and is not apparent from texture, or the adults have a mutual attraction which leads to their congregation in small areas. There is no evidence for mutual attraction, but work on the physical properties of submerged marine sands has shown that important and subtle differences, to which organisms living in the sand react, exist within deposits (Webb, 1969b).

A sedimentary deposit can be considered as a lattice of particles within which there is a network of void. The particles are arranged in groups or clusters as the section through sand in Fig. 8 shows. At

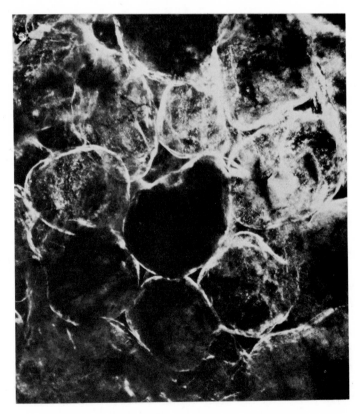

FIG. 8. A section through a sand column impregnated with resin and cut with a diamond wheel to show a cluster of regularly arranged grains. (From Webb, 1969b).

the centre of each cluster there is a cavity entered by the triangular or rectangular pores between the surrounding particles. The cavities and pores together form pathways of varying diameter which anastomose throughout the deposit (see Fig. 9). Natural

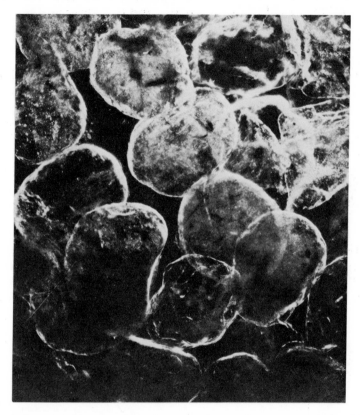

FIG. 9. A section through a sand column showing a pathway composed of cavities and pores.

sands differ in the size and shape of their particles and these give rise to a lattice that is characteristic in each individual case. Sands of nearly similar composition indistinguishable by granulometric analysis may differ considerably in the lattice they form and hence in the relative dimensions of the interstitial cavities and the pores between the grains. The ratio of void to particle volume, that is the porosity of the system, is one important characteristic. Another is the rate at which water will flow through the system under given conditions of temperature and pressure. This is the permeability (or drainage factor) of the deposit and depends on the size and shape of the pores linking the interstitial cavities. When a force is uniformly applied to a loosely packed sand, which can be done in the laboratory with ultrasonic radiation, consolidation takes place reducing the size of both the cavities and the pores and hence the porosity and the permeability of the deposit.

The porosity of a sand column is here defined as the volume in millilitres of interstitial space per 100 ml of sand particles and is calculated from the volume of the column ($\pi r^2 h$), the dry weight of the sand (w) and its mean relative density (D) by the formula ($\pi r^2 h - [w/D]$) $\times$ 100 D/w, the volume of the sand grains remaining unchanged for all degrees of consolidation. The permeability (drainage factor) is given by $[h/(t - K)] \times 10^2$ where h is the height of the column, t is the time in seconds for the water level above the sand to fall 10 cm due to drainage through the sand and K is a constant being the rate of flow of water through the apparatus in seconds in the absence of sand. This formula thus gives comparable values irrespective of column height.

If the process of consolidation of a sand with ultrasonic radiation is repeatedly interrupted and the porosity and permeability measured at different degrees of compaction, it is found that the relation between porosity and permeability is linear (see Fig. 10). It is true that, with some deposits, the slope of the graph so

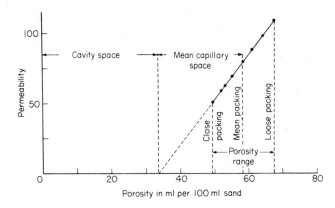

FIG. 10. Graph showing the permeability/porosity characteristics of a graded sand at 20°C with distilled water. Consolidation is single phase and the estimation of cavity space and capillary space is illustrated. Specific permeability is given by the slope of the graph. (From Webb, 1969b).

obtained suddenly changes during the course of consolidation. However, there is evidence that this is due to reorientation of elongated particles under pressure, to give a second phase of consolidation with an essentially new geometry in the lattice and new characteristics. The line of a graph of permeability against porosity when extended cuts the porosity axis at some point from the origin. This may be interpreted to mean that a part of the void varies as the permeability and a part remains constant within a single phase of consolidation. These parts of the void have been termed *capillary space* and *cavity space* respectively. The slope of the graph, which is constant within a phase of consolidation, gives the relative change in permeability with porosity (and hence capillary space since the cavity space by definition is constant) and has been termed the *specific permeability*. Specific permeability is expressed as the change in permeability (drainage factor) at a given temperature for a fall in porosity of 10 ml/100 ml of sand at the gradient of the phase of consolidation.

The passage of water through sand appears to take place along pathways of preferential flow at rates limited by the size of the pores entering the interstitial cavities. It is thought that, whether flow is laminar or turbulent, streams pass across the interstitial cavities, entering by one pore and leaving by another, with little or no mixing with the water occupying the peripheral parts of the cavities or with those parts of the system not in a direct line with the flow pathways. The nature of the capillary and cavity spaces is interpreted in these terms, the capillary space being held to be that of the flow pathways and the cavity space the remainder of the void which is stagnant (see Fig. 11).

Specific permeability expresses a relationship between the rate of flow through the deposit and the volume of the flow pathways. A high specific permeability indicates a high rate of change of the water in the flow pathways and a low specific permeability a correspondingly low rate. This can be represented in a model system by the inflow and outflow of a tank of water. Thus a system with a high rate of flow into and out of a large tank may change the water in the tank at the same rate as one in which there is a low rate of flow through a small tank. In both systems the specific permeability would be the same. A coarse sand with high permeability may have either a high or a low specific permeability according to the geometry of the lattice and hence the pore size and the interstitial volume occupied by flow pathways. The same is true of fine sands with low permeability. It is also true, however, that the majority of

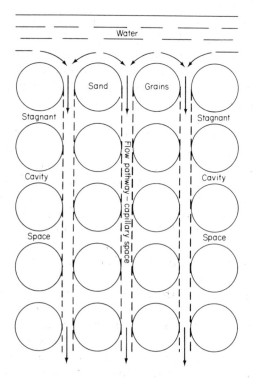

FIG. 11. Diagram to show an interpretation of the interstitial void of a sand deposit in terms of cavity and capillary space. Specific permeability expresses the relationship between the rate of flow of water through the deposit and the volume of the capillary space and hence the rate of change of the water in the flow pathway.

coarse sands have high specific permeability and most fine sands low specific permeability.

The relationship between capillary space, cavity space and specific permeability depends on the geometry of the lattice and alters not only with very small differences in particle size composition, but also with changes in the adhesiveness of the grains. Zeller (1967) pointed out that the characteristics of an aggregate alter with the mobile and adhesive tendencies of its constituent elements. Adhesiveness depends in part on the roughness of the grain surfaces, but much more and variably on the extent to which the grains are coated with micro-organisms. Thus although different deposits with minor differences of composition may differ widely in capillary and cavity space and in specific permeability, the same deposit carrying different growths of micro-organisms may differ

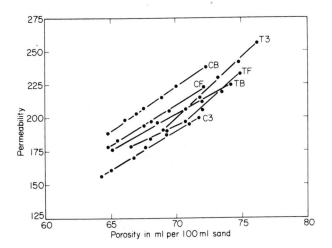

FIG. 12. Permeability/porosity graphs for sands from a ripple crest (C) and the adjacent trough (T) freshly removed from the seabed (F) after burning to remove the organic component (B) and after reconstitution of the organic component by standing in sea water for three days (3). These graphs show the wide differences in capillary and cavity space and in specific permeability arising from minor differences in granulometric composition and differences in the growth of micro-organisms on the grain surfaces. (From Webb & Theodor, 1972).

in these characteristics to an even greater extent (see Fig. 12). Organic films on the grains alter the size of both the pores in the sand and the volume of capillary space and can lead to an increase in specific permeability of over 70% compared with the same sand in which the organic component has been removed (Webb & Theodor, 1972).

THE IRRIGATION OF SAND DEPOSITS

Division of the interstitial void into capillary and cavity space and the concept of specific permeability, depend on a flow of water through the sand and cannot exist under stagnant conditions. Water currents in sand deposits at the sea bed may arise from several causes such as convection, tidal flow and the activity of the animals themselves that live in the sand. But perhaps the most widespread and continuous circulation of water in the sands in shallow seas is that induced by pressure waves passing across the bottom and generated by the surface waves (Webb & Theodor, 1968, 1972). Circulation takes place in the unconsolidated surface sand, which may be as much as 30 cm deep, and in one case at least is related to the sand ripples. Injection of dye has shown water passing into the sand in the troughs of the ripples and leaving it at the crests (see Fig. 13). Such currents carry into the sand finely divided organic particles and organic material in solution. These

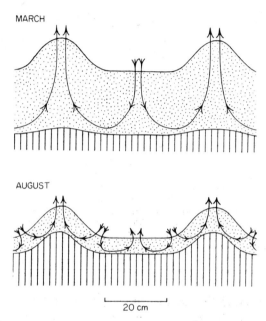

FIG. 13. Diagram showing the suggested circulation of water in the unconsolidated (stipple) layer of sand ripples at 3 m depth at Le Racou induced by pressure waves passing across the bottom. (From Webb & Theodor, 1972).

nourish the micro-organisms on the sand grains and lead to a rich but differential organic growth in different parts of the ripple system. Thus ripple trough and ripple crest sands of essentially similar granulometric composition may develop different organic components and correspondingly divergent properties (see Fig. 12).

THE SEASONAL DISTRIBUTION OF *BRANCHIOSTOMA LANCEOLATUM*

In the summer the number of amphioxus per unit area of deposit at a given depth of sea bears a direct linear relationship to specific permeability (see Fig. 14), other factors such as disturbance by wave action not being limiting. Agreement between the numbers of amphioxus and permeability of the deposit at maximum porosity was less good and no such relationship was

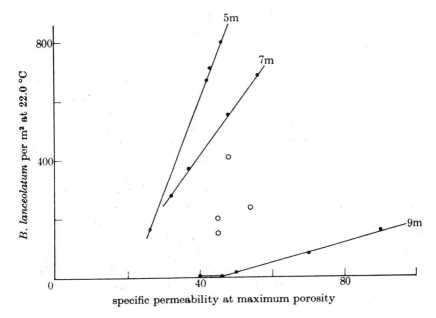

FIG. 14. Graph showing the relationship between the numbers of *Branchiostoma lanceolatum* and the specific permeability of the sands at Le Racou at 3, 5, 7 and 9 m in August 1968. Nonlinear arrangement at 3 m (○) was probably due to disturbance of the sand by wave action in shallow water. (From Webb, 1969b).

found with any other parameter of these sands at this time. The relationship with specific permeability was first discovered from collections of amphioxus from the volcanic sands of the Bay of Naples and subsequently confirmed in the mixed sands and gravels at Helgoland and in the highly uniform quartz sands from Le Racou near Banyuls-sur-Mer on the Mediterranean coast of France where the importance of the depth of the water was corroborated under conditions of low tidal rise and fall (Webb, 1969b).

In the course of these investigations numerous grab samples were taken of animals and their sand at various times in the year. But these showed a direct relationship between animal numbers and specific permeability in only about one half of the samples. This discrepancy was not resolved until systematic collection of animals and their deposits at Helgoland under warm water and cold water conditions showed that the distribution of amphioxus changed with the season (Webb, 1971). In the summer the animals live in greatest numbers in the coarse sands, which in general have high specific permeability, whereas in the winter they are absent from the coarse sands and are common in the fine sands with low specific permeability (see Fig. 3a, b & d). The sea temperature range at Helgoland is extreme. At the bottom, which is at 20 m over the amphioxus ground, it reaches a maximum of 16–18°C in August and September and falls to a minimum in a normal year of 3–4°C in February and March (Courtney & Webb, 1964). At Le Racou the temperature is higher and the seasonal range is less, being 21–22°C in August, falling to 12·5°C in December. Nevertheless here also amphioxus in the winter is found in sands with different properties from those it inhabits in the summer. Collections made at Le Racou in December showed that the numbers of animals recovered from each of ten van Veen grab samples bore an approximate inverse linear relation to specific permeability, but a very close direct linear relation to the capillary space of the sands in which the animals were living (see Figs 15 & 16). Thus, unlike the summer amphioxus sands which have a high specific permeability, the preferred winter sands have a high capillary space and incidentally a low specific permeability.

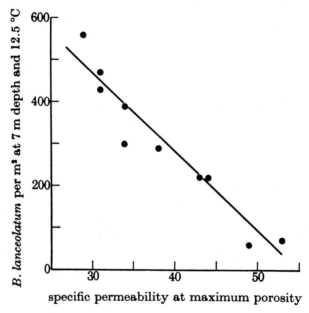

Fig. 15. Graph showing the relationship between the numbers of *Branchiostoma lanceolatum* and the specific permeability of the sands at Le Racou at 7 m in December 1967. (From Webb, 1969b).

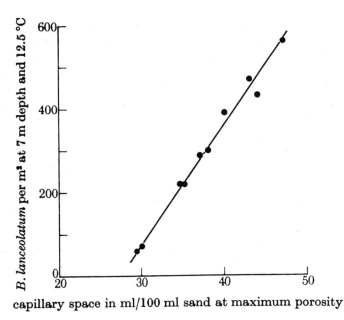

FIG. 16. Graph showing the relationship between the numbers of *Branchiostoma lanceolatum* and the capillary space of the sands at Le Racou at 7 m in December 1967. (From Webb, 1969b).

The high summer demand for a continuous stream of uncontaminated water for feeding and respiratory purposes (Courtney & Newell, 1965) and for the rapid removal of metabolic waste is evidently met by high specific permeability. Amphioxus reacts to falling temperature. For example in *B. nigeriense*, a tropical species accustomed to sea temperatures of 25–30°C, a reduction to 17°C induces continuous swimming. This does not happen in *B. lanceolatum* but a reaction to a lower temperature is to be expected in this species also (Webb & Hill, 1958; Courtney & Webb, 1964). It seems probable, therefore, that in the autumn, cooler water penetrates most rapidly into deposits with high specific permeability. This induces the amphioxus to move into the finer sands with high capillary space and low specific permeability where the rate of exchange of warm interstitial water with cold water above the sand would be less rapid. When the temperature falls to about 7°C the metabolic rate of *B. lanceolatum* is low and the animal's movements become sluggish (Courtney & Webb, 1964). In even colder water, such as at Helgoland in the winter, the animal moves very little and would be expected to remain buried in the fine sands. At temperatures below 7°C the ciliary current through the pharynx is minimal while at 3°C the pharyngeal mechanism cannot function normally. The respiratory demands are low and for this reason the animal would be able to tolerate the low permeability of the fine sands. In the spring, with rising temperature, the animal's metabolic rate increases and with it the ciliary beat of the pharynx. The specific permeability of the fine sands is insufficient to meet the demands for a fast stream of water through the pharynx and again the animal is stimulated to move until deposits with high specific permeability are found. Thus the environmental stimulus inducing amphioxus to move away from high specific permeability summer sands appears to be a fall in temperature. Continued residence in the high capillary space winter sands arises from the reduced activity of the animal at lower temperatures. Vacation of the winter sands, with rise in temperature in the late spring and early summer, follows an increase in activity and the corresponding demand for conditions of high specific permeability.

THE EFFECT OF WATER DEPTH ON SIZE DISTRIBUTION

The need for a copious stream of interstitial water will clearly vary with body size, a large amphioxus passing a greater volume through the pharynx than a small animal. As the depth of the sea increases the deposit becomes finer and less permeable and the rate of irrigation, from this cause and through diminution of pressure waves at the bottom for a given surface wave height, becomes correspondingly less. As depth increases, therefore, a point should be reached where the rate of flow of water through the deposit will be sufficient to maintain a small amphioxus but not a large one. This has been found to be the case in the Mediterranean at Le Racou where amphioxus occurs in a belt parallel to the shore delimited on the shoreward side by turbulence and seaward by the rate of irrigation of the sand. Here at the shoreward margin large amphioxus occur, at the centre of the belt all sizes are found together, but at the seaward margin only small animals are present. This is shown in Fig. 17 which gives the range in size of amphioxus collected by van Veen grab at depths from 3 to 12 m. Water depth will be expected to influence size distribution in other areas, but measurements such as these can only be made where tidal rise and fall is slight or absent. Under exceptional conditions, where sand deposits are associated with continually sheltered and calm water, amphioxus is found at depths less than 1 m, or even intertidally. Intertidal amphioxus of small size are known from very restricted

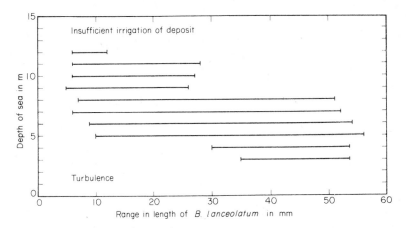

FIG. 17. The distribution of *Branchiostoma lanceolatum* of different size according to the depth of the water at Le Racou in November 1967 at 15°C.

localities in some of the creeks and lagoons at Lagos in Nigeria and in the Bassin d'Arcachon on the Atlantic coast of France (Salvat, 1962; Bouchet & Lasserre, 1965). In both cases they are associated with high water temperatures of about 30°C. It is perhaps significant that increase in permeability with temperature is relatively greater in fine than in coarse sands (Webb, 1969b), a factor that may lead to tolerance of fine intertidal sands by small amphioxus.

SHORT-TERM MOVEMENT OF AMPHIOXUS IN SAND

It is evident that sand deposits in shallow seas are a dynamic mosaic of small areas, many of less than a square metre, with widely different physical properties which are subject to change. The differences between one area and another arise partly from basic variation in particle size composition governed by the pattern of water flow over the bed and the consequent deposition or removal of particles of different shape and mass according to the rate of flow. But of greater importance are the differences in the organic component associated with ripple formation and with the patterns of irrigation within the deposit. Such sands are liable to change in the long term in particle size composition and in the short term through fluctuations in the growth of the organic component arising from changes in the input of particulate and dissolved organic matter such as glycollic acid from the photorespiration of marine autotrophs (James, 1971). It is difficult to see, therefore, how amphioxus could maintain itself in an optimal environment without continuously monitoring the changing substrate. Animals leaving the sand to swim in the sea above and then returning to burrow at random would be unlikely to produce the finely graded sequence of numbers in relation to sand property that evidently occurs. It has been shown, however, that amphioxus is extremely reluctant to leave the sand. Even when 150 large adults in a small basin of sand under seawater were vigorously stirred for several minutes only three animals appeared momentarily at the sand surface (Webb, 1971). A reason for this seems to be the reaction of the animal to light. Amphioxus, dark-adapted by the exclusion of light for upward of four minutes, react to exposure to light by rapid undulatory movements of the body which may continue for as much as 50 seconds in the absence of a deposit. Moreover movement may be either head-first or tail-first through the sand away from a source of stimulation (Webb, 1972, 1973). Rapid movement within the unconsolidated sand layer and continuous reaction to the changing physical properties of the deposit would seem to be essential for the selection of the best conditions available for life and for the aggregation of very large numbers within a small area that is known to occur.

THE CONGREGATION OF ADULTS AND SPAWNING

The congregation in the summer of large numbers of amphioxus in sands with high specific permeability would seem to be important for reproductive purposes. There is not known to be any attraction between the sexes and, as sperm and eggs are shed into the sand, it is essential that the mature males and females should be in close association. This is evidently achieved by the development of an avoidance reaction to all deposits except those satisfying a relatively extreme condition, that is high specific permeability, which although local in occurrence is at the same time widespread in a range of different deposits. Amphioxus is a warm water animal. *B. lanceolatum*, which is the most northerly of the species and ranges as far north as the Norwegian coast at Bergen (61°N), has an activity peak between 10 and 20°C in both the North Sea and the Mediterranean forms. *B. lanceolatum* can survive at temperatures between 3° and zero for several months, but it is doubtful if it can spawn in waters permanently below 10°C (Courtney & Webb, 1964; Courtney, 1975). This is probably the factor which establishes the limit of between about 60°N and 45°S latitudes for the world distribution of amphioxus of which some 20–30 species have been described.

THE DISTRIBUTION OF AMPHIOXUS 205

POPULATION AND SPECIFIC DIFFERENCES IN *BRANCHIOSTOMA*

Most of the diagnostic characters used to determine *Branchiostoma* species are quantifiable. They involve the number and size of the fin chambers and the distribution of the myotomes in the various regions of the body, and are independent of the size of the animal (Webb, 1955). Assuming a normal distribution of the values for a character within a sample, the probability that the differences given by the Means and the Standard Deviations between two samples in respect of that character could occur between random samples of one population can be calculated by the "t" test. This procedure, repeated for each diagnostic character, shows to what extent the differences between two samples may be held to characterize the populations from which they were taken and has been used for this purpose in taxonomic studies of the group (Webb, 1956a, b, c, 1957). Nevertheless, in spite of such statistical evaluations of the separate diagnostic features, determinations of species and populations usually involve some intuitive recognition of differences judged by the taxonomist to be sufficient. This process can be greatly assisted if a numerical value representative of the difference between pairs of species or populations of a species can be assigned to each character. A useful simplification less subject to intuitive bias is then achieved by combining the values for the various characters into a single figure.

Numerical evaluation of the difference between a pair of samples from a sequence of species, such as that from the west coast of Africa shown in Table I, should involve and give equal weight to all the characters subject to the following provisions (Sokal & Sneath, 1963; Clarke, 1968). The distribution of values for each character should be normal within a sample (which has been shown to be the case, Webb, 1955), the characters used should be independent of each other and the sample itself should be of a reasonable size (about 20). The six characters selected from those normally used for diagnostic purposes in the genus *Branchiostoma* and listed below have been shown graphically not to be correlated with each other and can be used in the calculation of a composite value of difference.

1. Number of dorsal fin chambers
2. Number of preanal fin chambers
3. Height/breadth of tallest dorsal fin chamber
4. Myotomes from anterior end to atriopore
5. Myotomes from atriopore to anus
6. Myotomes posterior to anus

Thus where samples from two populations A and B have the Means $\mu A1$ and $\mu B1$ and Standard Deviations σ_{A1} and σ_{B1} for the first of the six normally distributed characters, then a measure of the similarity (or difference) with respect to this first character is given by

$$\frac{(\mu_{A1} - \mu_{B1})^2}{2(\sigma_{A1}^2 + \sigma_{B1}^2)}$$

If the formula is applied to each of the six characters of the pair of samples under consideration, the six values obtained are dimensionless and are of comparable numerical magnitude. Summing the values over the six characters gives, in one sense, a measure of the difference between A and B which can be compared to the values obtained for other pairs of samples. The sum of the difference values for each of the six characters has been termed the *difference factor*. This factor is here used, first, to indicate whether differences between samples held to represent different populations of the same species and between those determined as distinct species are such as to warrant this recognition and, secondly, for comparison with the distances between the localities at which the samples were taken.

POPULATION AND SPECIFIC DIFFERENCES IN RELATION TO
DISTRIBUTION

An analysis was made by computer of a wide range (but not all) of the populations and species of *Branchiostoma* from Europe, Africa, Asia and the west coast of the Americas. Each sample was compared with all others by means of difference factors calculated from published data (Webb, 1956a, b, c, 1957, 1958b). A selection of these results is given in Tables I, II and III and Fig. 18. Table I compares the difference factors with the distances between the localities at which the samples were taken for a sequence of species from the English Channel around the coast of Africa as far as Mozambique. Table II gives the same information for five populations of *B. lanceolatum*. Two points are clear: first, each pair of samples determined as different species has a difference factor considerably in excess of 3·0 whereas for populations of the same species the value is less than 3·0; secondly, there is no relationship between the difference factor and the distance between localities. The greatest distance (5500 km) is between two populations of *B. lanceolatum* with a difference factor of 0·9, and the shortest distance (160 km) between two West African species with a difference factor of 20·6. The difference factors for each West African species

TABLE III

The distance and difference factor between samples of populations and species of Branchiostoma *widely separated by ocean*

Species	Locality	Distance (km)	Difference factor
B. belcheri	East Africa		
		16 000	0·8
	China		
		18 100	33·7
B. capense	South Africa		
		23 000	7·7
B. elongatum	Peru		
		6700	17·1
B. californiense	California		

(including the species from the Cape) and all of the others in that sequence are shown diagramatically in Fig. 18. The differences between the populations of B. *lanceolatum* drawn to the same scale are included for comparison. It is of course not possible to construct such a two-dimensional diagram in which the lines joining the species accurately represent the differences between them. A better approximation would be given by a three-dimensional model. There seems little doubt from this diagram that specific recognition of the West African forms is appropriate. They probably comprise a related group, which would be in accordance with the current patterns of the West African coast. In Table III difference factors and distances are given for widely separated species and populations of the same species in which, unlike the African sequence in Table I, any contact between the localities from which the samples were taken would seem to be via transoceanic currents. Here again populations of the same species have a difference factor less than 3·0 and there is no relation between difference factor and distance. The three species, B. *capense*, B. *elongatum* and B. *californiense*, have been the centre of taxonomic controversy and doubt has been expressed whether they are distinct (Webb, 1957). Certainly there are similarities between B. *capense* and B. *elongatum* as the low difference factor (7·7) testifies, but this could be due to chance or it may indicate recent or incomplete separation. In the latter case it is not the shortest distance between the two localities that is significant, but the possibility of eastward drift of larvae from the Cape to the coast

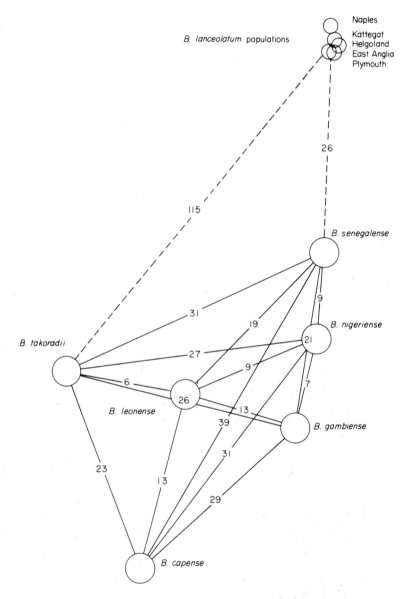

FIG. 18. Diagram to show the magnitude of the differences between the populations and species of *Branchiostoma* from the coasts of Europe and Africa. The full lines are approximately proportional to the difference factors between the species; the broken lines are abbreviated. Values for the difference factors are included in the diagram.

of S. America, a distance of some 23 000 km. Isolation leading to the formation of species and populations within the species must have arisen from the infrequency or absence of larval interchange. It is clear that this is not a function of distance of separation, but of the pattern of water flow between the localities.

AMPHIOXUS AS AN INDICATOR OF CURRENT PATTERNS

The distribution of amphioxus is determined by the currents dispersing the larvae and the physical nature of the substratum in which the adults live. The length of the pelagic phase of up to 200 days or more ensures prolonged drifting in circulating or reciprocal, coastal or transoceanic currents necessary for transmission of larvae between the widely separated localities at which the adults are established. The requirements of the adults for deposits that meet a number of physical conditions, the most important particularly in the temperate summer being high specific permeability, leads to their congregation in large numbers in small areas for spawning. It is evidently rare for the larvae they produce to return to their site of origin so that maintenance of the adult groups by recruitment depends on the arrival of larvae from other groups in a circulating or reciprocal current system. If this were not so all amphioxus grounds would be transient which is certainly not the case. Free larval exchange between the groups of adults in such a current system must provide the genetic unity characteristic of a population. Reduced, sporadic or intermittent larval exchange would lead to the development of genetic difference between the adult groups and hence to distinct populations and eventually species. Taxonomic recognition of the populations and species and the determination of their distribution therefore must at the same time establish the pattern of water flow on which that distribution depends. This being so, amphioxus would appear to be an appropriate organism to serve as an indicator of both coastal and oceanic currents, more particularly since the diagnostic features are quantifiable and the behaviour of the adult ensures their presence in large numbers in discrete areas from which good samples can be taken for statistical purposes. If the distribution of amphioxus were used in this way recognition of the physical characteristics that render a deposit suitable for the adult would become important. Without this information it is impossible to say whether a ground that does not harbour amphioxus is outside a current system and therefore never colonized by larvae or is barren through lack of some essential physical requirement.

Preliminary work is beginning to show the potential value of amphioxus as an indicator of current flow and the occurrence of isolated water masses. For example it may be inferred from the distribution of species and populations given here that there is or has been greater isolation of water masses along the apparently uniform West African coast than exists between the China Sea and East Africa, or again between the North Sea and the Bay of Naples. In another case population differences in the amphioxus of the Gulf of California indicate that the water of the Gulf north of about latitude 26°N is very largely isolated from that of the southern reaches (Webb, 1959). It is evident that much could be done with the adult, but a possible fruitful field for further work is the identification of the larvae collected from the high seas with their corresponding adult forms (Wickstead, 1964).

ACKNOWLEDGEMENTS

The author is grateful to Mr P. J. Webb for devising the formula used to determine the difference factor and for the computer analysis of the various species and populations of *Branchiostoma*. Thanks are also due to Professor S. J. Taylor for mathematical advice and to Dr W. A. M. Courtney and Professor J. P. Harding for helpful criticism of the typescript.

REFERENCES

Bouchet, J-M. & Lasserre, P. (1965). Note préliminaire sur les fonds à *Amphioxus* du Bassin d'Arcachon. *P.-v. Soc. linn. Bordeaux* **101**: 1–7.
Chin, T. G. (1941). Studies on the biology of the Amoy amphioxus *Branchiostoma belcheri* Gray. *Philipp. J. Sci.* **75**: 369–424.
Clarke, D. L. (1968). *Analytical archaeology.* London: Methuen.
Courtney, W. A. M. (1975). The temperature relationships and age-structure of North Sea and Mediterranean populations of *Branchiostoma lanceolatum* (Pallas). *Symp. zool. Soc. Lond.* No. 36: 213–233.
Courtney, W. A. M. & Newell, R. C. (1965). Ciliary activity and oxygen uptake in *Branchiostoma lanceolatum* (Pallas). *J. exp. Biol.* **43**: 1–12.
Courtney, W. A. M. & Webb, J. E. (1964). The effects of the cold winter of 1962/63 on the Helgoland population of *Branchiostoma lanceolatum* (Pallas). *Helgöländer wiss. Meeresunters.* **10**: 301–312.
Franz, V. (1922). Systematische Revision der Akranier. (Fauna et Anatomia ceylanica Nr. 10). *Jena Z. Naturw.* **58**: 369–452.
Gosselck, F. & Hagen, E. (1973). Vorkommen und Verbreitung der Larven von *Branchiostoma senegalense* (Acrania, Branchiostomidae) vor Nord-West-Africa. *Fisch.-Forsch. wiss. Schrift.* **11**: 101–106.

Gosselck, F. & Kuehner, E. (1973). Investigations on the biology of *Branchiostoma senegalense* larvae off the northwest African coast. *Mar. Biol.* **22**: 67–73.

Hartmann, J. & John, H. C. (1971). Planktische *Branchiostoma* nordwestlich der Doggerbank (Nordsee). *Ber. dt. wiss. Kommn Meeresforsch.* **22**: 80–84.

Hill, M. B. & Webb, J. E. (1958). The ecology of Lagos lagoon II. The topography and physical features of Lagos harbour and Lagos lagoon. *Phil. Trans. R. Soc.* (B) **241**: 319–333.

James, W. O. (1971). *Cell respiration.* London: English Universities Press.

Kuhl, H. (1972). Hydrography and biology of the Elbe estuary. *Oceanogr. Mar. Biol. A. Rev.* **10**: 225–309.

Meadows, P. S. & Campbell, J. L. (1972). Habitat selection and animal distribution in the sea: the evolution of a concept. *Proc. R. Soc. Edinb.* (B) **73**: 145–157.

Rice, H. J. (1880). Observations upon the habits, structure and development of *Amphioxus lanceolatus. Am. Nat.* **14**: 1–19 & 73–95.

Salvat, B. (1962). Faune des sédiments meubles intertidaux du Bassin d'Arcachon. Systématique et écologie. *Cahiers Biol. mar.* **3**: 219–244.

Sokal, R. R. & Sneath, P. H. A. (1963). *Principles of numerical taxonomy.* San Francisco & London: Freeman.

van Wijhe, J. W. (1927). Observations on the adhesive apparatus and the function of the ilio-colon ring in the living larva of Amphioxus in the growth-period. *Proc. K. ned. Akad. Wet.* **30**: 991–1003.

Webb, J. E. (1955). On the lancelets of West Africa. *Proc. zool. Soc. Lond.* **125**: 421–443.

Webb, J. E. (1956a). On the populations of *Branchiostoma lanceolatum* and their relations with the West African lancelets. *Proc. zool. Soc. Lond.* **127**: 125–140.

Webb, J. E. (1956b). Cephalochordata of the coast of tropical West Africa. *Atlantide Rep.* No. 4: 167–182.

Webb, J. E. (1956c). Cephalochordata. *Scient. Rep. "John Murray" Exped.* **10**: 121–128.

Webb, J. E. (1957). On the lancelets of South and East Africa. *Ann. S. Afr. Mus.* **43**: 249–270.

Webb, J. E. (1958a). The ecology of Lagos lagoon III. The life history of *Branchiostoma nigeriense* Webb. *Phil. Trans. R. Soc.* (B) **241**: 335–353.

Webb, J. E. (1958b). On a collection of lancelets from the Gambia, with a description of a new species of *Branchiostoma. Proc. zool. Soc. Lond.* **131**: 627–635.

Webb, J. E. (1958c). The ecology of Lagos lagoon I. The lagoons of the Guinea Coast. *Phil. Trans. R. Soc.* (B) **241**: 307–318.

Webb, J. E. (1958d). The ecology of Lagos lagoon V. Some physical properties of lagoon deposits. *Phil. Trans. R. Soc.* (B) **241**: 393–419.

Webb, J. E. (1959). Lancelets and the ocean currents. *Int. Oceanogr. Congr.* **1**: 356–358.

Webb, J. E. (1969a). On the feeding and behaviour of the larva of *Branchiostoma lanceolatum. Mar. Biol.* **3**: 58–72.

Webb, J. E. (1969b). Biologically significant properties of submerged marine sands. *Proc. R. Soc.* (B) **174**: 355–402.

Webb, J. E. (1971). Seasonal changes in the distribution of *Branchiostoma lanceolatum* (Pallas) at Helgoland. *Vie Milieu* (suppl.) **22**: 827–839.

Webb, J. E. (1972). *Amphioxus.* Cinefilm made by University of London Audio-Visual Centre.

Webb, J. E. (1973). The role of the notochord in forward and reverse swimming and burrowing in the amphioxus *Branchiostoma lanceolatum*. *J. Zool., Lond.* **170**: 325–338.

Webb, J. E. & Hill, M. B. (1958). The ecology of Lagos lagoon IV. On the reactions of *Branchiostoma nigeriense* Webb to its environment. *Phil. Trans. R. Soc.* (B) **241**: 355–391.

Webb, J. E. & Theodor, J. (1968). Irrigation of submerged marine sands through wave action. *Nature, Lond.* **220**: 682–683.

Webb, J. E. & Theodor, J. L. (1972). Wave-induced circulation in submerged sands. *J. mar. biol. Ass. U.K.* **52**: 903–914.

Wickstead, J. H. (1964). Acraniate larvae from the Zanzibar area of the Indian Ocean. *J. Linn. Soc. (Zool.)* **45**: 191–199.

Wickstead, J. H. (1973). in "Report of the Council for 1972–3". *J. mar. biol. Ass. U.K.* **53**: 1017.

Wickstead, J. H. & Bone, Q. (1959). Ecology of Acraniate larvae. *Nature, Lond.* **184**: 1849–1851.

Willey, A. (1894). *Amphioxus and the ancestry of the vertebrates.* New York and London: MacMillan.

Zeller, C. (1967). A geometric model with some properties of biological systems. *Gen. Syst.* **12**: 53.

Symp. zool. Soc. Lond. (1975) No. 36, 213–233.

THE TEMPERATURE RELATIONSHIPS AND AGE-STRUCTURE OF NORTH SEA AND MEDITERRANEAN POPULATIONS OF *BRANCHIOSTOMA LANCEOLATUM*

W. A. M. COURTNEY

Department of Zoology, Westfield College, Hampstead, London, England

SYNOPSIS

The recorded range of *Branchiostoma lanceolatum* (Pallas) extends from Bergen on the Norwegian coast to the Morrumbene Estuary on the east coast of Africa and this lancelet extends into higher latitudes than other species of the genus.

The age-structure of the *B. lanceolatum* population at Helgoland is different from that of the population at Le Racou on the French Mediterranean coast. This results from the colder winters of the North Sea having a sub-lethal effect upon the whole population by limiting growth when the temperature falls below 10°C. Severe winters are lethal with respect to the youngest and oldest individuals in the short term which, in the longer term, causes a paucity of mature animals up to several years later. Reproduction takes place in the summer and both the behaviour and oxygen uptake temperature profiles show that this animal is adapted to temperatures of 10–25°C. The data which are available indicate that the other species of *Branchiostoma* are suited to higher temperatures.

INTRODUCTION

Despite frequent references to *Branchiostoma* in general texts, many aspects of its life-history are poorly understood. The work of Chin (1941) on the sub-tropical population of *B. belcheri* at Amoy was interrupted by the Second World War and the life-cycle of the tropical lagoon population of *B. nigeriense* studied by Webb (1958) is terminated annually by gross fluctuations in salinity.

The distribution of *B. lanceolatum* extends from the North Sea through the Mediterranean Sea to the East African coast (Webb, 1956) and this species is active over a lower range of temperature than the sub-tropical and tropical forms (Parker, 1908; Webb, 1958). A comparison of a North Sea and a Mediterranean population of *B. lanceolatum* shows that further temperature-dependent factors operate within populations of this species towards the northern limit of its distribution.

MATERIAL AND METHODS

The population of *Branchiostoma lanceolatum* studied in greatest detail is found at a depth of 20 m in sandy gravel 5 km north-east of Helgoland in the German Bight. Twenty Van Veen samples were taken on one day at approximately 6-week intervals throughout 1963 together with samples during 2 days late in 1962 and a further 2 days in 1965. Each of the 0·1 m² samples taken with the Van Veen grab was repeatedly stirred with 2% formalin and the dead animals strained off from the supernatant fluid.

Larvae were taken during the daily plankton hauls of the Meeresstation Helgoland from the waters of the Rade between Helgoland and the nearby island, the Dune. During the spawning period further plankton nettings were made over the lancelet-grounds themselves at a variety of depths.

A frequency/body length histogram was constructed for each collection of 20 Van Veen samples which was analysed, where necessary using the methods of Harding (1949).

Monthly temperatures at 20 m depth over the grounds were measured during the greater part of the years 1960-1963 to augment the routine daily recordings of surface-water temperature.

A Mediterranean population of *B. lanceolatum* about 100 m offshore at Le Racou was sampled quantitatively during 1966-1968 using a Stramin dredge and Van Veen grab. The animals, which are found at depths of 3 to 11 m, were extracted from the sand samples using formalin.

Animals were separated by hand from shell-gravel of the Eddystone grounds during one day of dredging 40 km south-west of Plymouth.

GROWTH AND LONGEVITY OF HELGOLAND LANCELETS

In August 1962 the mean length of the youngest *B. lanceolatum*, the 0-group, at Helgoland was 13 mm (Fig. 1). These animals were spawned in 1961 and were almost one year old. The mean length of the 1-group animals adults (spawned in 1960) was 23 mm, a difference of 10 mm in a year. Three months later, in early November 1962, what was now the 1-group had grown a further 8 mm to a mean length of 21 mm (Fig. 1). Thus, growth is not continuous throughout the year and the greater part of the annual growth occurs during the months of August, September and October.

After a further settlement had taken place in September 1963, the 1-group was represented by one animal of length 20 mm (Fig. 1), the remainder of the 1962 settlement having been eliminated during the cold winter of 1962-1963 (Courtney & Webb, 1964).

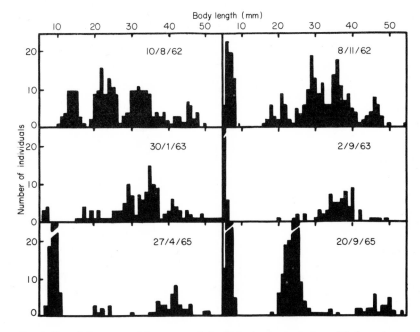

FIG. 1. Length/frequency histograms of the Helgoland population of B. *lanceolatum*.

The high mortality amongst the 1962 settlement caused a paucity of individuals of length 25 to 33 mm in the population three years later in April 1965 (Fig. 1) followed by a gap in the population at length 30 to 35 mm in September of the same year. This loss of an annual group provided invaluable confirmatory evidence of the growth and longevity of the Helgoland lancelet.

Interpretation of the polymodal-frequency histograms was difficult after January 1963 when the average number of individuals taken in 20 Van Veen samples fell by 50% so that in April and May 1963 and also in May 1965 the grounds were worked for two or three days a month. Samples for each month were amalgamated.

The population in August 1962 consisted of six annual groups of animals with mean lengths of 13, 23, 33, 38, 44 and 47 mm respectively (Table I). In November the new settlement varied in length from 4 to 9 mm. Meanwhile, the older animals had grown by amounts varying between 8 and 2 mm depending upon their age.

There was no change in the mean length of the 0-group during the period 5/11/62 and 30/1/63 and the mean length of each of the

TABLE I

Annual means of length (mm) in monthly samples of the Branchiostoma lanceolatum *population at Helgoland 1962–1963*

Year Month	1963 Jan.	1963 Apr.	1963 May	1963 June	1963 July	1962 Aug.	1963 Aug.	1962 Nov.
0-group	7	abs	abs	abs	abs	13	abs	7
1-group	20	20	20	21	24	23	23	21
2-group	29	29	29	29	31	33	33	30
3-group	36	36	36	36	37	38	37	36
4-group	41	41	41	42	43	43	44	41
5-group	46	45	45	46	46	47	47	46
6-group								50
*Max size	53	49	47	48	50	50	49	55
†No. of VV	20	40	40	20	20	20	20	20
‡Animals in 20 VV	143	41	68	94	54	115	48	213

(The maximum size of animal(*) and the number of Van Veen samples(†) each month, together with the average number of individuals in 20 Van Veen samples,(‡) is given.)

remaining annual groups in the population changed by 1 mm or less during the same period (Table I). Clearly, there was little or no growth within the population between November 1962 and June 1963 so that growth is confined to about six months of the year (Fig. 2). The percentage increase in growth, that is relative growth, decreased logarithmically with age (Fig. 3).

Both the length of the pharynx and the number of primary gill bars increased linearly with the body length (Fig. 4). However, the maximum length of the primaries failed to keep pace with increase in body length, though the direct logarithmic relationship between (length of the primaries)^2 and the body length suggests that the volume of the pharynx relative to the length of the body may remain constant.

Helgoland lancelets reach a length of about 47 mm after six years of adult life and the largest animal ever collected at Helgoland, which was 55 mm long, is presumed to have lived for about eight years.

RECRUITMENT MORTALITY AND POPULATION
STRUCTURE OF HELGOLAND LANCELETS

Gonads appear between September and November in animals which are two years old and all of the mature animals in the

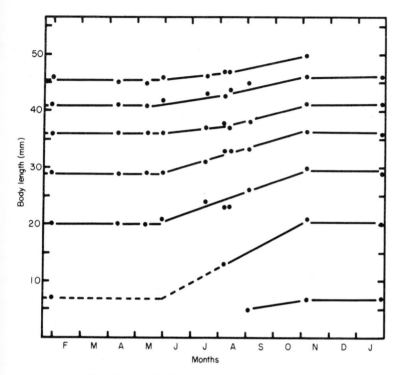

FIG. 2. Growth of *B. lanceolatum* at Helgoland.

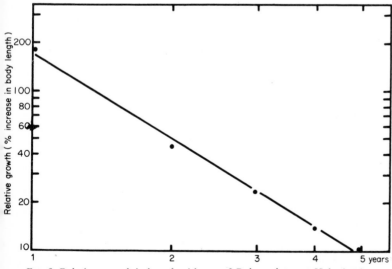

FIG. 3. Relative growth in length with age of *B. lanceolatum* at Helgoland.

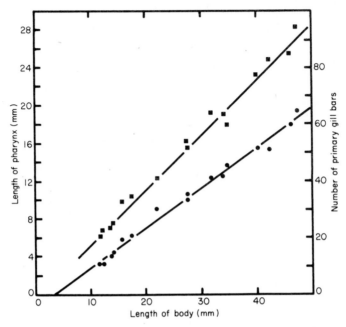

FIG. 4.The relationship of body length to pharyngeal length (●) and to number of primary gill bars (■) in *B. lanceolatum* taken at Helgoland in May 1963.

population spawn about seven months later. Spawning had not commenced on 7/6/63 but all of the animals with gonads had spawned by 15/7/63. Larvae were taken in plankton hauls on seven days between 10/8/63 and 6/9/63 with consecutive daily catches on and between August 20th and 24th. The larvae taken on 20/8 and 6/9 had begun to metamorphose. Newly metamorphosed adults were taken from the grounds in Van Veen samples on 2/9/63 and 20/9/65. These observations indicate a larval life of about two months duration and it is possible that the Helgoland grounds are repopulated from other populations of *B. lanceolatum* which are known to exist in the German Bight. A thorough search of the grounds was undertaken and mid-water plankton hauls were made during July and August of 1963 without yielding a single developing larva. There is no doubt that larval development is planktonic.

The larvae taken in the plankton at Helgoland on and before 20/8 had been spawned about two weeks earlier than the subsequent catches of larvae. However, the variation in the size of the 0-group (Table II), where in November 1962 adults of the minimum size of 4 mm were present together with animals 9 mm long, is not necessarily indicative of prolonged spawning. Hartmann & John (1971) caught planktonic larvae between

TABLE II

Size-range in mm of 0-group at Helgoland
within four months of settlement

	1962	1963	1965
September	4–6	—	5–7
November	4–9	6–8	—

12/9/69 and 19/9/69 of length 4 to 8 mm north-west of the Dogger Bank. Nevertheless, all of the larvae collected during the present work were less than 5 mm in length.

The time when *Branchiostoma lanceolatum* spawns in the North Sea probably varies by about two weeks with location and year. Larvae were taken about a fortnight later in 1969 off the Dogger Bank (Hartmann & John, 1971) than at Helgoland in 1963. Hagmeier & Hinrichs (1931) found that spawning was confined to a period of two weeks early in June during laboratory observations of the Helgoland population. In 1963, the same population spawned in late June or early July.

Fewer animals settled in 1963 than in the other years in the period 1962–1965 at Helgoland (Table III). Recruitment was probably greater in 1964 and 1965 than in 1962.

TABLE III

Annual density of settlement per m^2 at Helgoland

Settlement	Samples taken	Density/m^2
1962	5/11/62	32
1963	2/ 9/63	19
	4/11/63	6
1964	23/ 4/65	43
	27/ 4/63	51
1965	20/ 9/65	54

There were 22 animals per m^2 with spent gonads in August 1962, 17 per m^2 after spawning in July 1963 and an average of 20 per m^2 before spawning in 1965. Therefore, the recruitment was poorest following the cold winter and there were fewer mature individuals after 1962. But since death from senility probably follows the spawning period, the figure obtained, in August, for 1962 is probably depressed.

The highest rate of recruitment per mature adult of the population was in 1965 with a ratio of five newly-settled adults to two mature adults in the population. Therefore the mortality rate associated with larval life is extremely high.

The 0-group may be wiped out by extremely cold winters at Helgoland and this occurred in 1962–1963. It is probable that the 0-group is affected by winters which are less severe than the low temperatures of early 1963. The size of the annual groups in the August 1962 samples varied inversely with the length of time that the temperature was less than 5° C during the first winter of each settlement as seen in Fig. 5. There is also a relationship between the

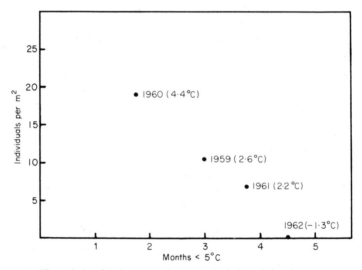

FIG. 5. The relationship between the numerical size of the three most recent annual settlements in the population at Helgoland in August 1962 and the length of time the temperature was less than 5°C during the first winter of life. The minimum temperature during the first winter is given in parenthesis. The data for the 1962 settlement is added.

minimum temperature of the first winter of life, since the duration of low temperature and temperature minimums are related in this area of the North Sea. Of course, such an analysis of the 1962 histogram takes no account of any variation in settlement density from year to year nor of mortality rates in later years of life.

The lowest rates of mortality are experienced in the second and third year of the life of the Helgoland lancelet. Animals of intermediate size in the population are least affected by severe winters (Courtney & Webb, 1964). When such winters occur during the fourth and fifth year of life the mortality-rate is higher

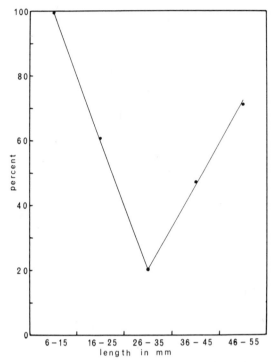

FIG. 6. Reproduced with permission from Courtney & Webb (1964). Percentage of dead lancelets in a catch taken April 1963 at Helgoland.

(Fig. 6). During severe winters, because of different rates of mortality operating within the population, a change takes place in the structure of the lancelet population at Helgoland. In August 1962 the size-distribution within the population was considerably greater than a year later when the 2- and 3-group animals form a substantial part of the population totalling 80% of the individuals present (Table IV).

Animals may grow to 47 mm in their sixth year at Helgoland (Fig. 2). However, it is evident from the size-distribution of the population in 1962 and 1965 (Fig. 1) that few members of the population actually do so, even in the absence of severe winters. In 1962 the 4- and 5-group individuals formed less than 10% of the total population in August and, in contrast to the size-distribution a year later, the 3-group was smaller than either the 1- or 2-group (Table IV). In 1965 the structure of the population was quite different and the proportion of 4- to 6-group individuals had doubled to 20%. The total numbers of animals present in 1962 and in September 1965 were similar.

TABLE IV

Percentage of the Helgoland population

Group–	0	1	2	3	4	5	6+
August 1962	16	34	29	12	7	2	0
*November	—	15	34	30	12	7	0
*January 1963	—	14	26	35	21	4	0
April	—	5	32	54	5	4	0
June	—	5	35	48	10	1	0
August	—	7	37	45	10	1	0
May 1965	66	6	0	9	13	5	1

* The 1962 settlement ignored.

Owing to the annual variation in the relative proportions of the annual groups each year, it was not possible to calculate the mortality-rate amongst the older animals in mild winters. Further, it is known that the distribution of older animals varies with season on the Helgoland grounds (Webb, 1971). Nevertheless, it is probable that the loss is about 70% during both the fifth and sixth year of life. Whether or not an animal lives for more than six years and

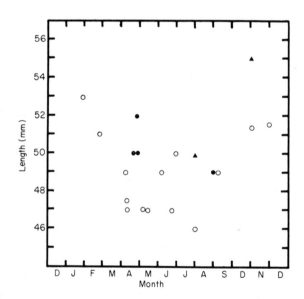

FIG. 7. Seasonal variation in maximum size of animal taken in 1962 (▲), 1963 (○) and in 1965 (●) at Helgoland.

reaches 50 mm in length depends upon the winter conditions late in life. The maximum length recorded in each set of daily samples is shown in Fig. 7. Animals of length 50 mm or longer were present in 1962 and in January 1963, but thereafter, the maximum length fell as the summer of 1963 approached. The maximum size of animal in the population increased again at the end of the year. Again, in 1965, the size-range was less in September than in April, despite an intervening period of suitable conditions for growth. Probably fatalities occur amongst the older animals after spawning.

THE EFFECTS OF HELGOLAND TEMPERATURES ON THE *BRANCHIOSTOMA LANCEOLATUM* POPULATION

Influence of temperatures of 10°C and above

Growth of the Helgoland population of *B. lanceolatum* takes place between June and November when the sea temperature is 10°C or higher. The temperature rose to 10°C at the beginning of June in both 1960 and 1961 and during the second week of June 1962. The temperature fell to 10°C during late November of these years, some five and a half to six months later. At these summer temperatures growth varied between 2 mm a month in the youngest animals and 1 mm per month in older animals. There is little or no growth in the winter although the growth of the 0-group was not recorded between February and July.

It is known that the respiratory rate of *Branchiostoma* at Helgoland and Naples varies considerably with temperature (Courtney & Newell, 1965). The pattern of oxygen consumption in these animals is a complicated one with three possible rates of respiration being shown by any animal of any weight at temperatures greater than 10°C (Fig. 8). It is not possible to distinguish three rates statistically at temperatures of less than 10°C. The fastest rate, rate *a* Fig. 9, is probably associated with feeding animals when all of the lateral cilia together with the frontal cilia are in use. A slower inhalent stream observed in the pharynx, set up by some of the lateral cilia and not involving the frontal cilia, may be the activity underlying the moderately fast respiratory-rate *b*. The respiratory-rate *c* is the result of the metabolic activities of the animal, excluding both ciliary activity in the pharynx and locomotory activity, since all the measurements were made with the animals in sterile gravel. The increased oxygen consumption associated with feeding is believed to be *a–b* and Fig. 9 shows the variation of this rate with temperature and size of animal. The rate rises steeply above 10°C but remains steady below this temperature. There is also a tendency, seen particularly in larger individuals, for the rate to level off above 15°C. The significance of this phenomenon is not understood. However, the rapid change in the level of oxygen consumption associated with feeding at temperatures above 10°C suggests that it is the intake of food which is limiting at lower temperatures when only the maintenance requirements of the animals are met and growth does not take place.

The swimming activity of this species is also markedly temperature-dependent (Courtney & Webb, 1964), with active swimming restricted to temperatures between 9° and 25°C. The upper limit is elevated in a regime of falling temperature and is reduced when temperatures are rising.

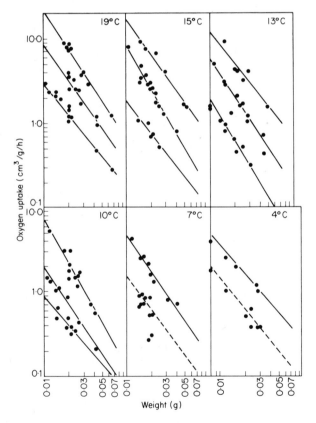

FIG. 8. A log. plot of respiratory rate against dry weight for a series of temperatures using *B. lanceolatum* at Helgoland. Reproduced with permission from Courtney & Newell (1965).

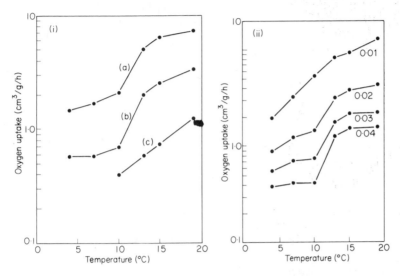

FIG. 9. Graphs (i) showing the relationship between oxygen uptake and temperature for an animal of 0·02 g and (ii) showing the effect of temperature on the respiratory level *a–b*, which is believed to be associated with feeding, in animals of different weights (g). Both graphs are compiled from the data illustrated in Fig. 8. Reproduced with permission from Courtney & Newell (1965).

The temperature had risen to 11°C in the first week of June 1963 when the population had not yet spawned and was 13·6°C when the next samples were obtained and it was found that the gametes had been shed. Therefore, the gonads start to develop in immature two-year-old adults in September when the temperature is above 15° and, after the intervening winter, spawning takes place when the temperature is between 11° and 13·6°C.

The summer temperatures at Helgoland rise above 15°C from the beginning of August to the middle of October when *Branchiostoma* larvae are in the plankton. In most years the temperature is 17°C during the first half of September when the larvae are undergoing metamosphosis.

The effects of temperatures of less than 10°C

During the period 1958–1965 (excluding 1963) the temperature fell below 5°C for an average of three months each winter and the mean minimum temperature was 2·3°C. These temperatures are at the critical lower limit for *Branchiostoma lanceolatum*. At 3°C the contraction of the muscles of the atrial wall is insufficient to drive water out of the mouth and at 2°C this "cough" mechanism, described by Dennell (1950) disappears altogether (Courtney & Webb, 1964). Also, the locomotory muscles cease to function at 3°C. It is clear that these animals are stressed at temperatures of less than 5°C and that the survival of the 0-group is dependent upon the duration of cold conditions and the minimum temperature of the winter (Fig. 5).

In the winter of 1962–1963 the bottom temperature was less than 0°C for seven weeks. The minimum temperature recorded was −1·3°C. What is surprising is that so many of the population survived these conditions, for it is known that the pharyngeal cilia stop when the temperature falls to 1°C, although the production of mucus continues (Courtney & Webb, 1964). In the absence of the "cough" mechanism and ciliary activity, the pharynx must have been congested for weeks. Furthermore, at these temperatures the animals are completely immobile and would have been unable to reburrow if displaced from the substrate.

The reason for the differential rates of survival operating within the population, with the young and old animals most affected by the severe conditions, is not known. There is some indication, in the oxygen profiles at low temperatures, that physiological differences exist between the small animals and the remainder of the population. There is a greater depression of oxygen uptake in smaller animals at lower temperatures (Fig. 9). It is noteworthy that more than one out of ten of the November 0-group individuals were still alive on 30/1/63 although the bottom temperature had fallen to −0·6°C six days earlier (Fig. 1).

COMPARISON OF THE HELGOLAND, LE RACOU AND
PLYMOUTH POPULATIONS OF *BRANCHIOSTOMA LANCEOLATUM*

The distribution of the Le Racou population of *B. lanceolatum* on the south coast of France, at a depth of between 3 and 11 m, varies with size. The larger individuals inhabit shallower water than the smaller animals in the population (Webb, 1975).

The 0-group had a mean length of 17 mm in August 1966 and in November 1967 a mean length of 25 mm, a difference of 8 mm. By February 1968 the mean length of these individuals was 29 mm, representing an overall growth of 12 mm in six months, with faster growth during the warmer months of the period.

During the years 1951–1964 the average minimum winter sea-temperature at Le Racou was 10·6°C with an annual variation in the range 8·6° to 11·6°C. At the time of sampling in February 1968 the temperature was 10°C.

A new settlement appeared in the November 1967 histogram of the Le Racou population (Fig. 10) with a mean length of 9 mm, by which time the 1966 settlement had a mean length of 29 mm, having grown by 20 mm in length during the year. This is a faster rate of growth than occurs in the Helgoland population.

TABLE V

Annual means in length (mm) of B. lanceolatum *at Le Racou*

Year Month	1966 Aug.	1967 Nov.	1968 Feb.
0-group	17	7	9
1-group	35	25	29
Older animals	45	46	45
*Max size	51	53	56

(The maximum size (*) in the sample.)

Although it is not possible to analyse completely the age-structure of the population in November and February it can be seen in the August histogram that the animals reach 50 mm in length soon after entering their fourth year of life (Fig. 10).

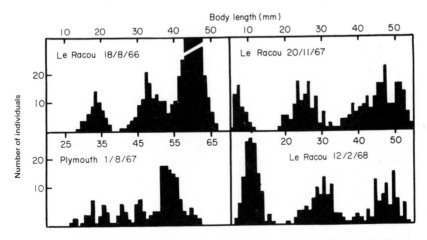

FIG. 10. Length/frequency histograms of the *B. lanceolatum* populations at Le Racou and Plymouth.

Therefore, *B. lanceolatum* grows twice as quickly at Le Racou as at Helgoland and there are twice as many annual groups in the population in the North Sea as there are in the Mediterranean. The maximum length of animal recorded at both sites is similar.

The average minimum summer temperature at Le Racou was 21·8°C during the years 1951–1964 with annual variation within the range 20·4° to 24°C. These temperatures are below the temperature at which active swimming ceases in this species and it is unlikely that the population is stressed at all by the summer temperatures of its environment. However, the rate of growth probably slows slightly when the winter temperatures drop to 10°C.

At Plymouth, where the temperatures do not fall to 5°C, the maximum size of lancelets in the Eddystone gravel at 20 m is 61 mm (Fig. 10). The absence of young animals in the sample may be the result of the method of collection; alternatively, the young animals may live elsewhere than on the adult grounds.

Three specimens of *B. lanceolatum* taken offshore at Bergen, and kindly lent by the University there, measured 37, 40 and 43 mm respectively so that there is no reason to suppose that the size range of this species is reduced at the northern limit of its range. A solitary specimen taken by the Port Erin Marine Station from the Irish Sea measured 56 mm (Dr Colman, personal communication) showing that the maximum size range of this species is fairly uniform.

GROWTH IN OTHER SPECIES OF LANCELET

The other populations where growth has been studied in any detail include *B. belcheri* off the coast of China at Amoy (Chin, 1941) and *B. nigeriense* off the west coast of Africa (Webb, 1958).

B. belcheri has a life span of about three years and commonly reaches 50 mm in length living between 14° and 28°C. Calculation of the growth from the data of Chin (1941) is complicated by asynchronous settlement of larvae. Also, these studies were interrupted by the outbreak of the Second World War. An approximate growth-rate in this species is shown in Fig. 11 and it is slightly faster than the growth of *B. lanceolatum* at Le Racou although the life-spans are similar.

A growth-rate of about 2 mm per month during the first few months of life was recorded by McShine (1971) for the populations of *B. caribaeum* in Kingston Harbour, Jamaica, where the temperature varies between 25° and 32°C. This is similar to the growth in

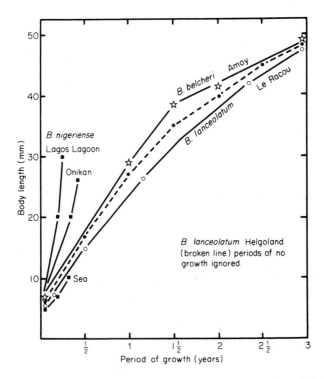

FIG. 11. Growth of lancelets. The data for *B. belcheri* at Amoy taken from Chin (1941) and for *B. nigeriense* in West Africa from Webb (1958).

other lancelets with the exception of *B. nigeriense* in Lagos lagoon and at nearby Onikan on the west coast of Africa where lancelets may grow 10 mm per month (Fig. 11).

Webb (1958) related the differences in growth-rate between the lagoon and sea forms of *B. nigeriense* shown in Fig. 11, to the inferior food supply in the sea compared with the lagoon. It is extremely unlikely that differences in temperatures of the two environments are responsible for the differences in rate of growth for the temperatures are similar, 25° to 27° in the sea and 25° to 29°C in the lagoon.

The maximum size of *B. nigeriense* is smaller than *B. belcheri* which may grow to 57 mm in length. In Lagos lagoon reduced salinity destroys the lancelet population when it has reached 35 mm in length. The maximum size of animal taken from the marine population offshore is also 35 mm. At Kingston, Jamaica, *B. caribaeum* may grow to 45 mm but in many months of the year the largest animal in the population is 32 mm in length. Seventy per cent of this population is lost between 10–18 mm and 90% of the remainder at lengths 18–34 mm according to McShine (1971). Salinity fluctuations drastically reduce the size of this population in June and July although densities of $5 \cdot 10^3/m^2$ are quoted by this author.

DISCUSSION

Millar (1952) has shown that there is no growth in length in any part of the population of *Ascidellia aspersa* and *Ciona intestinalis* during the winter around the British Isles. The same author (1954) has shown that the growth of *Dendrodoa grossularia* is confined to the months of April to September not only off the Essex coast but also in the Clyde where it is warmer in the winter and it would seem that the growth of this species is associated with temperatures of 10°C or more.

B. lanceolatum grows throughout the year in the Mediterranean although the growth rate may be slightly slower in the winter months, when the temperatures approach 10°C, than in the summer months, when the temperature rises above 20°C. At Le Racou animals grow to 50 mm early in their fourth year. In sharp contrast, this species takes six or seven years to reach this length at Helgoland, where the mean annual temperatures are 2° to 17°C. Here, *B. lanceolatum* grows for only half of the year and winter temperatures have a sub-lethal effect upon the population. However, if growth is plotted as a continuous process for this population in the North Sea, and the periods in the life when growth is not occurring are ignored, then the growth-rate is not dissimilar from the Mediterranean population (Fig. 11). Thus, when suitable temperatures permit, this species grows at 2 mm per month in early life and at less than 1 mm per month when it is approaching its maximum size of about 55 mm. Both the growth-rate and the maximum size are similar in the population of *B. belcheri* at Amoy (Chin, 1941) though slightly larger *B. lanceolatum* are taken from the Eddystone shell-gravel off Plymouth.

The growth of *B. nigeriense* in Lagos lagoon is accelerated when compared with the marine forms off the West African coast owing to a better supply of food in the lagoons. The lagoon populations of this species perish when the salinity falls and the maximum size recorded for this species is 35 mm. Salinity reduces the maximum size of *B. caribaeum* in Kingston Harbour although individuals have reached 45 mm length.

Despite the tendency for the warmer-water species to grow to a smaller size than individuals from populations of colder climates, Hubbs (1922) describes many species with maximum lengths greater than 50 mm. Also, *B. belcheri*, with a distribution from Japan through Ceylon to East Africa, grows to a similar size to *B. lanceolatum*.

The temperature range at which active swimming occurs in both *B. caribaeum* (Parker, 1908) and *B. nigeriense* (Webb & Hill, 1958) is 13° to 37°C, whereas *B. lanceolatum* from both Helgoland and Naples swim actively between 10° and 25°C (Courtney & Webb, 1964).

B. lanceolatum certainly extends into cooler waters than other species of this genus with populations in Loch Ewe on the west coast of Scotland, the North Sea, Kattegat and Bergen. The Norwegian populations at 60°N are at the highest latitude recorded for this genus. Its presence there is in part the result of the moderating influence of the warm North Atlantic Drift, and in part due to the tolerance of this species to low winter temperatures. Elsewhere in the world the genus is generally restricted to the seas lying between 40°N and 40°S where temperatures rarely fall below 10°C.

The distribution studies of Cory & Pierce (1967) on the Atlantic coast of the United States showed that the northward limit of *B. caribaeum* is associated with a winter minimum of 9°C. Earlier, the associate author in that investigation had reported that the same species is found where an annual range of surface temperature is 5°–33°C in the Gulf of Mexico off the west coast of Florida (Pierce, 1965). In the southern hemisphere lancelet populations appear to be associated with local winter minimums of about 10°C, for the Peru and River Plate populations off South America, and about 11° off the coast of Southern Australia and the northern island of New Zealand.

The winter temperatures at Helgoland affect the structure of the population of lancelets there. Since growth is curtailed for six months in the year, then there are twice as many annual groups in the population compared with the Mediterranean population at Le Racou (Figs 1 and 10) and the animals take twice as many years to reach their maximum size in the North Sea.

There are short-term and long-term effects of an extremely severe winter at Helgoland when temperatures are below average. In the short term these temperatures are lethal for the youngest and the oldest animals of the population. In the longer term, the loss of mature animals results in a reduction in the reproductive stock the following summer. However, the loss in the reproductive elements of the population is more marked three to six years later when the lost settlement would have matured.

The doubling of the number of annual groups which occurs in populations of lancelets at this latitude tends to smooth the effects of the loss of a spawning since the reproductive members of the population come from several annual spawnings. If the population was composed of fewer annual groups then the loss of the 0-group would have a proportionately greater deleterious effect.

The larval development of *B. lanceolatum* in the North Sea is planktonic. In contrast, Wickstead & Bone (1959) found larvae of *B. belcheri* in the benthos, although, more recently, Gosselck & Kuehner (1973) found that the larvae of *B. senegalense* avoided direct contact with the sediment despite being present in all depths of water.

ACKNOWLEDGEMENTS

I am indebted to Professor O. Kinne for the generous facilities and excellent boat-work provided at the Meeresstation Helgoland of the Biologische Anstalt Helgoland. Dr M. Gillbricht kindly provided the temperature profiles over the lancelet grounds and the technical administration of Herr A. Holtmann was invaluable. For the collecting and laboratory facilities at the Stazione Zoologica, Naples, A am indebted to Dr P. Dohrn and to the Royal Society of London. The samples from Le Racou were kindly collected by Professor J. E. Webb using facilities at the Laboratoire Arago at Banyuls-sur-Mer generously provided by Professor P. Drach. Further thanks are due to Professor Webb for invaluable discussion and encouragement and to Miss C. Earle who assisted with the diagrams.

Part of this work was supported by a DSIR grant.

REFERENCES

Chin, T. G. (1941). Studies on the biology of the Amoy amphioxus *Branchiostoma belcheri* (Gray). *Philipp. J. Sci.* **75**: 369–424.
Cory, R. L. & Pierce, E. (1967). Distribution and ecology of lancelets (Order Amphioxi) over the continental shelf of the Southeastern United States. *Limnol. Oceanogr.* **12**: 650–656.
Courtney, W. A. M. & Newell, R. C. (1965). Ciliary activity and oxygen uptake in *Branchiostoma lanceolatum* (Pallas). *J. exp. Biol.* **43**: 1–12.
Courtney, W. A. M. & Webb, J. E. (1964). The effect of the cold winter of 1962/63 on the Helgoland population of *Branchiostoma lanceolatum* (Pallas). *Helgoländer wiss. Meeresunters.* **10**: 301–312.

Dennell, R. (1950). Note on the feeding of Amphioxus (*Branchiostoma bermudae*). *Proc. R. Soc. Edinb.* (B.) **64**: 229–234.

Gosselck, F. & Kuehner, E. (1973). Investigations on the biology of *Branchiostoma senegalense* larvae off the Northwest African coast. *Mar. Biol.* **22**: 67–73.

Hagmeier, A. & Hinrichs, J. (1931). Bermerkungen über die Ökologie von *Branchiostoma lanceolatum* (Pallas) und das Sediment seines Wohnortes. *Senckenbergiana* **13**: 255–267.

Harding, J. P. (1949). The use of probability paper for graphical analysis of polymodal frequency distributions. *J. mar. biol. Ass. U.K.* **28**: 141–153.

Hartmann, von J. & John, H. C. (1971). Planktische *Branchiostoma* nord-westlich der Doggerbank (Nordsee). *Ber. dt. Wiss. Kommn Meeresforsch.* **22**: 80–84.

Hubbs, C. L. (1922). A list of lancelets of the world with diagnosis of five new species of *Branchiostoma*. *Occ. Pap. Mus. Zool. Univ. Mich.* **105**: 1–16.

McShine, A. H. (1971). *Some aspects of the biology of* Branchiostoma caribaeum *Sundevall (Cephalochordata) in Kingston Harbour, Jamaica.* Ph.D. Thesis, University of West Indies.

Millar, R. H. (1952). The annual growth and reproductive cycle in four ascidians. *J. mar. biol. Ass. U.K.* **31**: 41–62.

Millar, R. H. (1954). The annual growth and reproductive cycle of the ascidian, *Dendrodoa grossularia* (van Beneden). *J. mar. biol. Ass. U.K.* **33**: 33–48.

Parker, G. H. (1908). The sensory reactions of Amphioxus. *Proc. Am. Acad. Arts Sci.* **43**: 415–455.

Pierce, E. L. (1965). The distribution of lancelets (Amphioxi) along the coasts of Florida. *Bull. mar. Sci. Gulf Caribb.* **15**: 480–494.

Webb, J. E. (1956). On the populations of *Branchiostoma lanceolatum* and their relations with the West African lancelets. *Proc. zool. Soc. Lond.* **127**: 119–123.

Webb, J. E. (1958). The ecology of Lagos Lagoon. Part III. The life history of *Branchiostoma nigeriense* Webb. *Phil. Trans. R. Soc.* (B.) **241**: 335–353.

Webb, J. E. (1971). Seasonal changes in the distribution of *Branchiostoma lanceolatum* (Pallas) at Helgoland. *Proc. Europ. mar. Biol. Symp.* **3**: 827–839.

Webb, J. E. (1975). The distribution of amphioxus. *Symp. zool. Soc. Lond.* No. 36: 179–212.

Webb, J. E. & Hill, M. B. (1958). The ecology of Lagos Lagoon. Part IV. On the reactions of *Branchiostoma nigeriense* Webb to its environment. *Phil. Trans. R. Soc.* (B.) **241**: 355–391.

Wickstead, J. H. & Bone, Q. (1959). Ecology of acraniate larvae. *Nature, Lond.* **184**: 1849–1851.

Symp. zool. Soc. Lond. (1975) No. 36, 235–251.

FINE STRUCTURE AND PHYLOGENY OF THE POGONOPHORA

EVE C. SOUTHWARD

The Laboratory, Citadel Hill, Plymouth, England

SYNOPSIS

It is now known that Pogonophora are not three-segmented coelomates, as Ivanov thought when he assigned them to the Deuterostomia, but multi-segmented coelomates. The segmented posterior end was discovered fairly recently and its septa and setae have reinforced earlier speculations that pogonophores might be more closely related to annelids than to hemichordates. The fine structure of the cuticle is similar in pogonophores and annelids; the setae are also very similar, but Echiura and Brachiopoda have setae of the same type. The septa which separate the segments in the hind end of a pogonophore are three-layered as in annelids. The central nervous system of this part of the body has a segmental arrangement which recalls that of some annelids, though it is not exactly equivalent to any known annelid system. Other cell details are common to many animal groups.

Pogonophores have some unique characteristics: the lack of an internal digestive system and the consequent dependence on epidermal absorption of food; the biochemical peculiarity of the chitinous tube and the specialized cells for its secretion; and among finer details of the anatomy, the single-celled pinnules on the tentacles and the cuticular plaques on the body surface. Because of these features the Pogonophora may still be regarded as a separate phylum but, from their body form, septa and setae, their nearest relatives seem to be annelids or the ancestors of the annelids.

INTRODUCTION

Pogonophores are worm-shaped benthic animals which completely lack a gut. Perhaps because most of them live in the deep sea, they were discovered later than most of the other groups of Metazoa, and the earliest description of a pogonophore was published by Caullery in 1914. The first thorough anatomical studies appeared in papers by A. V. Ivanov between 1949 and 1959, and these papers formed the basis of a monograph on Pogonophora (Ivanov, 1960) the English edition of which was published in 1963. Ivanov's studies of anatomy and embryology of pogonophores led him to believe that they should be classified among the Deuterostomia and that their nearest relatives would be Hemichordata.

Though Ivanov did not realize it, at the time he was supervising the English translation of his monograph no one had yet seen a complete pogonophore. A tiny but important posterior region was always missing. This region was discovered by Webb in 1963, when he was working on the Norwegian *Siboglinum fiordicum* (Webb,

1964). Ivanov soon found the hind end of *Siboglinum caulleryi* (Ivanov, 1964, 1965), and the two species proved to have very similar hind ends. They are both segmented, with septa between the segments and serially arranged bristles. Naturally this discovery stimulated speculation about the systematic position of the Pogonophora (Liwanov & Profirjewa, 1967; Webb, 1969a; Ivanov, 1970; Nørrevang, 1970a, b; Southward, 1971), but Ivanov himself stressed that the similarities to the hind end of an annelid could be coincidental and need not imply homology.

In this paper I shall first outline our present knowledge of the anatomy of pogonophores, and then describe some fine structural details which may help us to understand their phylogenetic position.

ANATOMY

Pogonophores live in narrow individual tubes made of chitin and protein, which they secrete themselves. The tubes are positioned more or less vertically in the sea-floor sediment, with a short anterior part projecting above the sediment surface. The living animal is about the same length as its tube and can move up and down inside it, but cannot turn round. The middle part of the body is held in position by two rings of small bristles with toothed heads, which can grip the tube wall and allow the two halves of the body to extend and contract independently. The body is bilaterally symmetrical and there is a considerable difference between the dorsal and ventral sides. However, it is difficult to decide which side is ventral; Ivanov (1963 and earlier) has always described the side of the body containing the nerve trunk as dorsal, since the nerve trunk is dorsal in Deuterostomia. In most worm-shaped animals we can decide which side is ventral by observing the way they orientate themselves when crawling, but pogonophores live in a vertical position and if one is removed from its tube it just coils up and does not attempt to crawl. A second essential clue should come from the position of the mouth, but pogonophores have no mouth. For objectivity, I shall therefore not use the terms dorsal and ventral, but shall call the side of the body with the main nerve trunk the neural side and the opposite side the antineural side.

At the anterior end of the body is a small cephalic lobe with some long tentacles attached to the antineural side (Fig. 1). The number of tentacles depends on the species and the size of the animal. The cephalic region and the following short muscular region are together termed the forepart. The very long main part

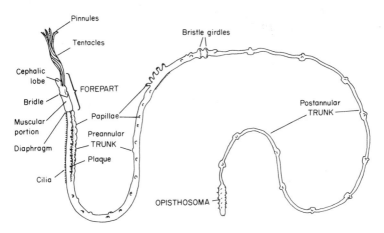

FIG. 1. Diagram of a pogonophore, much shortened, showing regions of body.

of the body is called the trunk and the small segmented region behind the trunk is called the opisthosoma. There is a muscular diaphragm between the forepart and trunk. The trunk is sub-divided into preannular and postannular regions, before and behind the two girdles of bristles mentioned. Two short longitudinal bands of cilia are usually present on the neural side in the preannular region (Fig. 1), while there are papillae arranged in various ways along the antineural side (Ivanov, 1963; Southward & Southward, 1966; Gupta & Little, 1970; George & Southward, 1973). The postannular part of the trunk is thin and delicate, which is why the opisthosoma is so often lost when the animals are collected. This opisthosoma is thicker than the preceding part of the postannular region and is divided into several segments, most of which bear bristles. It ends in a small depression or pit (Fig. 2a) (Webb, 1964; Ivanov, 1965; Southward, in press).

The body wall is made up of cuticle, epidermis, basement membrane, circular muscle and longitudinal muscle. The coelom is lined by the nucleate bodies of muscle cells and non-muscular epithelial cells in the same layer, for there is no separate peritoneum. The number of coelomic compartments in the anterior part of the body is still slightly uncertain. Ivanov describes one median anterior coelom close to the region where the tentacles are attached, with branches into the tentacles and a pair of coelomoducts which are thought to be excretory. In some species a second median space, doubtfully coelomic in origin, called the pericardium by Ivanov, is present between the blood vessels in the

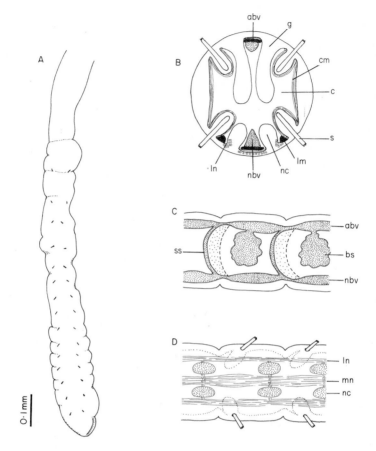

FIG. 2. A. Opisthosoma of *Siboglinum ekmani*. B. Diagrammatic transverse section of opisthosoma segment: abv, antineural blood vessel; c, coelom; cm, circular muscle; g, gland; lm, longitudinal muscle; ln, lateral nerve tract; nbv, neural blood vessel; nc, nerve cells; s, seta. C. Blood vessels of opisthosoma, side view: abv, antineural vessel; bs, blood sinus (paired); nbv, neural vessel; ss, septal sinus (paired). D. Nervous system of opisthosoma: ln, lateral nerve tract; mn, median nerve tract; nc, nerve cells (ganglion).

"heart" region, just behind the tentacle region. Behind this, in the muscular section of the forepart, is a pair of long coelomic spaces, separated by a median mesentery. Another pair of coeloms extend right down the trunk, separated only partially by an incomplete median mesentery. The diaphragm between forepart and trunk appears to form a barrier between the coeloms of the two body regions, but a doubt as to its completeness has been expressed recently (Nørrevang, 1970b) and it needs more investigation. The coelom of each opisthosome segment is separated completely from the next by a muscular septum, but there is no median mesentery (Southward, in press). Coelomoducts are present only in the cephalic region, already mentioned, and in the trunk, where they act as gonoducts.

The blood, which contains haemoglobin, is carried in a neural and an antineural vessel throughout the body. At the anterior end blood is able to flow from one to the other through fine capillaries which run in loops within the pinnules on the tentacles (Nør-revang, 1965a; Gupta, Little & Philip, 1966). In the opisthosoma the longitudinal vessels are again connected, by lateral vessels or sinuses within the septa (Figs 2c, 4, 5a). In the cephalic region there is a section of the antineural vessel surrounded by strong circular muscles, which is termed the heart, but it may perhaps act as a valve rather than a pump.

The gonads are in the trunk and the trunk coelomoducts act as gonoducts, but the arrangement is different in the two sexes. The testes are in the postannular region, from where the spermatozoa are carried forward in two sperm ducts, formed into sper-matophores, and stored in the anterior parts of the capacious sperm ducts. They are eventually released through paired pores at the anterior end of the trunk. In the female, a pair of ovaries are at the anterior end of the trunk, the oocytes grow in size and move back until they enter the oviducts, which then curve forward and open to the exterior a little way in front of the girdle region (Ivanov, 1963). There are details which still need clarifying about both male and female reproductive systems.

The nervous system is completely intraepidermal. Small axons can be found close to the basement membrane under most of the epidermal cells (Gupta & Little, 1970), but the main concentration of axons forms a median tract on the neural side of the body. The "brain" is a concentration of axons and neurons in and behind the cephalic lobe, encircling the body but thickest on the neural side (Ivanov, 1963). The nervous system of the opisthosoma is com-paratively elaborate; there are three longitudinal nerve tracts, one median and two lateral, with a pair of ganglion-like concentrations of nerve cells in each segment (Fig. 2d); axons from these cells run into the median and lateral nerve tracts (Southward, in press).

Ivanov's attribution of Pogonophora to Deuterostomia and his theory of their relationship to hemichordates were based mainly on the apparent tricoelomate condition of the adult. That is, before the opisthosoma was known, it seemed that the animal had an anterior median coelom followed by two pairs of coeloms. Ivanov also studied an incomplete series of pogonophoran embryos (which are incubated by the female in her tube) and he found in the oldest stage available a three-segmented condition of the coelom, which he compared with that of the adult (Ivanov, 1957, 1963). Naturally, he thought that the trunk developed from the third segment of the embryo, and that the second embryonic segment represented the muscular region of the forepart, while the first became the cephalic region and tentacles. Since the discovery of the opisthosoma our ideas about segmentation have been revised (Webb, 1964; 1969b; Nørrevang, 1970a, b; Southward, 1971) and it now seems clear that the second embryonic segment gives rise to the trunk as well as to the muscular region of the forepart, the diaphragm between them developing later than the septum between trunk and opisthosoma. The third embryonic segment represents the first opisthosome segment plus a posterior growth region which starts to produce more segments before the embryo leaves the maternal tube.

Our understanding of the cleavage and early developmental stages is still very incomplete, and relies too much on studies of two species of *Siboglinum* with very yolky and somewhat distorted eggs. The type of cleavage, in these species, is not clearly radial or spiral, but somewhat intermediate (Bakke, in press). There is no blastula stage and a gastrula is formed by epiboly (Ivanov, 1963; Nørrevang, 1970a). The coelom develops by splitting of mesodermal strands (Nørrevang, 1970a) but the number of compartments and their arrangement in the anterior region is still not clear.

It is obvious, now, that the Pogonophora have much less in common with the hemichordates or echinoderms than was originally supposed, and they should not be included in the Deuterostomia. We have, therefore, to find a place in the animal kingdom for a group of animals with more than three coelomate segments; no mouth, gut or anus; tentacles on the first segment, and an intraepidermal nervous system whose median longitudinal tract becomes ladderlike at the posterior end. We need to know more about the embryology, but should begin to compare the ciliated stage with the trochophore type rather than with the dipleurula or tornaria.

FINE STRUCTURE

I hope that study of fine structure will help to elucidate the relationships of pogonophores with other animal groups. The electron microscope has already proved useful in clarifying some problems. Accounts have been published of the anatomy of the tentacles and their pinnules (Nørrevang, 1965a; Gupta *et al.*, 1966; Gupta & Little, 1969); the epidermis and cuticle (Gupta & Little, 1970); the bristles and the cells which produce them (Gupta & Little, 1970; Orrhage, 1973; George & Southward, 1973); the distribution of glycogen in the tissues (Southward, 1973); and the fine structure of the spermatozoa (Franzén, 1973). Gupta & Little (in press) have discussed the relevance of fine structural comparisons to phylogeny, and have questioned the coelomic status of the pogonophore body cavity.

Epidermis

The cuticle covering the body is made up of criss-cross layers of fibres. Microvilli protrude through the fibre mesh and through a surface layer of mucus. An outermost layer of tiny dark bodies seems to be formed by the tips of the microvilli (Fig. 3a,b) (Gupta *et al.*, 1966; Gupta & Little, 1970). The thickness of the cuticle varies greatly in different parts of the body, the pinnules and the postannular region of the trunk having particularly thin cuticles with few fibres; and the forepart having a particularly thick cuticle with many layers of fibres. It is common for soft-bodied invertebrates to have a surface layer of microvilli, often with cilia, but only annelids and sipunculids are known to have a combination of microvilli and layers of fibres like the cuticle of pogonophores (Moritz & Storch, 1970; Coggeshall, 1966; Storch & Welsch, 1970), and a distal layer of surface particles similar to those of pogonophores seems to occur only in oligochaetes and some polychaetes (Potswald, 1971; Boilly, 1967); the cuticle of echiurids has not been investigated.

Pogonophores have characteristic surface plaques and bridle keels, which are disc- or ridge-shaped thickenings of the cuticle (Ivanov, 1963; Gupta & Little, 1970). They appear to help the animal to grip the tube wall and move about inside the tube. Such external cuticular structures are not found in annelids or sipunculids, unless the paragnaths of the mouth region of some polychaetes may be comparable.

The bristles or setae of pogonophores, though produced by epidermal cells, have a different chemical composition from the cuticle. The latter is probably collagenous (Southward & Southward, 1966; Gupta & Little, 1970), but the bristles certainly contain chitin and are chemically similar to annelid setae (George &

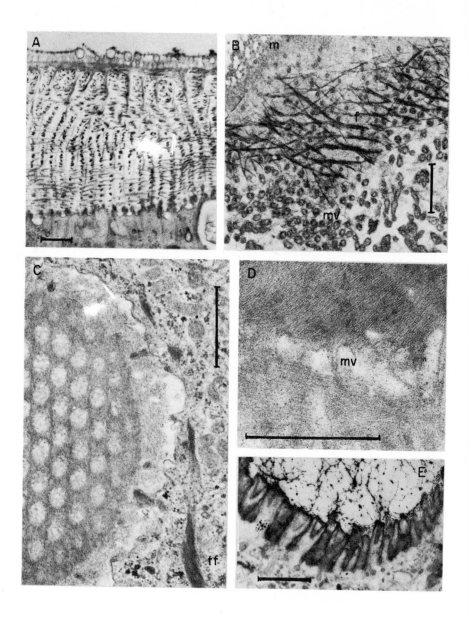

FIG. 3. Products of epidermis (electron micrographs): A. Cuticle of *Siboglinum atlanticum*, anterior trunk region. B. Cuticle of opisthosoma of *S. fiordicum*, oblique section: f, fibres; m, surface mucus layer; mv, microvilli. C. part of T.S. of seta from opisthosoma of *S. fiordicum*, basal cell on right: tf, tonofibrillae. D. part of lumen of pyriform gland, containing fibrous secretion: mv, microvilli. E. part of ampulla of pyriform gland, with lining of short, concave-tipped, microvilli. Scale = 1 μm.

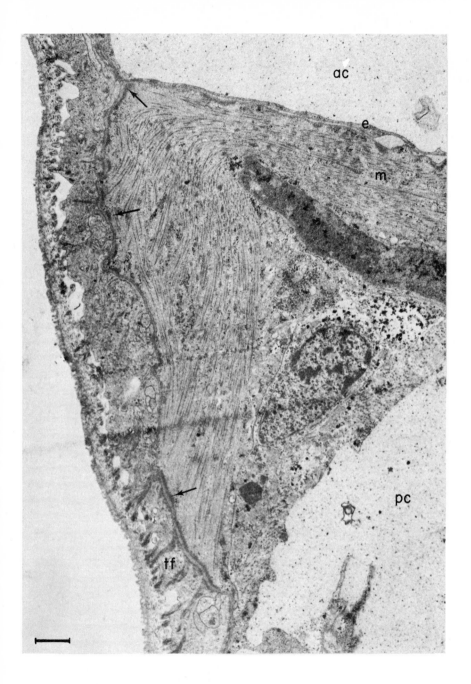

Fig. 4. Part of L.S. of opisthosoma of *Siboglinum fiordicum* (electron micrograph), showing attachment of septum to body wall; epidermis and cuticle on left; ac, coelom of anterior segment; e, non-muscular epithelium; m, muscle layer; pc, coelom of posterior segment; tf, tonofibrillae; arrows, hemidesmosomes. Scale = 1 μm.

Southward, 1973). They are also similar in fine structure to annelid setae, consisting as they do of many longitudinally arranged cylinders with coalescing fibrous walls (Fig. 3c). Each bristle is secreted by microvilli in the bottom of a pocket in a single basal cell, which is part of a multicellular setal pouch. The setae of annelids are also secreted inside epidermal pouches by single basal cells (Bouligand, 1967). Setae of the annelid type have also been described in echiurids (Orrhage, 1971) and in brachiopods (Storch & Welsch, 1972; Gustus & Cloney, 1972; Orrhage, 1973). Brachiopods are not regarded as related to annelids, so their possession of the same kind of setae might open up some interesting new speculations. Other groups of invertebrates have setae, spines and hooks which are cuticular in origin, and not multicylindrical like the annelid type.

In pogonophores there is a type of gland cell (possibly comparable to the basal cell of the seta sac), which secretes the chitinous component of the tube (Southward & Southward, 1966). The pyriform glands, which are abundant in the forepart and trunk, consist of groups of these cells. Each cell contains a deep cleft, or lumen, with a small ampulla at the base, both lined with short microvilli (Fig. 3d,e). A finely fibrous extracellular secretion is produced. The secretion of the ampulla is modified in the upper part of the lumen, and the threadlike material is stored as a lamella. The lamellae, when released, uncoil to produce threads (Southward & Southward, 1966), which the animal can incorporate into the wall of its tube (Webb, 1965). This kind of cell is unlike any known annelid tube-secreting cell, and annelid tubes do not contain chitin (Jeuniaux, 1963). The only hemichordate with a tube that looks a little like a pogonophore tube is *Rhabdopleura*; however, its tube, or coenecium, is not chitinous, but contains keratin-like fibres (Dilly, 1971).

Another peculiar epidermal cell, very characteristic of Pogonophora, is the tentacle pinnule cell. Most species have a large number of pinnules on their tentacles, and each pinnule is a single epidermal cell. It is greatly elongated but still attached at the base to the basement membrane and its nucleus is in the basal part, level with the ordinary epidermal cells. A fine blood vessel, which is really a projection of the basement membrane, pushes up inside the cell, forming a loop. Each of these capillaries forms a connection between the two longitudinal blood vessels of the tentacle. The pinnule cell has a thin cuticle and small amount of cytoplasm (Gupta *et al.*, 1966), and seems only to form the thinnest possible

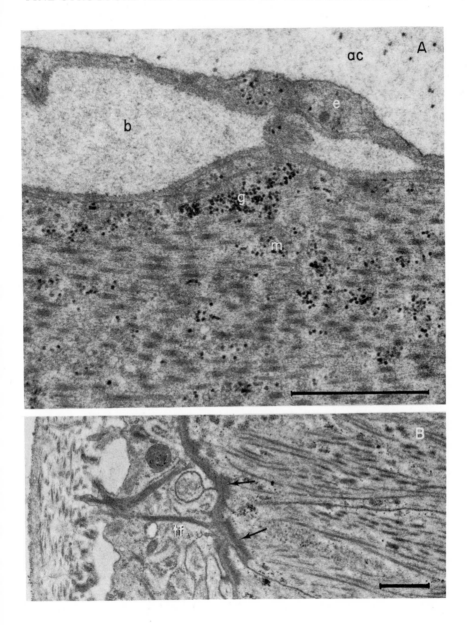

FIG. 5. Septa of *Siboglinum fiordicum* (electron micrographs): A. Anterior side of septum (see Fig. 4): ac, coelom of anterior segment; b, blood in septal sinus; e, non-muscular epithelial cell; g, glycogen particles; m, muscle layer. B. Part of septum seen in T.S. opisthosoma; cuticle on left, muscles on right; tf, tonofibrillae; arrows, hemidesmosomes. Scale = 1 μm.

covering for the enclosed capillary. These pinnule cells do not, apparently, have any equivalent in other groups of animals.

However, most of the other epidermal cells, such as nerve cells and their axons, mucus glands and granule-containing cells are very similar to cells found in various other animal groups.

Mesoderm

Under the thin basement membrane of the epidermis lies the musculoepithelial layer (Fig. 6a). The cells of this layer form elongated muscle fibres and the nucleus of each is contained in a large cell-body which projects into the body cavity. The cell bodies form a lining to this cavity, or if they are very large they almost fill it (Southward, 1973). In the forepart and trunk there are two muscle layers, circular outside longitudinal, but in the postannular region of the trunk both are reduced, and the most obvious feature is a band of longitudinal muscle fibres along the neural side. The opisthosoma has a thin layer of circular or oblique fibres round the sides of the segments and around the setal sacs (Southward, in press), and four longitudinal bands of fibres running continuously from segment to segment (Figs 2b, 6d). The longitudinal blood vessels have a thin lining of extracellular material similar to the epidermal basement membrane and a wall of epithelial cells, some of which contain muscle fibrils (Fig. 6e). There is no separate peritoneum. In some parts of the body the cells of the blood vessel epithelium almost fill the body cavity, but in the opisthosoma the cells of blood vessel and body wall provide only a thin lining to a large cavity (Southward, in press).

It may be argued that, since there is no separate peritoneal epithelium, the pogonophore body cavity is not a true coelom (Gupta & Little, in press). However, most of our ideas about the coeloms of invertebrates are based on light microscopy, much of it done long ago. If the fine structure of the coelomic lining of other invertebrates is investigated, it may be found that the so-called peritoneum is part of the muscle layer in many cases. For example, Hama (1960) has examined the blood vessels of an earthworm and found a single-layered wall, with extracellular lining, very like that of pogonophore blood vessels. The coelomic lining in an enteropneust was found by Nørrevang (1965b) to consist of a layer of musculoepithelial cells. I have made a brief survey of sections of two polychaetes (*Oridia armandi* and *Nereis diversicolor*) and the hemichordate *Rhabdopleura*. The polychaetes have a single cell layer over the blood vessels and round the thinnest parts of the

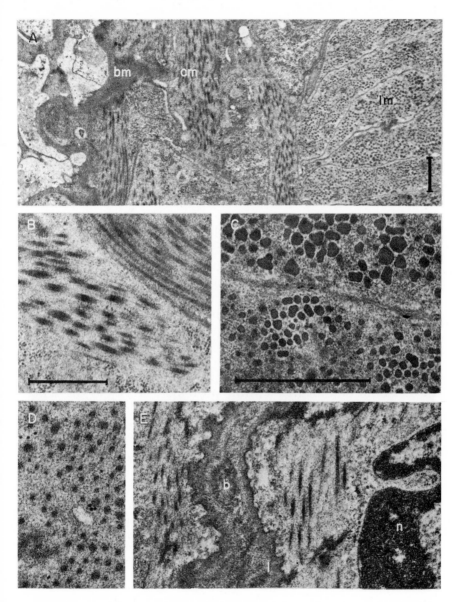

FIG. 6. Muscles of *Siboglinum* (electron micrographs): A. part of T.S. anterior trunk of *S. atlanticum*, showing two muscle layers: bm, basement membrane; cm, circular muscle; lm, longitudinal muscle. B. Circular muscles, cut longitudinally; C. Longitudinal muscle fibres, cut transversely. D. T.S. of longitudinal muscle fibre in opisthosoma of *S. fiordicum*. E. Blood vessel of *S. atlanticum*, from T.S. anterior trunk; note almost empty lumen in centre, with thick extracellular lining (l) and a little blood (b); striated muscle fibrils on right; n, nucleus of musculoepithelial cell. Scale = 1 μm; C, D and E same scale.

body wall, but the situation is more complicated where there are thick blocks of muscle, and more careful study will be needed. *Rhabdopleura*, like the enteropneust *Harrimania*, has a single layer of musculoepithelial cells lining the coelom. Obviously, more species, and more phyla, must be investigated, but I am inclined to think that the body cavity of a pogonophore is not very different from the coeloms of other invertebrates, whether protostomian or deuterostomian.

Septa

The septa of the opisthosoma are thick domeshaped sheets of muscle between the segments (Southward, In press). The muscle fibres run transversely and are inserted directly on to the basement membrane of the segment behind the septum. In front of the muscle layer is a thin sheet of extracellular material similar to, and continuous with, the basement membrane (Figs, 4, 5b). Anterior to this is a thin, non-muscular epithelium which is continuous with the body wall muscle layer of the segment in front of the septum. In the middle of this three-layered septum there are blood-filled spaces which are simply splits in the septal membrane (Fig. 5a). The sinuses form lateral links between the longitudinal blood vessels.

I have found a similar three-layered structure in the septa of the sabellid polychaete *Oridia armandi*, which also has muscle fibres on the posterior side of the septum and blood sinuses in the septal membrane, but I do not know how typical this is of polychaetes in general. In many polychaetes there is specialization of the septal muscles for control of the parapodia and partial, or complete, loss of septa. Clark (1964) has suggested that in primitive annelids the septal muscles may have been inserted directly on the basement membrane of the epidermis (which is just the situation in pogonophores) though in modern annelids they are inserted on the body-wall muscles. Though not enough is known about the fine structure of annelid septa, it does seem that the muscular septum is only found in annelids and pogonophores.

DISCUSSION

Despite the preceding emphasis on the features which pogonophores share with annelids and the peculiarities they do not share with other animal groups, they do contain many cell types which are almost universally found in other animal groups, such as nerve cells, sensory cells, mucus cells and the non-cellular lining of the blood vessels; but these do not help in the classification of the Pogonophora.

Most of the features shared with annelids are concentrated in the opisthosoma, for example: muscular septa between segments; segmentally arranged chitinous setae; segmentally arranged ganglia and blood vessels; these might be a consequence of the segmentation rather than an indication of affinity. I have argued in earlier papers (Southward, 1971; George & Southward, 1973) that the function of the opisthosoma must be a burrowing one, and that in any worm-shaped animal a contractile bulbous organ is necessary for burrowing. Many animals can burrow efficiently without segments or setae (Clark, 1964; Trueman & Ansell, 1969); therefore, though segments and setae are useful in burrowing, it seems unlikely that pogonophores would have developed septa and setae of annelid type specially for this purpose if they had evolved from a non-annelidan ancestor. The fibrous cuticle, shared by pogonophores, annelids and sipunculids, is well adapted for its functions. The way in which the fibres wind round the body in criss-crossing layers is ideally suited to keeping the volume of the animal constant during changes in length or diameter (Clark, 1964). Among other elongated animals, nematodes have a different type of fibrous cuticle, without microvilli, and nemerteans have developed a fibre system in the basement membrane which has the same function in volume control and so compensates for the lack of any cuticle. Again, if pogonophores evolved separately to annelids it seems unlikely that they should have developed such a very similar cuticle.

The most obvious ways in which pogonophores differ from present-day annelids are the lack of a gut and the production of a chitinous tube. In my view, these justify the placing of Pogonophora in a separate phylum, about as closely related to Annelida as are Echiura and Sipuncula, both of which have affinities with Annelida or the ancestors of Annelida.

ACKNOWLEDGEMENTS

I am very grateful to Dr Q. Bone, Mr A. C. G. Best and Mr K. Ryan for their help with electron microscopy and advice on problems of fine structure.

REFERENCES

Bakke, T. (In press). Early cleavage in embryos of *Siboglinum fiordicum* Webb. *Z. Zool. Syst. Evol.*

Boilly, B. (1967). Contribution à l'étude ultrastructurale de la cuticle épidermique et pharyngienne chez une annélide polychète (*Syllis amica* Quatrefages). *J. Microsc.* **6**: 469–484.

Bouligand, Y. (1967). Les soies et les cellules associées chez deux annélides polychètes. Étude en microscopie photonique à contraste de phase et en microscopie electronique. *Z. Zellforsch. mikrosk. Anat.* **79**: 332–363.

Caullery, M. (1914). Sur les Siboglinidae, type nouveau d'Invertébrés recueilli par l'expédition du Siboga. *C. r. hebd. Séanc. Acad. Sci., Paris* **158**: 2014–2017.

Clark, R. B. (1964). *Dynamics in metazoan evolution.* Oxford: Clarendon Press.

Coggeshall, R. E. (1966). A fine structural analysis of the epidermis of the earthworm, *Lumbricus terrestris* L. *J. Cell Biol.* **28**: 95–108.

Dilly, P. N. (1971). Keratin-like fibres in the hemichordate *Rhabdopleura compacta. Z. Zellforsch. mikrosk. Anat.* **117**: 502–515.

Franzén, A. (1973). The spermatozoon of *Siboglinum* (Pogonophora). *Acta zool., Stockh.* **54**: 179–192.

George, J. D. & Southward, E. C. (1973). A comparative study of the setae of Pogonophora and polychaetous Annelida. *J. mar. biol. Ass. U.K.* **53**: 403–424.

Gupta, B. L. & Little, C. (1969). Studies on Pogonophora. II. Ultrastructure of the tentacular crown of *Siphonobrachia. J. mar. biol. Ass. U.K.* **49**: 717–741.

Gupta, B. L. & Little, C. (1970). Studies on Pogonophora. 4. Fine structure of the cuticle and epidermis. *Tissue Cell* **2**: 637–696.

Gupta, B. L. & Little, C. (In press). Ultrastructure, phylogeny and Pogonophora. *Z. Zool. Syst. Evol.*

Gupta, B. L., Little, C. & Philip, A. M. (1966). Studies on Pogonophora. Fine structure of the tentacles. *J. mar. biol. Ass. U.K.* **46**: 351–372.

Gustus, R. M. & Cloney, R. A. (1972). Ultrastructural similarities between setae of brachiopods and polychaetes. *Acta zool., Stockh.* **53**: 229–233.

Hama, K. (1960). The fine structure of some blood vessels of the earthworm *Eisenia foetida. J. biophys. biochem. Cytol.* **7**: 717–723.

Ivanov, A. V. (1957). [Materials on the embryonic development of Pogonophora]. *Zool. Zh.* **36**: 1127–1144. [In Russian].

Ivanov, A. V. (1960). [Pogonophora] *Fauna S.S.S.R.* N.S. **75**. [In Russian].

Ivanov, A. V. (1963). *Pogonophora.* London: Academic Press.

Ivanov, A. V. (1964). [On the structure of the hind region of the body in Pogonophora]. *Zool. Zh.* **43**: 581–589. [In Russian].

Ivanov, A. V. (1965). Structure de la région posteriéure sétigère du corps des Pogonophores. *Cah. Biol. mar.* **6**: 311–323.

Ivanov, A. V. (1970). Verwandtschaft und Evolution der Pogonophoren. *Z. Zool. Syst. Evol.* **8**: 109–119.

Jeuniaux, C. (1963). *Chitin et chitinolyse.* Paris: Masson.

Liwanow, N. A. & Profirjewa, N. A. (1967). Die Organization der Pogonophoren und deren Beziehungen zu den Polychäten. *Biol. Zentr.* **86**: 177–204.

Moritz, K. & Storch, V. (1970). Über den Aufbau des Integumentes der Priapuliden und der Sipunculiden (*Priapulus caudatus* Lamarck, *Phascolion strombi* Montagu). *Z. Zellforsch. mikrosk. Anat.* **105**: 55–64.

Nørrevang, A. (1965a). Structure and function of the tentacle and pinnules of *Siboglinum ekmani* Jägersten (Pogonophora). *Sarsia* **21**: 37–47.

Nørrevang, A. (1965b). Fine structure of nervous layer, basement membrane and muscles of the proboscis in *Harrimania kupfferi* (Enteropneusta). *Vidensk. Meddr dansk naturh. Foren.* **128**: 325–337.

Nørrevang, A. (1970a). On the embryology of *Siboglinum* and its implications for the systematic position of the Pogonophora. *Sarsia* **42**: 7–16.

Nørrevang, A. (1970b). The position of Pogonophora in the phylogenetic system. *Z. Zool. Syst. Evol.* **8**: 161–172.

Orrhage, L. (1971). Light and electron microscope studies of some annelid setae. *Acta zool., Stockh.* **52**: 157–169.

Orrhage, L. (1973). Light and electron microscope studies of some brachiopod and pogonophoran setae, with a discussion of the "annelid seta" as a phylogenetic character. *Z. Morph. Tiere* **74**: 253–270.

Potswald, H. E. (1971). A fine structural analysis of the epidermis and cuticle of the oligochaete *Aeolosoma bengalense* Stephenson. *J. Morph.* **135**: 185–212.

Southward, E. C. (1971). Recent researches on the Pogonophora. *Oceanogr. mar. Biol. A.* **9**: 193–220.

Southward, E. C. (1973). The distribution of glycogen in the tissues of *Siboglinum atlanticum* (Pogonophora). *J. mar. biol. Ass. U.K.* **53**: 665–671.

Southward, E. C. (In press). A study of the structure of the opisthosoma of *Siboglinum fiordicum*. *Z. Zool. Syst. Evol.*

Southward, E. C. & Southward, A. J. (1966). A preliminary account of the general and enzyme histochemistry of *Siboglinum atlanticum* and other Pogonophora. *J. mar. biol. Ass. U.K.* **46**: 579–616.

Storch, V. & Welsch, U. (1970). Über die Feinstruktur der Polychaeten-Epidermis (Annelida). *Z. Morph. Tiere* **66**: 310–322.

Storch, V. & Welsch, U. (1972). Über Bau und Entstehung der Mantelstacheln von *Lingula unguis* L. (Brachiopoda). *Z. wiss. Zool.* **183**: 181–189.

Trueman, E. R. & Ansell, A. D. (1969). The mechanisms of burrowing into soft substrata by marine animals. *Oceanogr. mar. Biol.* **7**: 315–366.

Webb, M. (1964). The posterior extremity of *Siboglinum fiordicum* (Pogonophora). *Sarsia* **15**: 33–36.

Webb, M. (1965). Additional notes on the adult and larva of *Siboglinum fiordicum* and on the possible mode of tube formation. *Sarsia* **20**: 21–34.

Webb, M. (1969a). An evolutionary concept of some sessile and tubicolous animals. *Sarsia* **38**: 1–8.

Webb, M. (1969b). Regionation and terminology of the pogonophoran body. *Sarsia* **38**: 9–24.

Symp. zool. Soc. Lond. (1975) No. 36, 253–318.

FOSSIL EVIDENCE CONCERNING THE ORIGIN OF THE CHORDATES

R. P. S. JEFFERIES

Dept. of Palaeontology, British Museum (Natural History), London, England

SYNOPSIS

The ancestors of all living chordates are to be sought among a group of strange fossils found in marine rocks of Cambrian to Devonian age. These show definite echinoderm affinities, but are best regarded as primitive chordates constituting the subphylum Calcichordata Jefferies 1967. Chordate features for which there is evidence among some or all calcichordates include: gill slits; a post-anal tail with notochord, muscle blocks and dorsal nerve cord; a fish-like brain and cranial nervous system; a tunicate-like, filter-feeding pharynx; and various asymmetries which recur in cephalochordates and tunicates. Among calcichordates certain forms are likely to be particularly closely related to vertebrates, since they possessed the rudiments of a lateral line. Another group, represented by the genus *Lagynocystis*, was ancestral to cephalochordates.

INTRODUCTION

The animals discussed here are a group of peculiar fossils known from Middle Cambrian to Middle Devonian marine rocks, which thus range in known age from about 550 to about 370 million years (Fig. 1). They are traditionally placed in the carpoid echinoderms, and this is still the view of most palaeontologists closely involved with them (Ubaghs, 1967, 1968, 1970, 1971 and other earlier papers; Caster, 1972; Chauvel, 1971). In my view, however, they should certainly be regarded as chordates with echinoderm affinities and can be referred to as the subphylum Calcichordata of the chordates (Jefferies, 1967, 1968, 1969, 1971, 1973; Jefferies & Prokop, 1972; Eaton, 1970; Bone, 1972). The name Calcichordata alludes to their skeleton which is echinoderm-like in that each of the many plates is a meshwork formed from a single calcite crystal. The living chordate subphyla, of vertebrates, of tunicates and of amphioxus and its allies, are probably descended from the calcichordates.

Gislén (1930) was the first to argue in detail that the calcichordates have to do with the origin of the chordates, in an extremely erudite and original paper. Matsumoto (1929) was the first to suggest it in print.

The subphylum Calcichordata is coextensive with the class Stylophora, Gill & Caster 1960. It is divided into two orders: the Cornuta, which are more primitive, and the Mitrata which evolved from Cornuta.

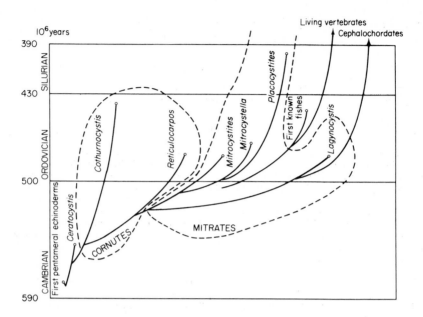

FIG. 1. The stratigraphy and phylogeny of the forms discussed in the text, and their probable relationships with other chordates and with echinoderms.

In this account I shall briefly describe three cornutes and four mitrates. The cornutes are *Cothurnocystis, Ceratocystis* and *Reticulocarpos,* and the mitrates are *Mitrocystites, Mitrocystella, Placocystites* and *Lagynocystis.* In the descriptions the words anterior and posterior, dorsal and vertical and right and left are defined to correspond with a fish.

Before beginning the descriptions I wish to emphasize a change of mind. All calcichordates consist of two parts which hitherto I have called theca and stem, by supposed homology with crinoids. I have always maintained, and indeed it is now virtually certain, that these parts are homologous with the body and tail of other chordates.* I now think, however, that they are probably not homologous with the theca and stem of stemmed echinoderms but represent a parallelism to these structures. My new opinion agrees with a suggestion by Eaton (1970). To emphasize my change of

---

* *Note added in proof.* By strict homology with vertebrates it might be better to speak of the head and the tail of mitrates.

opinion I shall henceforth refer to the parts of calcichordates as body and tail instead of theca and stem.

The reasons for my change of view are palaeontological. The genus *Gogia* is an eocrinoid from the Middle Cambrian of Utah, and as such is representative of the group from which crinoids are probably descended. It has the stem only weakly distinct from the theca (Robison, 1965; Sprinkle, 1973), as if in process of arising.* If *Gogia* represents the phylogenetic origin of the stem in echinoderms, then the appendage of calcichordates must have arisen separately. For *Gogia*, with its pentameral symmetry and lack of gill slits, cannot be ancestral to the calcichordates. It follows that other supposed homologies between stems and tails, which I have maintained previously, are also only parallelisms. Any two appendages arising in a posterior position from the body or theca are likely to contain tumid endodermal or mesodermal hydraulic skeletons (notochord, chambered organ). The nervous system, primitively ectodermal in position, will concentrate round these hydraulic skeletons to form the dorsal nerve cord or peduncular nerve. And a ganglion is likely to arise at the anterior end of the appendages to coordinate the activities and sense impressions of the theca or body with those of the stem or tail (brain, aboral nerve centre).

There is functional-morphological evidence that the tail of calcichordates contained a notochord, and direct evidence of a dorsal nerve cord overlying this notochord. The conclusion, therefore, that stems and tails are parallels rather than homologues does not greatly alter the arguments for connecting calcichordates with other chordates, which are now confirmed by detailed resemblances of their pharyngeal structures with those of tunicates.

COTHURNOCYSTIS (ORDER CORNUTA)

I describe this genus from *Cothurnocystis elizae* Bather from the Upper Ordovician of Scotland. A fuller but somewhat outdated account is given in Jefferies (1968).

Cothurnocystis (Fig. 2a,b) has an anterior body joined to a posterior tail. The body is shaped like a mediaeval ankle boot (*cothurnos* = boot). The skeleton of the body is made up of a frame of marginal plates to which dorsal and ventral, flexible, plated integuments are attached. Spikes on the ventral surface, adapted to

*This was pointed out to me by Professor J. Wyatt, Durham.

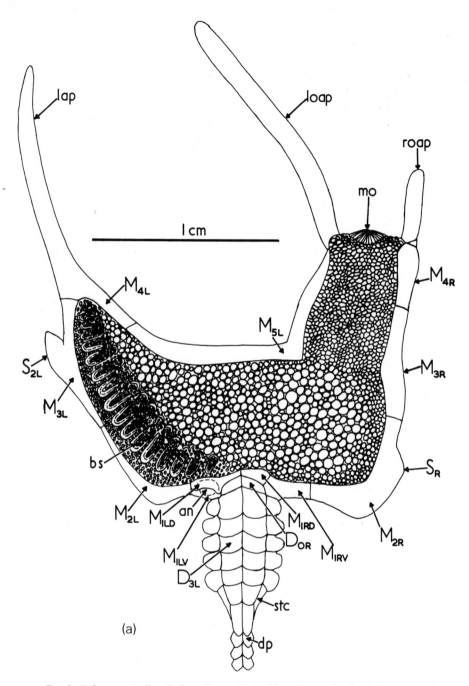

FIG. 2. *Cothurnocystis elizae* Bather, Upper Ordovician, Girvan, Scotland. Reconstruction of a) dorsal and b) ventral aspect. Most of hind tail omitted. an = Gonopore-anus; bs = branchial slit; dp = dorsal plate of hind tail; lap = left body appendage; loap = left oral appendage; mo = mouth; stc = stylocone; str = strut; roap = right oral appendage;

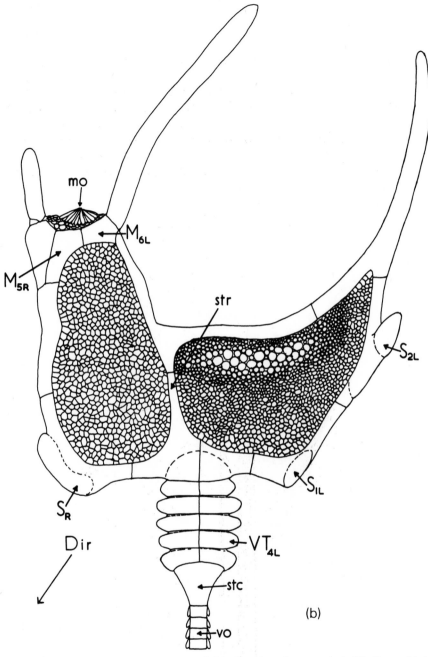

vo = ventral ossicle of hind tail; Dir = probable direction of movement in life; D_{3L} = third left dorsal plate of fore tail; M_{4R}, M_{1LV}, etc. = marginal plates; S_{1L}, S_{2L}, S_R = ventral spikes of body; VT_{4L} = fourth left ventral plate of fore tail. Details of plate notations in this and other figures are explained in Jefferies (1973: 423) (Reproduced with permission of the British Museum (Natural History)).

support the body on the sea floor, indicate that this surface was indeed ventral, and downwards in life.

Openings in the wall of the body are anatomically crucial. The most obvious are a series of about 16 elliptical openings. Each of these holes is surrounded by a frame consisting of an anterior and posterior u-shaped calcite plate (Fig. 3). Attached to the smaller, anterior u-plate was an outwardly convex flexible tongue, stiffened with tiny ossicles. Gislén (1930) suggested that these openings were

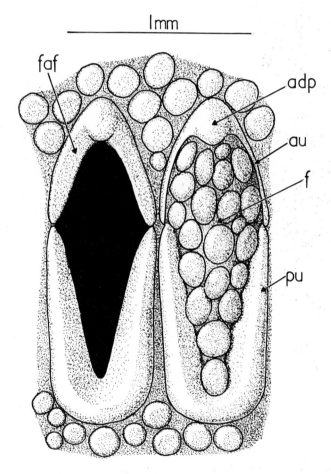

FIG. 3. *Cothurnocystis*. Reconstruction of two branchial slits, external aspect. The left slit has had the flap or tongue (f) removed. Other features: adp = antero-dorsal process; au = anterior u-plate; faf = flap attachment facet; pu = posterior u-plate. Reproduced with permission of the British Museum (Natural History).

branchial slits, and in this he was almost certainly right, for each one is well adapted as an outlet valve. High pressure beneath the slits would inflate the dorsal integument, cause the two u-plates of each slit to bend downwards towards each other and thus lift up the tongue, so letting water out. When the pressure inside fell, the tongue would fall down again, and prevent water from re-entering. The fact that *Cothurnocystis* had left gill slits only is not strange, for so has larval amphioxus.

Another opening is situated just behind the most median gill slits, just left of the tail. Grooves in the skeleton indicate that two tubes ended at this hole—a larger which was probably the rectum, and a smaller which was probably the gonoduct. The hole can therefore be called the gonopore-anus. This identification is confirmed by its position downstream of the gill slits which would allow faeces and gametes to be washed away from it. Also the grooves inside the skeleton lead rightwards from it and end just right of the tail, in the position where the gonopore and anus were present in *Ceratocystis*. And, in this the two holes can be identified by analogy with primitive echinoderms (see below).

The situation in the cornute *Scotiaecystis curvata* (see *Cothurnocystis curvata* in Jefferies, 1968) confirms that the gonopore-anus was indeed an outlet of some sort, for in that form it actually opens into the most median gill slits, rather than downstream of them.

Finally there was a large opening at the anterior end of the body in *Cothurnocystis*. This was the mouth, as is suggested by its size and position and confirmed by the conclusion that all the other openings were outlets.

Four chambers inside the body of *Cothurnocystis* can be deduced on direct evidence and a fifth on circumstantial evidence (Fig. 4). The first chamber filled the "ankle" part of the "boot", or buccal lobe, behind the mouth and can be called the buccal cavity. It is probably homologous with the buccal cavity of other chordates. A second chamber lay just anterior to the stem and the gonorectal groove to the gonopore-anus ran across its floor. I call this chamber the posterior coelom. There is evidence, taken from mitrates, that the posterior coelom of cornutes was homologous with the left epicardium of tunicates.

Two other chambers are indicated by direct evidence in the "foot" part of the "boot". One of these lay above the other, and the two are separated by an undulating line on the inner face of the marginals. Above this line the marginals are sometimes striated both vertically and longitudinally, which suggests that the upper

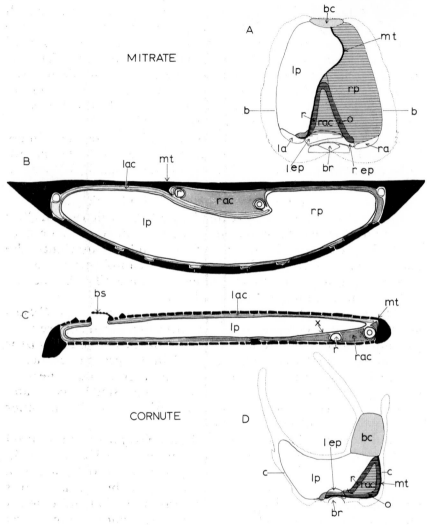

FIG. 4. The body chambers of a mitrate (*Mitrocystites*) and a cornute (*Cothurnocystis*). B is a transverse section of A through b–b; C is a transverse section of D through c–c. The mitrate condition can be derived from the cornute condition by the growth of a right pharynx (rp) out of the left pharynx (lp) at the point marked x so as to lift up the cavity and contents of the right anterior coelom (rac), squash them against the roof of the body, and force them in a medial direction. Other features: bc = buccal cavity; br = brain; bs = branchial slit; la = left atrium; lac = left anterior coelom (purely virtual), l ep = left epicardium (= posterior coelom of cornutes, part of posterior coelom of mitrates); lp = left pharynx; mt = mesenteric trace, between the left and right anterior coeloms (oblique ridge of mitrates); o = oesophagus; r = rectum; ra = right atrium; rac = right anterior coelom; r ep = right epicardium (left & right epicardia = posterior coelom of mitrates); rp = right pharynx.

chamber was indeed a unity. Its major part lay in the anterior left portion of the body and the gill slits opened through its roof. It is therefore likely to have been the pharynx.

The chamber beneath the pharynx was most voluminous near the posterior, right corner of the body. The gonorectal groove passed out of this part of it to run beneath the posterior coelom to the gonopore-anus. This lower chamber therefore probably contained the gonads and all the non-pharyngeal gut except the distal part of the rectum running beneath the posterior coelom. Evidence in *Ceratocystis* suggests that it also contained the heart and remnants of the axial gland of echinoderms, the latter probably being related to the neural gland of tunicates and therefore, to the hypophysis of vertebrates. There is a large spindle-shaped depression inside the marginals of *Cothurnocystis* which probably contained the heart. Judging by its contents, therefore, this lower chamber was a coelom. I have previously called it simply the anterior coelom but I now propose to call it the right anterior coelom.

The reason for this change of name is a purely comparative argument which suggests that there was probably also a left anterior coelom in *Cothurnocystis* and all other calcichordates. Thus there is evidence, discussed later, that the ventral surface of calcichordates, and chordates in general, corresponds to the right surface of hemichordates. In these latter animals the gut is suspended between right and left metacoels. The right anterior coelom of *Cothurnocystis*, lying on the floor of the body beneath the pharynx, would correspond to the right metacoel of hemichordates. And so, by analogy, *Cothurnocystis* would be expected to have a homologue of the hemichordate left metacoel dorsal to the pharynx. This homologue of the left metacoel could only have had a purely virtual cavity, and can be called the left anterior coelom. The right, and presumed left, anterior coeloms of *Cothurnocystis* would be homologous with the right and left body coeloms of amphioxus and the vertebrates.

Wherever the pharynx of *Cothurnocystis* came in contact with the integuments, or more precisely was separated from them only by the membranes of the right and left anterior coeloms, the skeleton of the integuments is made up of circular plates with gaps between them. Everywhere else the plates are polygonal without gaps. The gaps between the circular integument plates probably housed muscles serving to contract the pharynx in a coughing action.

The course of the post-pharyngeal gut in *Cothurnocystis*, as shown in Fig. 4, can be deduced from the situation in mitrates and will be discussed later.

The tail of *Cothurnocystis* has three portions, as in all known calcichordates. In my earliest papers I called these portions the anterior, medial and posterior parts of the stem (Jefferies, 1967, 1968, 1969). Later I called them the fore, mid and hind stem (Jefferies & Prokop, 1972; Jefferies, 1973). I now call them the fore, mid and hind tail. Use of the informal terms fore, mid and hind is intended to imply that like-named parts are not homologous in cornutes and mitrates.

The fore tail of *Cothurnocystis* had a large lumen surrounded by a skeleton made of about five rings of calcite plates. In each ring there were left dorsal, left ventral and right dorsal and right ventral plates. Successive rings were in contact dorsally and ventrally, but not at the sides, in a manner which would have allowed the fore tail to flex vigorously from side to side, but scarcely at all up or down. This flexion would wag the mid and hind tail sideways as a unit.

The lumen of the fore-tail presumably contained muscles, and there would also need to be an anticompressional notochord to prevent telescoping and dislocation. The evidence for this notochord is in fact clearer in other calcichordates where the skeleton of the fore-tail is very loose, for in *Cothurnocystis* the compressional stress caused by the muscles would partly be taken up by the contact between successive rings of plates in the dorsal and ventral mid-line.

The skeleton of the hind tail of *Cothurnocystis* (Fig. 5) consists of a ventral line of hemicylindrical ossicles, each ossicle being roofed over dorsally by a pair of dorsal plates. The hind tail ends abruptly, as if part had broken off, and the same is true of all known calcichordates. The sculpture of the dorsal surface of the ossicles includes a median longitudinal groove from which a pair of transverse grooves go out in each ossicle to lateral longitudinal grooves. These latter are deepened into ill-defined lateral pits, opposite the transverse grooves.

The median groove presumably contained a backward extension of the notochord that can be postulated on mechanical grounds in the fore tail. In the hind tail the notochord would again have helped to keep the skeleton in alinement. Each pair of dorsal plates imbricates inside the next pair anterior. This suggests that the hind stem could flex dorsally, presumably by the contraction of muscles in the tunnel-like lumen underneath the arched dorsal

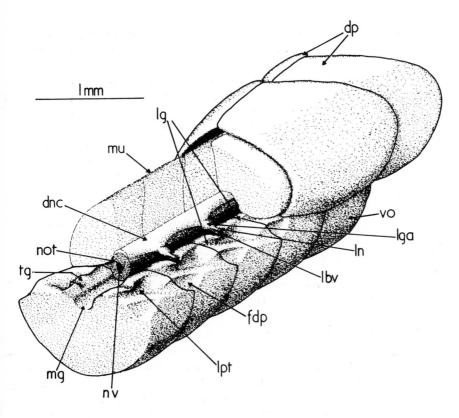

FIG. 5. *Cothurnocystis*. Anatomy of hind tail, partly based on analogy with mitrates: dp = dorsal plate; d nc = dorsal nerve cord; fdp = facet for dorsal plate; lbv = lateral blood vessel; lg = lateral groove; l ga = lateral ganglion; ln = lateral nerve; lpt = lateral pit; mg = median groove; mu = muscle block; not = notochord; nv = notochordal vessel; tg = transverse groove; vo = ventral ossicle. Reproduced with permission of the Trustees of the British Museum (Natural History).

plates and above the ventral ossicles. The contacts between the ventral ossicles suggest that the terminal portion of the hind tail was able to bend downwards. The transverse grooves probably carried blood vessels. There is circumstantial evidence that a main blood vessel passed down the middle of the notochord.

The skeleton of the mid tail includes a massive ventral ossicle called the stylocone, roofed over by two pairs of dorsal plates. The stylocone resembles several hind tail ossicles fused together and deeply excavated anteriorly. It would act as a socket in which the fore-tail muscles were inserted and by which they could move the mid and hind tail as a unit.

The brain of *Cothurnocystis* was situated at the anterior end of the tail in a deep cerebral depression. It was thus at the anterior end of the notochord like the brain of any other chordate. There is evidence of: a pair of ganglia anterior to the brain (pyriform bodies, trigeminal ganglia); of two nerves arising from the brain on either side of the mid line; and of pits in the skeleton which may represent places where olfactory fibres left the skeleton.

Ubaghs (1961, 1971 and other papers) has proposed a thoroughly different explanation of the calcichordate tail. He sees it as an echinoderm arm with a water-vascular system. The median groove of the hind tail, which in my view carried the notochord, is thought by Ubaghs to have carried a radial water vessel. The latter was supposedly connected by the transverse grooves in each segment to tube feet situated in the lateral pits of the lateral grooves. Ubaghs also regards the dorsal plates as cover plates of an echinoderm pinnule or brachiole.

Ubaghs' interpretation has the result that the dorsal surface of cornutes, which all workers agree upon, was equivalent to what I call the ventral surface of mitrates. The ventral ossicles of the hind tail of cornutes would be equivalent to what I call the dorsal ossicles of the mitrate hind tail, and the dorsal hind-tail plates of cornutes to the ventral hind-tail plates of mitrates. The arguments against Ubaghs' interpretation are three-fold. Firstly the imbrication of the hind-tail plates with each other and with the ossicles in the cornutes *Scotiaecystis* and the mitrates *Mitrocystella, Chinianocarpos* (Fig. 12) and *Placocystites* strongly suggests that these plates could not open outwards in life. Secondly the internal chambers of cornutes and mitrates were fundamentally the same and support the mutual orientation which I adopt. If, in consequence, what I call the dorsal side of mitrates is truly equivalent to the dorsal side of cornutes, then Ubaghs' interpretation of the tail cannot be correct. And thirdly the comparison of the pharyngeal structures of mitrates with those of tunicates is now so strong and detailed that there can be little doubt of the correctness of the basic chordate interpretation of these animals. Comparing a calcichordate tail with a crinoid stem, though probably dependent on parallelism rather than homology, was less misleading than comparing it with a crinoid pinnule. For tail and stem both arise from the same posterior or aboral pole of the animal.

As regards habits, *Cothurnocystis* lived on a fine sandy sea bottom in marine, probably rather shallow, water, for associated fossils include starfishes, bryozoans, crinoids, brachiopods and trilobites. Spikes on the ventral surface of the body, and the fixed left oral and left body appendages, are pointed and slope downwards anteriorly. They would tend to prevent the body from moving forwards, but would have allowed it to slip obliquely backwards, pulled by the tail, as indicated by the arrow in Fig. 2b. The posterior distal tip of the tail, specially adapted for vertical flexion, would probably be stuck into the sea bottom to secure purchase. Evidence suggests that the habitual direction of crawling in all other calcichordates was also backwards. It is easier to pull an asymmetrical object than to push it, being directionally stable. This is the probable reason why such an ungainly method of locomotion was adopted.

The horizontally motile right oral appendage of *Cothurnocystis* probably had the function of stirring up the surface layer of sediment so that particles of food, accompanied no doubt by much silty sand, could be drawn into the mouth.

CERATOCYSTIS

I describe this genus from its only known species which is *Ceratocystis perneri* Jaekel from the Middle Cambrian of Bohemia. A fuller account is given in Jefferies (1969) and, with a completely different interpretation, in Ubaghs (1967). *Ceratocystis* is the most primitive cornute known and one of the oldest.

The body is boot-shaped (Fig. 6a, b, c). Its skeleton, unlike *Cothurnocystis*, is made up largely of big plates with no distinction into frame and dorsal and ventral integuments.

The body openings of *Ceratocystis* are again crucial. The gill slits are seven regular elliptical holes in the left part of the body. Three of these holes have u-shaped plates behind them confirming their homology with the gill slits of *Cothurnocystis*. A number of irregular holes (ag) between the plates in the general region of the gill slits were probably filled with muscle in life. They probably represent the beginnings of a slight flexibility which culminated in the very flexible dorsal integument of *Cothurnocystis*. This slight flexibility probably allowed *Ceratocystis* to cough a little. There are also signs of flexibility, in the form of rounded plate contacts, on the right side of the body.

There are a number of openings on the right side of the body of *Ceratocystis*. The mouth (m in Fig. 6b) was the largest of these. It lay

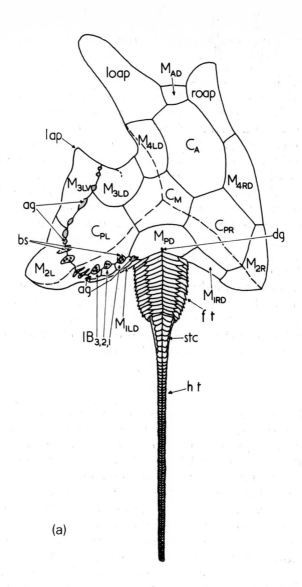

(a)

FIG. 6. *Ceratocystis perneri* Jaekel. Reconstruction; a) dorsal, b) ventral, c) posterior aspect: ag = accessory gap, between body plates; an = anus; bs = branchial slit; dg = dorsal groove for median eye; ft = fore tail; g = gonopore; h = hydropore; ht = hind tail; lap = left appendage; loap = left oral appendage; lpvp = lateral, non-overlapping, postion of a postero-ventral plate of the fore tail; m = mouth; ng = narrow groove (lateral line); or in = oral integument; roap = right oral appendage; stc = stylocone; vo = ventral ossicle; v pr = ventral process of ventral fore-tail plate; C_A etc. = centro-dorsal plates; $IB_{1,2,3}$ = infrabranchial or posterior u-plates; M_{AD} etc. = marginal plates; $S_{1R, 2R, 3R}$ = ventral spikes of body (copied with the permission of The Paleontological Association from the original figure in *Paleontology*).

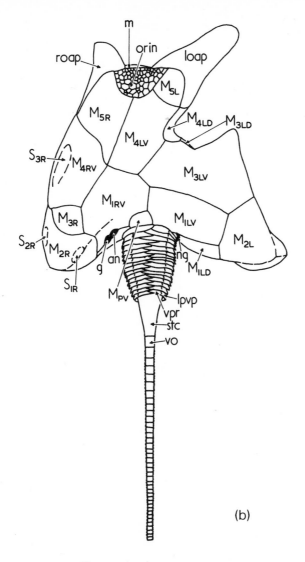

(b)

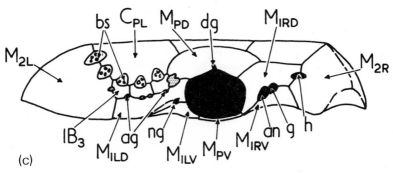

(c)

at the anterior end of the body with a flexible integument behind it
forming the lower lip. The other openings, marked h, g and an in
Fig. 6c are situated in the "heel" region of the body. They can be
identified by comparison with primitive echinoderms such as the
diploporitan cystoid shown in Fig. 7. Such echinoderms commonly

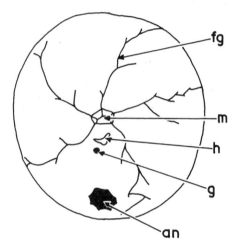

FIG. 7. Oral surface of an early echinoderm—the diploporitan cystoid *Glyptosphaerites
leuchtenbergi* (Volborth)—to show the usual openings and their arrangement: an = anus;
fg = food groove; g = gonopore; h = hydropore; m = mouth (copied with the permission
of The Paleontological Association from the original figure in *Paleontology*).

have three openings arranged roughly in a line. Going away from
the mouth these openings are the hydropore, gonopore and anus.
By analogy the openings h, g and an of *Ceratocystis* are also
hydropore, gonopore and anus.

These identifications can be confirmed by other arguments.
Firstly the hydropore is known among calcichordates only in
Ceratocystis. It disappeared as the cornutes evolved towards a more
customary chordate condition, as would be expected. The gonopore and anus migrated to left of the tail in *Cothurnocystis*, so as to be
behind the gill slits, where faeces and gametes could be washed by
the branchial current. This confirms that the gonopore and anus
were indeed outlets.

The position of the tail of *Ceratocystis*, relative to mouth,
hydropore, gonopore and anus, indicates that it was located in a

position on the body very like that of the stem of stemmed echinoderms.

The chambers inside the body of *Ceratocystis* (Fig. 9) were much as in *Cothurnocystis*. There is direct evidence of buccal cavity, pharynx, posterior coelom and right anterior coelom. There was probably also a purely virtual left anterior coelom as in *Cothurnocystis*.

The contents of the right anterior coelom of *Ceratocystis* are indicated by the fact that the hydropore, gonopore and anus opened directly out of it. It would therefore be expected to contain organs associated with the hydropore of echinoderms and also the gonads, opening at the gonopore, and the non-pharyngeal gut, opening at the anus.

The hydropore of modern echinoderms is associated with several organ systems (Fedotov, 1924). Into it open the stone canal of the water vascular system and the axial sinus. The axial sinus is partly surrounded by the axial gland, whose head process is included in the so-called dorsal sac or madreporic vesicle, situated just beneath the hydropore. Strands of tissue pass from the axial gland to the gonads. The head process and dorsal sac are pulsatile and probably homologous with the heart of hemichordates (Narasimhamurti, 1931; Boolootian & Campbell, 1964).

It is therefore possible that there was a homologous pulsatile structure near the hydropore of *Ceratocystis*. Being near the gonads and non-pharyngeal gut this structure could well be homologous with the heart of tunicates and vertebrates. There are direct indications of such a structure, spindle-shaped like the heart of salps, in a depression on the internal face of the right marginal plates in *Cothurnocystis*.

A stone canal would presumably be absent from the *Ceratocystis*, since there is no sign of any other part of the water vascular system. The hydropore probably functioned, therefore, mainly as an outlet for the axial sinus. Through it, as in modern echinoderms, dead coelomocytes from the axial gland could have been released to the outside.

The axial gland of echinoderms resembles, and is probably homologous with, the neural gland of tunicates. Both mainly function in expelling the degeneration products of coelomocytes (Godeaux, 1964; Pérès, 1943; Millar, 1953, for the neural gland, Millott, 1966, for the axial gland). Both probably have an endocrine function (Godeaux, 1964; Carlisle, 1953, for the neural gland, Millott, 1967, for the axial gland). Both are connected by strands of tissue to the gonad (dorsal strand of ascidians, genital strand of echinoderms). The principal anatomical difference between them is that the axial gland of echinoderms opens to the outside through the hydropore, whereas the neural gland of tunicates opens into the roof of the pharynx by way of the ciliated organ. There is evidence of a ciliated organ in the mitrate *Placocystites* opening out of the right anterior coelom. This implies that at least the duct of the gland was in that coelom, though the gland itself probably lay ventral to the brain. It seems very likely that the axial gland of *Ceratocystis* transformed itself into the neural gland by losing its old duct to the outside, acquiring a new opening into the pharynx, and itself migrating out of the right anterior coelom to lie beneath the brain. The loss of the duct to the outside, and therefore the acquisition of a pharyngeal opening, probably happened in evolution between *Ceratocystis* and the primitive *Cothurnocystis*-like genus *Nevadaecystis*, which had no hydropore.

The tail of *Ceratocystis* was much like that of *Cothurnocystis* but the skeleton of the fore-tail was looser. A notochord would be absolutely required for the fore tail to function.

The brain and cranial nerves (Fig. 8a, b) can be interpreted more fully than in other cornutes, mainly by comparison with mitrates. There is evidence of a nerve (nng, png) passing round the left trigeminal ganglion (pyriform body) to end at a groove in the surface. This probably represents the beginning of the acusticolateralis system, since its position is exactly comparable with the left acoustic nerve of *Mitrocystites*. In *Ceratocystis* this nerve, since it ends externally, would function as lateral line. A dorsal process (dp) of the brain goes up onto the dorsal surface of the body. By analogy with mitrates this process arose from the optic part of the brain and is likely to have been a sort of median eye. This however was not homologous with the pineal or parapineal eyes of vertebrates, since it was absent in the group of cornutes, represented by *Reticulocarpos*, from which mitrates and therefore all modern chordates were descended.

The peculiar shape of *Ceratocystis* can be explained if it was descended from a bilateral ancestor lying down on one side. More precisely it can be compared with the modern pterobranch *Cephalodiscus* resting on its right side. Figure 9 represents *Ceratocystis* in dorsal aspect and *Cephalodiscus* in left aspect. Passing round the two diagrams in a clockwise direction gives the following possible homologies.

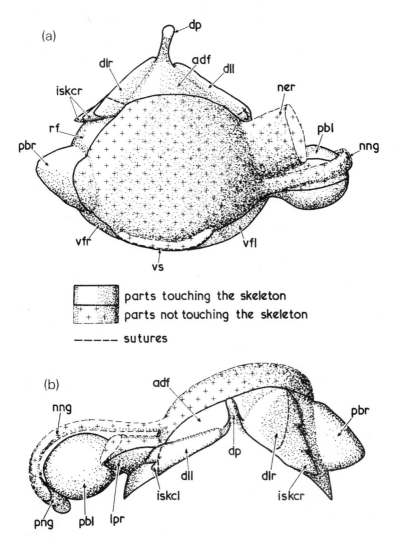

FIG. 8. *Ceratocystis*. Reconstruction of brain and associated nerves and ganglia; a) anterior, b) dorsal aspect; adf = antero-dorsal face of brain; dll, dlr = left and right dorsal lobes, corresponding to anterior part of brain, or olfactory bulbs, of mitrates; dp = dorsal process or median eye; iskcl, iskcr = left and right intra-skeletal cones, probably represent-ing places where olfactory nerve fibres from the skeleton entered the dorsal lobes; l pr = left process, representing large nerve going off to left; ner = nerve corresponding to left process; nng = nerve to narrow groove or lateral line, homologous with left auditory nerve of mitrates; pbl, pbr = left and right pyriform bodies (trigeminal ganglia); png = part of lateral-line nerve indicated by direct evidence in natural mould; rf = right face of brain; vfl, vfr = left and right ventral faces of brain; vs = ventral swelling of brain (copied with the permission of The Paleontological Association from the original figure in *Paleontology*).

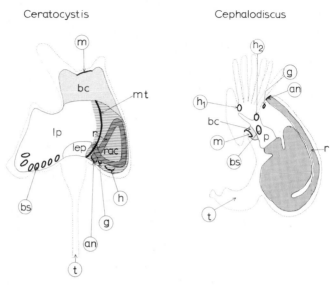

FIG. 9. Comparison between *Ceratocystis*, in dorsal aspect, and the modern hemichordate *Cephalodiscus*, in left lateral aspect. Compare *Ceratocystis* with *Cothurnocystis* in Fig. 4. The coeloms of *Cephalodiscus* have not been indicated. Note the similarity of sequence, passing round the figures, of the structures whose index letters have been included in circles: an = anus; bc = buccal cavity; bs = branchial slit; g = gonopore; h = hydropore; h_1 = protocoel pore; h_2 = mesocoel pore; lep = left epicardium, i.e. posterior coelom; lp = left pharynx; m = mouth; mt = mesenteric trace between left and right anterior coeloms; p = pharynx of *Cephalodiscus*; r = rectum; rac = right anterior coelom; t = tail of *Ceratocystis*, stem of *Cephalodiscus*.

Ceratocystis	*Cephalodiscus*
tail	stem
gill slits	left gill slit
mouth	mouth
hydropore	left axocoel pore, or hydrocoel pore, or both fused
gonopore	left gonopore
anus	anus
tail	stem

It is curious that echinoderms are also believed to be descended from a form like *Cephalodiscus* resting on its right side and which, like *Ceratocystis*, had acquired a calcite skeleton of echinoderm type (Grobben, 1924). Indeed it is likely that these hypothetical ancestors were one and the same, so that echinoderms and chordates are more closely related to each other than either are to hemichordates (Fig. 34).

Ceratocystis lived on a sandy bottom like *Cothurnocystis* and was probably similar to that genus in most of its habits.

I describe this genus from its only known species which is *R. hanusi* Jefferies & Prokop from the Lower Ordovician (Llanvirn) of Bohemia. For a fuller account see Jefferies & Prokop (1972).

Reticulocarpos is important because it represents the group of cornutes from which mitrates evolved. Unlike more primitive cornutes, but like mitrates, it is found in siliceous nodules in shales. It seems to have lived on very soft mud, using only the strength of the mud to support its weight (Fig. 13). This involved striking adaptations for weight reduction and load spreading. Mitrates, by contrast, supported their weight on soft mud by a much more efficient method resembling buoyancy.

The body of *Reticulocarpos* (Fig. 10a, b) was smaller than that of other cornutes, being only about 10 mm long. The small size would reduce its weight. The body was boot-shaped but bilaterally symmetrical like a mitrate. It had a frame, like *Cothurnocystis*, with dorsal and ventral flexible integuments. However the frame, and therefore the body, was almost bilaterally symmetrical, like a mitrate. The dorsal surface was probably smoothly convex in life while the ventral surface was flat, as a way of spreading the weight of the body evenly over the sea floor. Dorsal and ventral surfaces met at a peripheral flange which followed the edges of the marginal plates, as in mitrates, though this flange, because of the convex dorsal and flat ventral surface, was ventro-lateral, not dorso-lateral. The marginal plates were very loose in texture and the integument plates were two-dimensional networks of calcite, both of which features would help to reduce weight. The texture of the marginal plates was densest near the sutures between them, which suggests that the animals as found had almost stopped growing, and were adult despite their small size. The ventral strut, which in *Cothurnocystis* connects the anterior and posterior parts of the frame, was present in *Reticulocarpos*, but stopped short of the frame anteriorly. This would reduce weight and is mitrate-like, since the mitrates had no ventral strut. Again *Reticulocarpos*, unlike *Cothurnocystis*, had no bar ventral to the mouth. This would reduce weight and can be compared with the condition in such mitrates as *Mitrocystites* or *Peltocystis*. Functionally the ventral mouth bar was replaced in *Reticulocarpos* by a dorsal bar which connected the right

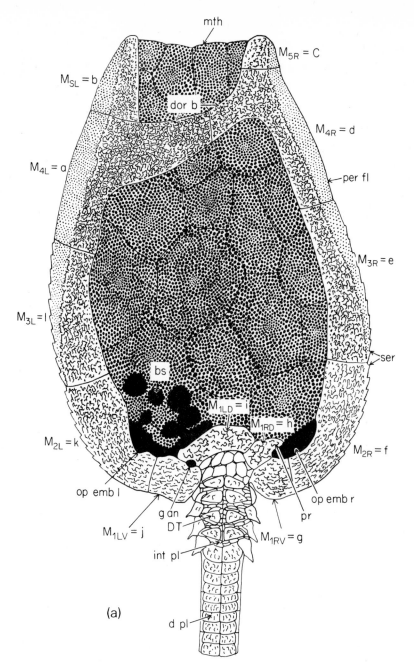

FIG. 10. *Reticulocarpos hanusi* Jefferies & Prokop, Lower Ordovician, Bohemia. Reconstruction of a) dorsal, b) ventral aspect. The whole of the hind tail is included: bs = branchial slit; dor b = dorsal bar; d pl = dorsal plate of hind tail; ft = fore tail; g an = gonopore anus; ht = hind tail; int pl = intercalary plate of fore tail; lat spk = lateral spike of ventral plate of fore tail; mt = mid tail; mth = mouth; op emb l, r = left and right optic embayment for transpharyngeal eyes; oss = ventral ossicles of hind tail; per fl = peripheral flange;

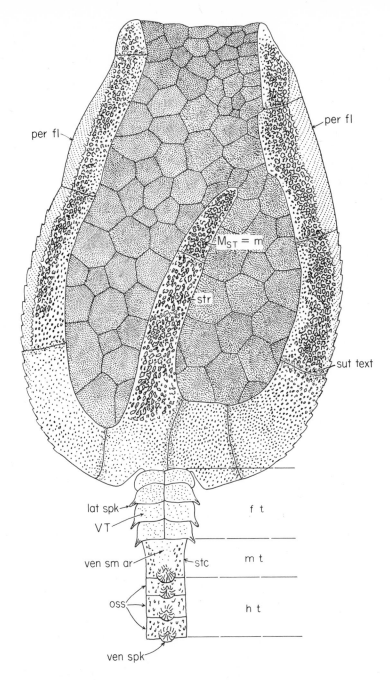

pr = prong in front of right transpharyngeal eye; ser = serrations of peripheral flange; stc = stylocone; str = ventral strut; sut text = dense texture of plates near to suture; ven sm ar = ventral smooth area of stylocone; ven spk = ventral spikes of mid and hind tails, M_{5R} = C etc. = marginal plates (two notations); VT = ventral major plate of fore tail; DT = dorsal major plate of fore tail.

and left parts of the frame posterior to the buccal cavity. This dorsal bar can be compared plate for plate, and in its oblique orientation, more posterior on the left, with the front edge of the dorsal shield of the primitive mitrate *Peltocystis*. It was perhaps the most striking mitrate-like feature which *Reticulocarpos* possessed. It can be calculated that, by combined weight reduction and load spreading, *Reticulocarpos* imposed a load of only about 10 mg/cm^2 on the sea bottom.

As concerns body openings, *Reticulocarpos* had external gill slits in the posterior left part of the dorsal integument like any other cornute. These had the form of mere round holes, however. The characteristic outlet-valve skeleton seen in *Cothurnocystis*, which had presumably existed in some of the ancestors of *Reticulocarpos*, had disappeared. Loss of this outlet-valve skeleton, again, would have reduced weight. Perhaps the slits could be closed by soft lips during life. The mouth was at the front end of the body, between dorsal and ventral integuments. The oral plates, however, were unspecialized. The gonopore-anus was situated just right of the tail, just as in *Cothurnocystis*, and between the same two plates. Two gaps in the dorsal skeleton (op emb 1 and op emb r) coincided in position with vesicles of soft tissue, indicated by excavations in the inner surfaces of the marginal plates. The vesicles were probably transpharyngeal eyes as in the mitrate *Mitrocystites*.

The chambers of the body were much as in *Cothurnocystis*. There is direct evidence of the posterior coelom, right anterior coelom, buccal cavity and pharynx. A purely virtual left anterior coelom probably also existed.

The tail was typically cornute in having fore, mid and hind portions with ventral ossicles and dorsal plates in the hind tail and a ventral stylocone in the mid tail (Fig. 10a, b, Fig. 11). However there were four ossicles or fewer in the hind tail, instead of about 56 as in *Cothurnocystis*, for example. This shortening of the tail would have reduced weight. The hind tail ossicles and the stylocone were each equipped with a big ventral spike, conical but curved anteriorly, which would have gripped the bottom mud in ventral flexion. Another indication of ventral flexion is seen in the ornament of the stylocone whose ventral surface is smooth, unlike that of the hind-tail ossicles. This ventral smooth area would have allowed the ventral surface of the stylocone to slide inside the posterior ventral plates of the fore tail (Fig. 11). This movement would have caused the mid and hind tail to flex ventrally, pushing downwards and forwards against the bottom mud. There were plane surfaces of

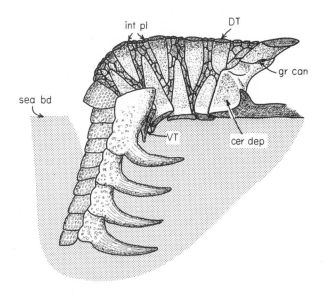

FIG. 11. *Reticulocarpos*. Reconstruction of tail in flexed condition. Fore tail shown as split sagittally and empty, to indicate how stylocone slid inside it in ventral flexion. Compare Fig. 12: cer dep = cerebral depression; gr can = gonorectal canal; int pl = intercalary plate; sea bd = sea bed; DT = dorsal major plate of fore tail; VT = ventral major plate of fore tail.

contact between the hind-stem ossicles and between the most anterior ossicle and the stylocone. The mid and hind tail would therefore have acted together as a rigid rod.

The skeleton of the fore tail consisted of major plates connected by a membrane plated with intercalary plates. The major plates were arranged in left and right ventral and left and right dorsal series, but, unlike *Cothurnocystis*, did not form rings. There were four pairs of ventral plates so shaped as to give the fore tail a broad ventral surface, flat when not flexed. All but the most anterior pair of ventral fore-tail plates had little spikes at their outer postero-ventral angles. The fore tail, like the mid and hind tail, seems to have been adapted to act by ventral flexion, pushing by its flat ventral surface downwards and forwards against the bottom mud. In this way the whole tail would have pulled the body backwards over the surface of the sea floor.

In acting by ventral flexion, instead of lateral flexion, the tail of *Reticulocarpos* differed from that of *Cothurnocystis* and resembled the mitrates. The change in direction of flexion was probably an adaptation to life on very soft mud, for lateral flexion required anti-yaw devices on the body—such as the ventral spikes and anterior appendages of *Cothurnocystis*. These would not have worked well on soft mud, where it was better to avoid yaw by developing a symmetrical body and a ventrally flexing tail.

A comparison of the tail of *Reticulocarpos* with that of the primitive mitrate *Chinianocarpos* (Fig. 12) is instructive. The latter

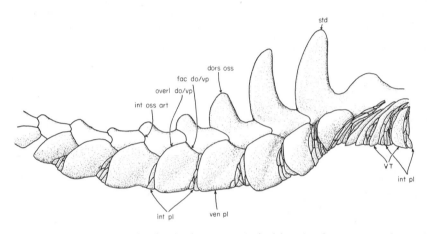

FIG. 12. *Chinianocarpos thorali* Ubaghs, Lower Ordovician, Southern France. Camera lucida drawing of the tail. The whole tail of this primitive mitrate should be compared with the homologous fore tail of *Reticulocarpos* in Fig. 11: dors oss = dorsal ossicle of hind tail; fac do/vp = facet at contact of a dorsal ossicle and a ventral plate; int oss art = interossicular articulation; int pl = intercalary plate; overl do/vp = overlap of ventral plate over dorsal ossicle; std = styloid; ven pl = ventral plate of hind tail; VT = ventral plate of fore tail.

was divided, as in all calcichordates, into fore, mid and hind portions; there were dorsal ossicles in the hind tail of *Chinianocarpos* and a massive dorsal styloid in the mid tail. It is likely, however, that the fore, mid and hind tail of cornutes were not homologous with those of mitrates. As already mentioned, the fore tail of *Reticulocarpos* was flexible in a vertical plane and its skeleton consisted of the major plates alternating with intercalary plates. The mid and hind tail of *Reticulocarpos*, on the other hand, was a rigid rod whose skeleton was made up of dorsal plates and ventral ossicles and stylocone. Now the tail of *Chinianocarpos* was ventrally flexible throughout its length and the ventral plates of the mid and hind tail alternate with intercalary plates (Fig. 12). It is therefore likely that the whole tail of *Chinianocarpos* is homologous only with the fore tail of *Reticulocarpos*. This implies that, in the early evolution of mitrates, the old cornute mid and hind tail had been lost. The mitrate fore, mid and hind tail were regionated from the remaining stump. The ventral plates of the hind tail of the mitrate *Lagynocystis* confirm this conclusion, for, with considerable differences in proportion, they show the same chief features as the ventral plates of the fore tail of *Reticulocarpos*, including a flat or concave ventral bearing surface and postero-lateral spikes (Fig. 29b).

The brain of *Reticulocarpos* was situated in a depression at the anterior end of the tail as in *Cothurnocystis*. There is evidence of the same two ganglia and the same nerves anterior to it.

MITROCYSTITES (ORDER MITRATA)

I describe this genus on the basis of *Mitrocystites mitra* Barrande from the Lower Ordovician of Bohemia. Like most mitrates, and like *Reticulocarpos*, it is found in a shale, preserved in siliceous nodules.

The body, as in nearly all mitrates, was flat dorsally and convex ventrally—opposite to the *Reticulocarpos* condition. Such a body would sink down in soft mud, displacing the mud sideways and thereby compensating the weight of the animal (Fig. 13). This

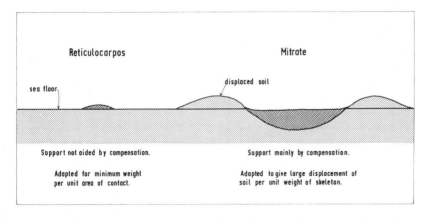

FIG. 13. The modes of support of *Reticulocarpos* amd of a mitrate on soft mud.

method of support on soft substrates was much more reliable than the "snow-shoe" system employed by *Reticulocarpos*, since mud will easily lose all of its weak strength by stirring, without changing appreciably in density. The dorsal surface of the body of

Mitrocystites (Fig. 14a) was formed of a shield made up of marginal plates surrounding centro-dorsal plates. The ventral surface (Fig. 14b) was largely made of a somewhat flexible plated integument, except for two large ventral marginal plates posteriorly. The anterior ventral integument plates imbricated with each other so

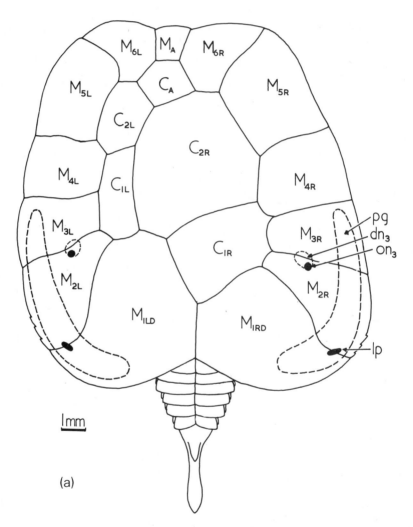

(a)

FIG. 14. *Mitrocystites mitra* Barrande, Lower Ordovician, Bohemia. Reconstruction in a) dorsal and b) ventral aspect: bo = branchial openings (paired); cu rib = cuesta-shaped rib; dn_3 = depression at end of transpharyngeal optic nerve n_3, i.e. for transpharyngeal eye; lp = opening of $n_{4\&5}$ onto dorsal surface; mo = mouth; ng = narrow groove, i.e. lateral

that each one was overlapped by its more posterior neighbours. There was a well marked peripheral flange round the marginals separating the dorsal from the ventral surface of the body.

The body openings of *Mitrocystites* included a large mouth anteriorly, with a flexible lower lip. There were no external gill slits,

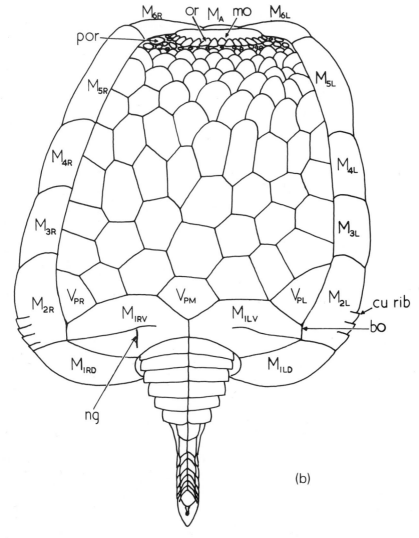

(b)

line; or = oral plate; pg = peripheral groove; por = post-oral plate; C₂ₗ, etc. = centro-dorsal plates; M_A etc. = marginal plates; V_{PR} etc. = ventral plates. Reproduced with permission of the British Museum (National History).

but there is evidence of a left and a right gill opening posteriorly, between the marginal and ventro-marginal plates. The gill openings would gape when the ventral marginal plates rocked relative to the other marginals about a transverse hinge-line. In this way the gill openings were guarded by outlet valves. There was a long groove on the right postero-ventral plate. This probably represented the beginnings of the lateral-line system, as discussed below. On the dorsal surface there were two pairs of openings. The more posterior were the openings of the paired nerves n_{4+5} onto the dorsal surface of the body, connected with peripheral grooves on the surface which were probably touch sensory areas. The more anterior openings represented the positions of transpharyngeal eyes.

Inside the body a number of chambers can be recognized (Fig. 4a, b). These were arranged fundamentally as in cornutes, but with important modifications.

There was a small buccal cavity at the anterior end of the body, homologous with that of cornutes. Anterior to the right and left gill openings were paired atria which presumably had gill slits in their front walls, i.e., there were right gill slits as well as left ones.

The posterior coelom lay just anterior to the tail and the brain. There is evidence that the posterior coelom in the mitrate *Placocystites* was equivalent to the right and left epicardia of a tunicate, and evidence in *Mitrocystella* that it was indeed formed from two chambers. The same presumably applies to *Mitrocystites* also. The right epicardium is formed in tunicates as an outpouching from the right pharynx and the left epicardium as an outpouching from the left pharynx. Since the left pharynx existed in both mitrates and cornutes, while the right pharynx existed in mitrates alone, it follows that the posterior coelom of cornutes can be homologous only with the left epicardium of tunicates. It is equivalent, therefore, to only part of the posterior coelom of mitrates.

The rest of the body, between the buccal cavity anteriorly and the atria and posterior coelom posteriorly, was divided up in a complicated way. There is evidence of a left, more anterior, more ventral chamber separated from a right, more posterior, more dorsal one. The division between these chambers shows as an obvious oblique groove (mt in Fig. 4a) in the dorsal surface of the internal natural moulds. The anterior left chamber corresponds in gross position to the pharynx of cornutes such as *Cothurnocystis*, and the posterior right chamber to the right anterior coelom. The basic comparison that this suggests is almost certainly correct.

However, there is also evidence of a right pharynx extending out of the primary or left pharynx towards the posterior right corner of the body. It is likely that in the early ontogeny of mitrates the cavity of the right anterior coelom, with its contents, lay on the floor in the posterior right part of the body as in cornutes. Later the right pharynx pouched out from the left pharynx, underneath the cavity of the right anterior coelom, and so lifted the latter upwards, squashed it against the ceiling of the body and pushed its contents leftwards into a median position. This suggested ontogeny explains the fact that the oblique groove—which represents the basic division between left pharynx and right anterior coelom—has clearly been affected in its course by the right pharynx and why the cavity of the right anterior coelom is dorsal in mitrates but ventral in cornutes.

The way in which the right pharynx appeared in ontogeny, deduced from the observed geometry of the parts, seems to have been parallel to its mode of appearance in phylogeny. Thus the mitrate situation shown in section in Fig. 4b can be obtained from the cornute situation (Fig. 4c) by expansion of the right pharynx from out of the left pharynx, underneath the right anterior coelom, starting from the point marked x. The origin of the right pharynx and right gill slits is an all-or-nothing process which could only have happened between one generation and the next. The first mitrate, defined as the first chordate with right gill slits and right pharynx, would have been "hopeful monster" and the child of cornute parents.

The left anterior coelom, clothing the dorsal surface of the left pharynx, can be postulated in mitrates on the same grounds as in cornutes. The course of the non-pharyngeal gut, within the right anterior coelom, as shown in Fig. 4a, is deduced from the condition in *Mitrocystella* and *Placocystites*.

There are sometimes longitudinal striations on the internal mould of the right pharynx in *Mitrocystites*. These are very significant for a comparison with the tunicate pharynx as discussed below in dealing with *Mitrocystella* and *Placocystites*. Another important feature in this respect is the short mid-dorsal process of the skeleton extending backwards and rightwards from the oblique ridge (i.e., from the skeleton filling the oblique groove) near the mid-line at the junction of left and right pharynxes.

The tail of *Mitrocystites* consists of fore, mid and hind portions. As already suggested it is likely that the whole of this tail is equivalent to only the fore tail of cornutes. The skeleton of the fore tail was rather loose in *Mitrocystites*, being made up of six imbricating rings of four plates each. The lumen of the fore tail presumably contained muscles and an anticompressional notochord.

The hind tail had dorsal ossicles and paired ventral plates. The ossicles were hinged to each other so that they could flex ventrally. In this flexed position the posterior, flat surfaces of the ossicles would be directed forwards and could be used as bearing surfaces to push forwards against the mud and pull the body backwards (Fig. 15). Cuesta-shaped, sediment-gripping ribs on the body

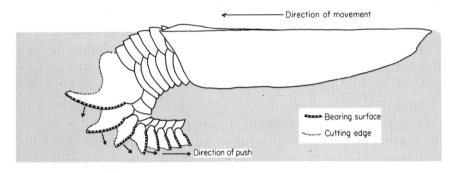

FIG. 15. *Mitrocystites*. Suggested mode of action of the tail in crawling.

(cu rib in Fig. 14b) whose steep slopes face anteriorly, and the direction of imbrication of the belly plates, would have aided backward movement of this sort.

The internal ventral surface of the hind-tail ossicles had a longitudinal median groove. This appears to have housed the dorsal surface of the notochord—in the form of a broad cylindrical organ—with a strap-like dorsal nerve cord on its back (Fig. 16). The notochord in the hind tail was a posterior extension of the notochord postulated for functional reasons in the fore tail. The dorsal nerve cord was connected in each segment with a pair of ganglia, housed in pits in the dorsal ossicles. The main lumen of the hind tail would carry muscles, presumably segmented, serving to flex the hind tail ventrally. These ventral muscles would be opposed by dorsal muscles and ligaments situated in cavities, between the ossicles, dorsal to the interossicular articulations. A blood supply to these dorsal muscles would probably be carried dorsalwards by the interossicular canals passing through the nerve cord, and there was also a dorsal longitudinal canal, presumably also carrying a blood vessel, running backwards through the series

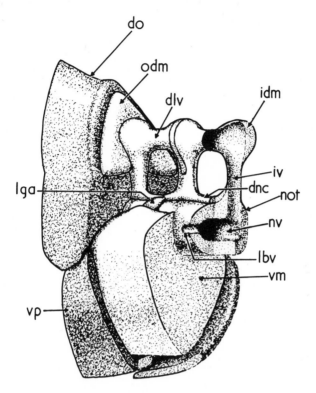

FIG. 16. *Mitrocystites.* Anatomy of a portion of the hind tail. Note the evidence, based on the sculpture of the ventral surfaces of the dorsal ossicles, for the notochord (not), the dorsal nerve cord (dnc) and the lateral ganglia (lga). The dorsal blood system, with the interossicular vessels (iv) and dorsal longitudinal vessel (d lv), is a specialization peculiar to *Mitrocystites* and *Mitrocystella* among mitrates. Other structures: do = dorsal ossicle; idm = inner dorsal muscle; lbv = lateral blood vessel; nv = notochordal vessel; odm = outer dorsal muscle, or perhaps ligament; vm = ventral muscle; vp = ventral plate. Reproduced with permission of the Trustees of the British Museum (Natural History).

of dorsal ossicles. This dorsal blood system was a specialized feature of *Mitrocystites* and *Mitrocystella*. In primitive mitrates it was absent.

The mid tail included a massive dorsal element, the styloid, serially homologous to two hind-tail ossicles. The anterior surface of the styloid was deeply excavated. It would have served as a socket, by which the muscles of the tail would move the mid and hind tail as a unit. It was analogous, but not homologous, with the stylocone of cornutes.

The brain and cranial nerves of *Mitrocystites* were exceedingly complicated. The solid lines in the reconstruction shown in Fig. 17

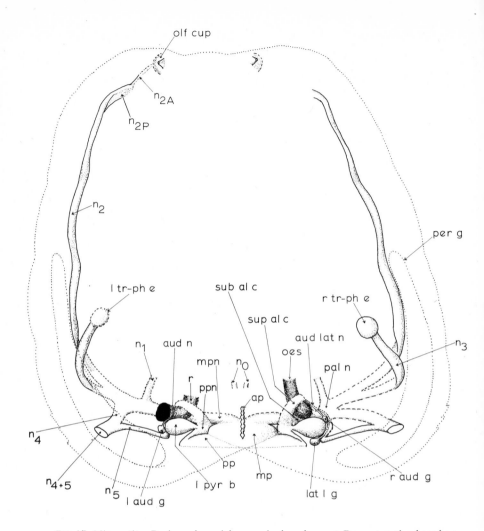

FIG. 17. *Mitrocystites*. Brain and cranial nerves in dorsal aspect. Reconstruction based on direct evidence from *Mitrocystites* (continuous lines) and also on indirect evidence (broken lines), partly derived from *Mitrocystella* and *Placocystites*: ap = anterior part of brain (olfactory bulbs); aud lat n = common auditory and lateral-line nerve of right side; aud n = auditory nerve; lat l g = lateral-line ganglion; l aud g = left auditory ganglion; l pyr b = left pyriform body (left trigeminal ganglion); l tr-phe = left transpharyngeal eye; mp = medial part of brain (diencephalon + optic lobes); mpn = medial part nerves (bases of optic nerves); n_0 = nerves n_0 (endostylar nerves); n_1 = first nerves of palmar complex (mandibular trigeminal); n_2 = second nerves of palmar complex (maxillary trigeminal) divided anteriorly, on the left, into n_{2A} and n_{2P}; n_3 = third nerves of palmar complex (transpharyngeal optic nerves); n_4, $n_{4\&5}$, n_5 = fourth and fifth nerves of palmar complex to dorsal surface; oes = oesophagus; olf cup = olfactory cup, in buccal cavity; pal n = palmar nerve; per g = peripheral groove; pp = posterior part of brain = (medulla oblongata); ppn = posterior part nerves (nerves from medulla oblongata); r = rectum; r aud g = right auditory ganglion; sub al c = subalimentary component of palmar nerve, going beneath rectum or oesophagus; sup al c = supraalimentary component of palmar nerve, going over rectum or oesophagus; r tr − ph e = right transpharyngeal eye.

are based on *Mitrocystites* itself. The dashed lines are largely by analogy with *Placocystites* and *Mitrocystella*. There is not space in this summary to justify the reconstruction in detail nor to do more than state, without detailed argument, the likely homologies with fishes.

The brain was divided into anterior, medial and posterior parts. The anterior part filled some small cavities in the skeleton near the posterior end of the median dorsal suture. It probably received fibres travelling through the skeleton from paired olfactory openings in the buccal cavity.The anterior part of the brain would represent the olfactory bulbs of fishes which likewise are the organs where olfactory fibres come together.

The medial part of the brain was inflated. A large medial-part foramen connected it with the body cavity and gave passage to medial part nerves. There was also a slit in the skeleton in the mid-line ventral to the medial part of the brain. The medial part of the brain was probably optic dorsally and hypophyseal ventrally, and corresponded to the diencephalon *plus* optic lobes of a fish. The suggested optic function is confirmed by the fact that the medial brain was more inflated in *Mitrocystites* and *Chinianocarpos*, which had external, transpharyngeal eyes, than in *Mitrocystella* or *Placocystites* where external eyes were lacking. The median slit beneath the medial brain probably connected its ventral hypophyseal portion with the neural gland or hypophysis.

The posterior part of the brain probably corresponds to the medulla oblongata of fishes. Posterior part nerves (= medullary nerves such as the trigeminal, facial etc) went off from it ventrally and admedian.

Turning to the cranial nervous system in general, just anterior to the posterior part of the brain were big pyriform bodies (trigeminal ganglia) which on their admedian sides were connected with the posterior-part nerves. The most prominent parts of the cranial nervous system were the two big palmar complexes, so called because each was arranged rather like a hand, with a central palmar nerve giving rise to five "fingers", i.e. the nerves n_1, n_2, n_3, n_4, and n_5. There is now evidence, mainly drawn from *Placocystites* and *Mitrocystella*, that the palmar nerves were formed by components going both over and under the oesophagus, on the right, and over and under the rectum, on the left. I refer to these as the supra- and sub-alimentary components. Evidence from *Placocystites* suggests that the sub-alimentary component was connected with the pyriform body (trigeminal ganglion), whereas the supra-alimentary component was not. This suggests, in turn, that the supra-alimentary component was probably purely motor, while the sub-alimentary component was sensory, and the trigeminal ganglia contained the cell bodies of sensory nerves as in fishes. The medial part or optic nerves of *Mitrocystites*, as judged by grooves in the skeleton, also passed under the rectum or the oesophagus and formed part of the sub-alimentary component of the palmar nerves.

The peripheral branches of the palmar nerve were: n_1, to the ventral wall of the body—this probably corresponds to the mandibular branch of the trigeminal of agnathan fishes; n_2, to the buccal cavity—this probably corresponds to the maxillary branch of the trigeminal; n_3, which ascends to the dorsal surface of the body and ends there in a bulb—this was probably an optic nerve of the type which I now call transpharyngeal; and n_4 and n_5 which join together and run to the peripheral grooves on the dorsal surface of the body—these were probably touch-sensory, and perhaps correspond to the dorsal branches of the trigeminal.

Transpharyngeal eyes are not the only types of paired eyes in mitrates. *Peltocystis*, *Mitrocystella* and *Placocystites* yield evidence of cispharyngeal eyes climbing up directly from the medial-part foramen and ending in bulbs in the dorsal part of the body cavity, separated from the outside world by presumably partly transparent calcite. It is possible that cispharyngeal eyes existed in *Mitrocystites* also. Contrary to what I believed previously, the transpharyngeal eyes, on the far side of the pharynx from the brain, cannot be homologous with the paired eyes of vertebrates. The cispharyngeal eyes, however, could well be so (Jefferies & Prokop, 1972: 90; Jefferies, 1973: 451).

The acustico-lateralis system of *Mitrocystites* is very complicated. There is evidence of a left auditory ganglion in the left atrium supplied by a nerve round the left pyriform body. This left auditory system is homologous with the lateral-line system of *Ceratocystis* which presumably existed in some form in *Cothurnocystis* and *Reticulocarpos* as well. On the right side *Mitrocystites* had a lateral line, with its ganglion and supplying nerve, and probably a right auditory ganglion also, by analogy with *Placocystites* and *Mitrocystella*. The mitrate *Peltocystis* is important with regard to the acoustic system. It had nerves going direct from the dorso-lateral angles of the posterior part of the brain to the right and left atria. The parts of the brain from which these nerves arise would be acustico-lateralis centres by analogy with fishes. And *Peltocystis* had no lateral line, so these nerves would supply auditory ganglia in both atria. The right acustico-lateralis nerves in *Mitrocystites* probably reached the ganglia after passing round the front of the oesophagus from the ventral to the dorsal side, i.e. in the opposite sense to the supra-alimentary component of the palmar nerve.

The paired nerves n_0, whose central connections are obscure, probably supplied the endostyle. This statement is based on evidence drawn from *Placocystites*.

The relationship of the cranial nerves to the oesophagus and rectum of *Mitrocystites* is evidently different to that in a fish. This is not surprising, since no modern fish has the brain posterior to the whole of the gut. Embryological experiment has shown that nerves will grow out from the central nervous system to their correct end organs, finding their way round obstacles artificially put in their path (Balinsky, 1960: 426). Accordingly I think that, in deciding homology with fishes, the central relations and end organs of the nerves of mitrates are relevant, but their intermediate courses are not.

MITROCYSTELLA

I describe this genus on the basis of *Mitrocystella incipiens* Barrande *miloni* Chauvel from the Middle Ordovician (Llandeilo) of Brittany. *Mitrocystella* is closely related to *Mitrocystites*, so it is not necessary to describe both in the same detail.

The body is more elongate than in *Mitrocystites* (Fig. 18a, b). There were extensive, transverse, cuesta-like ribs across the posterior ventral surface. The steeper faces of these ribs were anterior, as always in mitrates, and would have resisted forward movement.

The body openings consist of a mouth, with a flexible lower lip, right and left gill openings and the lateral line, just right of the tail. There are no openings on the dorsal surface of the body.

Inside the body the same chambers can be recognized as in *Mitrocystites*, but with some differences (Fig. 19a, b). The posterior coelom is, in some specimens, clearly divided into two chambers which, as shown by the situation in *Placocystites*, represent the right and left epicardia of tunicates. The left and right atria are much as in *Mitrocystites* except that the course of the rectum, passing out of the posterior coelom to open into the left atrium, is easier to follow. The position of the rectum can be compared on the one hand with that of *Cothurnocystis* and on the other hand, in opening into the left atrium, with that of a modern tunicate tadpole. The buccal cavity was a weakly defined chamber like that of *Mitrocystites*. The rest of the internal cavity of body, as in all mitrates, was taken up by the left

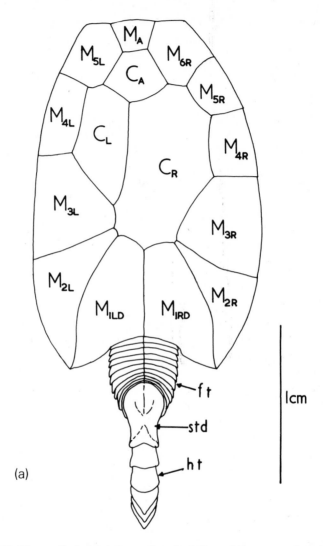

FIG. 18. *Mitrocystella incipiens* Barrande *miloni* Chauvel. Reconstruction of a) dorsal, b) ventral aspect: bo = branchial opening (paired); cu rib = cuesta-shaped rib; ft = fore tail; ht = hind tail; mo = mouth; ng = narrow groove (lateral line); or = oral plates; std = styloid; vp = ventral plates of hind tail; C_A etc. = centro-dorsal plates; M_A etc. = marginal plates; V_{PL} etc. = ventral plates. Reproduced with permission of the British Museum (Natural History).

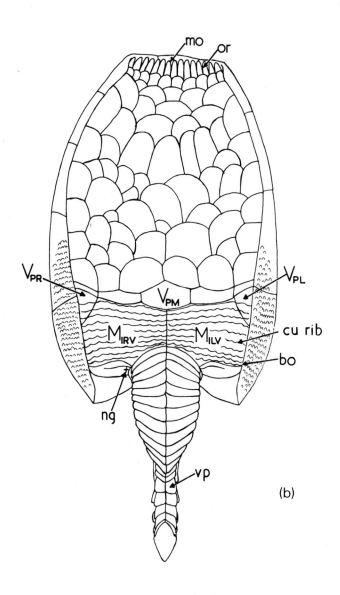

(b)

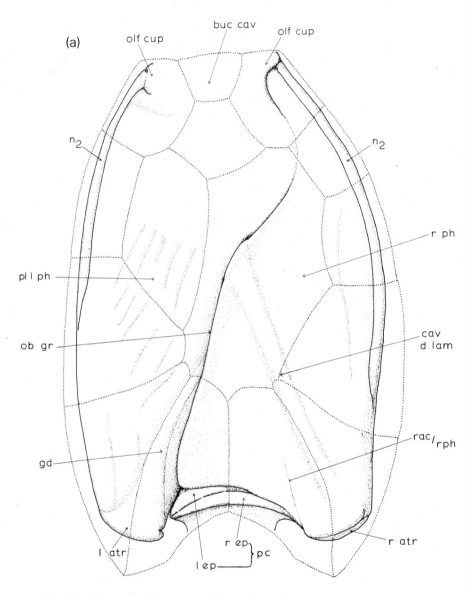

FIG. 19. *Mitrocystella*. Reconstructed internal mould, based on many different specimens, representing the soft parts of the body in positive. (a) Dorsal and (b) posterior aspects:
b l pyr b, b r pyr b = Broken-off base of left and right pyriform bodies; buc cav = buccal
cavity; cav d lam = cavity of dorsal lamina; gd = gonoduct; hyp cer p = hypocerebral
process; lat l n = lateral line nerve; l atr = left atrium; l aud g = left auditory ganglion; l
cisph op n = left cispharyngeal optic nerve; l ep = left epicardium; l pal n = left palmar
nerve; n_2 = nerve n_2 of palmar complex; ob gr = oblique groove; olf cup = olfactory cup;
pc = posterior coelom, made up of left and right epicardia (l ep, r ep); pl l ph = pleated left

pharynx, the right anterior coelom, the right pharynx and a presumed, virtual, left anterior coelom. The division between the left pharynx and the right anterior coelom is an oblique groove running across the dorsal surface of the internal mould from anterior right to posterior left as *Mitrocystites*.

The sculpture of the internal mould of the left pharynx shows two peculiarities (Fig. 19a). The first of these is a low ridge gd which leaves the right anterior coelom and passes back into the left atrium, widening as it does so. This ridge probably represents the course of the gonoduct. In *Cothurnocystis* this duct came out of the right anterior coelom along with the rectum, while its terminal portion followed the rectum to end left of the anus. In *Mitrocystella*, therefore, the origin and termination of the duct retained their primitive positions, but the intermediate course had straightened out, so that the gonoduct had come to be separated from the rectum.

A more important peculiarity of the left pharynx is a series of striae, roughly parallel to the length of the left pharynx, in the part marked pl 1 ph in Fig. 19a. These striae terminate anteriorly as if against a precise boundary. They are therefore absent in the anterior portion of the left pharynx. They are comparable with the striae sometimes observed in the right pharynx of *Mitrocystites*. The pharyngeal wall was probably longitudinally pleated where the

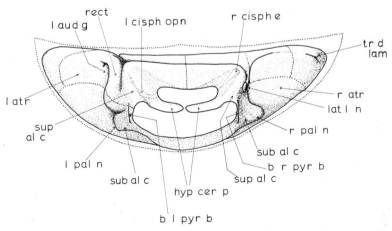

pharynx; r atr = right atrium; rac/r ph = boundary between residual right anterior coelom and right pharynx; r cisph e = right cispharyngeal eye; rect = rectum; r ep = right epicardium; r ph = right pharynx; r pal n = right palmar nerve; sub al c, c = sub- and supra-alimentary components of palmar nerve; tr d lam = posterior trace of dorsal lamina.

striae exist. There was probably a functional distinction between:
1) on the one hand the right pharynx observed to be striated in
Mitrocystites, together with the striated posterior part of the left
pharynx as seen in *Mitrocystella*; as against 2) the always unstriated,
anterior part of the left pharynx.

This distinction can be interpreted by comparison with the
pharynx of tunicates. These feed by means of a mucous trap
described more fully by Carlisle & Carlisle (in press) and Werner &
Werner (1954). Mucus is secreted in tunicates by the endostyle in
the ventral mid-line of the pharynx (Fig. 20a, b, c). The anterior
ends of the sheets or strands of mucus are held and carried dorsally
by the peripharyngeal bands which pass from the anterior end of
the endostyle up to the dorsal mid-line of the pharynx (Fig. 20a, c).

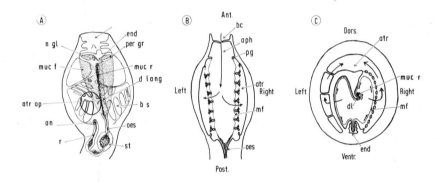

FIG. 20. Filter feeding in the living tunicate *Clavellina* (redrawn after Werner & Werner,
1954). The drawings represent the mechanism used under conditions of fast feeding, with
many particles in the water, when continuous mucous sheets are produced. In slow feeding
strands of mucus are produced instead (Carlisle & Carlisle, in press). A) sketch of animal
feeding; B) longitudinal section through pharynx; C) transverse section through pharynx:
an = anus; a ph = anterior part of pharynx; atr = atrium; atr op = atrial opening;
bc = buccal cavity; bs = branchial slit; d lang, d l = dorsal languet; end = endostyle; muc f,
mf = mucous filter; muc r = mucous rope; n gl = neural gland; oes = oesophagus; per gr,
pg = peripharyngeal grooves; r = rectum; st = stomach.

In the dorsal mid-line the strands or sheets are rolled up into a
mucous rope by one or more hooks called dorsal languets or a fold
of tissue called the dorsal lamina. The mucous rope is pulled
backwards into the opening of the oesophagus by oesophageal
cilia. There is therefore a functional distinction between the
pharynx anterior to the peripharyngeal bands, which is not involved in the transport of mucus, and the pharynx posterior to
those bands, which is so involved. Applying this distinction to

Mitrocystella and *Mitrocystites* it is likely that the left peripharyngeal band ran along the anterior margin of the left pharyngeal striae, while the right peripharyngeal band ran backwards along the anterior part of the oblique groove. The parts of the pharynx sometimes observed to be striated would thus be posterior to the peripharyngeal bands and involved in the transport of mucus, while the smooth, anterior part of the left pharynx would not be involved in it. It follows that the left peripharyngeal band of mitrates was not originally symmetrical to the right one although both presumably had the same function. And it is interesting that in many tunicates the left peripharyngeal band appears earlier than the right one in ontogeny (D. B. Carlisle, personal communication). The mucous filter trap occurs inside the pharynx of amphioxus, of tunicates and of the ammocoete larva among vertebrates. This in itself suggests that it would have existed in the latest common ancestor of cephalochordates, tunicates and vertebrates, which would have been a mitrate.

The right pharynx was limited by a weak line on the internal mould (rac/ph in Fig. 19a). More obvious than this weak line is a straight gentle ridge passing backwards from the oblique groove at the mid line to the posterior right corner of the body (cav d lam in Fig. 19a). This ridge probably overlay a dorsal lamina, i.e. a fold of tissue involved in rolling up the mucous rope. There is more definite evidence of such a dorsal lamina in *Placocystites*. Its condition in *Mitrocystella*, apparently attached by its posterior end to the inside of the body wall, was probably not primitive.

The left anterior coelom would probably have clothed the left pharynx as in *Mitrocystites*. The oblique groove, on this assumption, would represent the meeting place of the left and right anterior coeloms and would be a mesenteric trace.

The tail of *Mitrocystella* was like that of *Mitrocystites* in all important respects.

The cranial nerves of *Mitrocystella* (Fig. 21) were very similar to those of *Mitrocystites* (Fig. 17) but buried to a greater extent in the skeleton, which makes reconstruction easier, as indicated by the hard lines in the figure. Particularly noteworthy is the optic system. The transpharyngeal optic nerves (n_3) do not reach the dorsal surface but stop short inside the body. There is also direct evidence of cispharyngeal optic nerves going directly upwards from the medial part foramen. Nerves n_4 and n_5 did not issue onto the dorsal surface but joined a presumably touch-sensory peripheral canal just below the dorsal surface inside the skeleton.

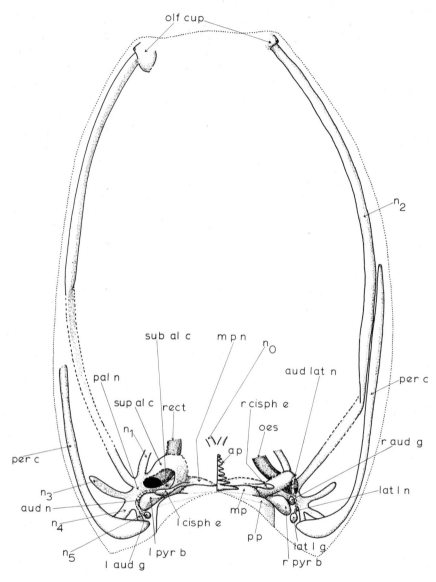

FIG. 21. *Mitrocystella*. Reconstruction of brain and cranial nerves in dorsal aspect. Only right half of anterior brain (ap) is shown. Legend as for Fig. 17, with the following additions: l cisph e = left cispharyngeal eye; lat l n = lateral-line nerve; per c = peripheral canal; r cisph e = right cispharyngeal eye; rect = rectum; r pyr b = right pyriform body.

PLACOCYSTITES

I describe *Placocystites* on the basis of *Placocystites forbesianus* de Koninck from the Middle Silurian near Dudley, Worcs. A fuller account of this species will be published elsewhere. *Placocystites* is preserved differently to the other forms discussed here, for it still retains the calcite plates. To study its internal structure it was therefore necessary to make two enlarged polystyrene models based on serial sections. These were constructed in the British Museum (Natural History) by Mr D. N. Lewis.

The body of *Placocystites* (Fig. 22a, b) is, in its main features, very like that of *Mitrocystella* which in most ways represents an ancestral condition. The most striking externally visible peculiarities of *Placocystites*, distinguishing it from *Mitrocystella*, are the presence of cuesta-like ribs on the dorsal as well as the ventral surface, the rigidity of the belly armour, the transformation of two marginal plates to form blade-like oral spines right and left of the mouth, and the reduction in size of the posterior surface of the body. These differences suggest that *Placocystites* pulled itself backwards by its tail just beneath the surface of the sea bed, instead of on the surface like *Mitrocystella*. As a result mud rode up over the ribbed areas of the dorsal surface of the body. The fore tail of *Placocystites* is adapted for very vigorous ventral flexion by having enormous, plated, dorsal imbrication membranes. It would probably have been a more powerful motor than the fore tail of *Mitrocystella*, as an adaptation to a shallow-burrowing mode of life.

The interior of the body of *Placocystites* is highly informative (Figs 23, 24, 25, 26). The buccal cavity existed anteriorly, as in all mitrates. Well developed olfactory cups opened into it dorsally, and its posterior border is defined by a weak groove in the dorsal skeleton (Fig. 23).

There is an oblique ridge crossing the internal surface of the dorsal skeleton from anterior right to posterior left. This would correspond to an oblique groove in an internal mould, like the oblique groove in the mitrates already discussed. Near where the oblique ridge crosses the mid-line of the body a prominent mid-dorsal process extends backwards towards the right corner of the body. This mid-dorsal process is like that of *Mitrocystites* and *Mitrocystella*, but much longer. By analogy with these two genera the left and right peripharyngeal bands would have met each other dorsally near to the front end of the mid-dorsal process. The right peripharyngeal band would have followed that portion of the oblique ridge anterior to the mid-dorsal process. The left

298   R. P. S. JEFFERIES

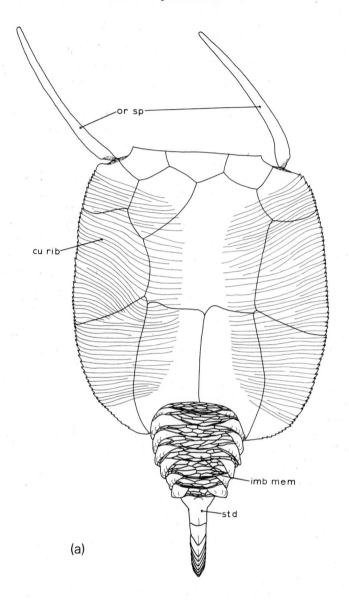

(a)

FIG. 22. *Placocystites forbesianus* de Koninck, Middle Silurian, nr. Dudley, England. Reconstruction of a) dorsal, b) ventral aspects: bo = branchial opening; cu rib = cuesta-shaped rib; ft = fore tail; ht = hind tail; imb mem = dorsal imbrication membranes of fore tail; or int = oral integument (omitted in Fig. 22a); or sp = oral spines; p surf = posterior surface of body; std = styloid.

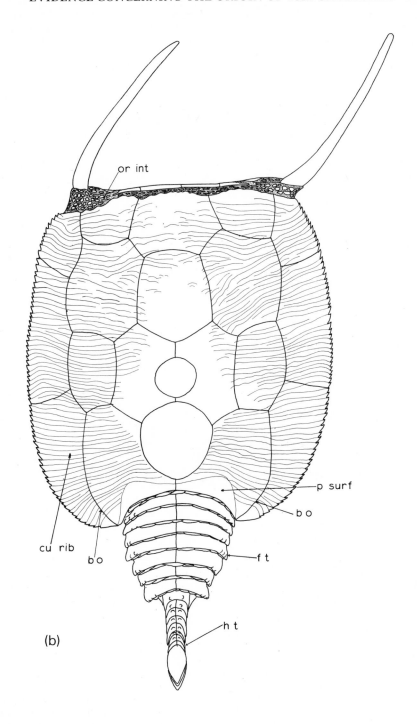

(b)

FIG. 23. First polystyrene model of *Placocystites*; ventral aspect of dorsal armour. a m pl l
ph = anterior margin of pleated left pharynx; hyp pr = hypocerebral process; hyp
sl = hypophyseal slit, connecting brain with hypophysis beneath it; m d pr = mid-dorsal
process; m p f = medial part foramen, anterior to brain; ob r = oblique ridge; op
hyp = opening of hypophysis into the pharynx; pm buc cav = posterior margin of buccal
cavity; r ac/r ph = boundary between right anterior coelom and right pharynx; rect
pr = process supporting rectum; r olf cup = right olfactory cup; r ph = right pharynx.

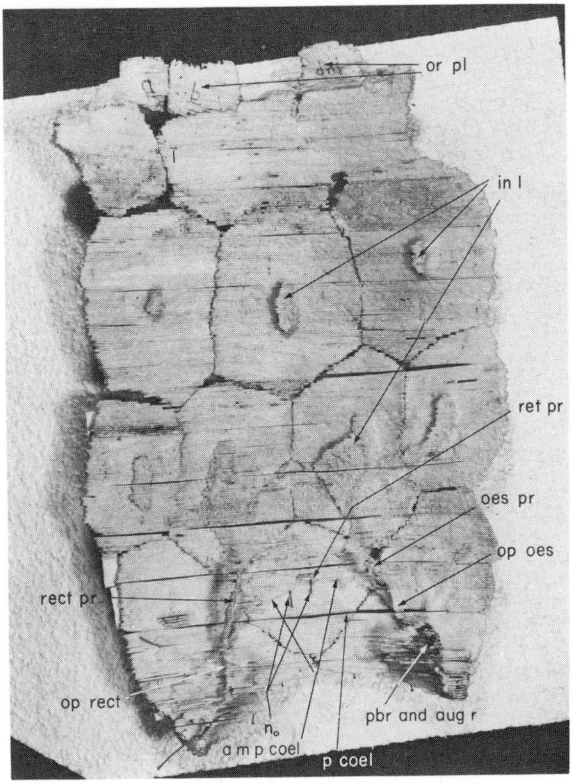

FIG. 24. First polystyrene model of *Placocystites*; dorsal aspect of ventral armour; a m p coel = anterior margin of posterior coelom; in l = inner calcite layer of ventral skeleton − the nerve canals run between the inner and outer calcite layers; n_0 = pieces of string passing into canals of endostylar nerves (n_0); oes pr = process supporting oesophagus; op oes = opening of oesophagus into pharynx; op rect = opening for rectum in wall of posterior coelom; or pl = oral plates; pbr & aug r = pit for right pyriform body and auditory ganglion; p coel = posterior coelom; rect pr = process supporting rectum; ret pr = retropharyngeal process.

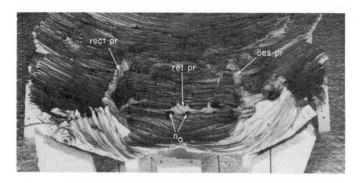

FIG. 25. First polystyrene model of *Placocystites*. Postero-dorsal aspect of posterior coelom. Note the canals for the endostylar nerves (n_0) and the retropharyngeal process (ret pr) positioned asymmetrically more over the right than the left n_0. Other structures: oes pr = process supporting the oesophagus; rect pr = process supporting the rectum.

FIG. 26. Second polystyrene model of *Placocystites*. Postero-dorsal aspect of the retropharyngeal process (ret pr). Note how the process leans rightwards, i.e., towards the opening of the oesophagus into the pharynx, and also the nicks in its base representing the pharyngo-epicardial openings (ph ep op): a m p coel = anterior margin of posterior coelom.

peripharyngeal band would have crossed the left pharynx transversely.

The left boundary of the right pharynx, where it came in contact with the dorsal skeleton, is represented by a low skeletal ridge. Throughout its length this ridge is situated some distance leftward of the dorsal process. If an attempt is made to reconstruct the chambers inside *Placocystites*, therefore, as in the transverse section in Fig. 27, the result is a fold of tissue, hook-shaped in section and curled rightwards and downwards, with the mid-dorsal process stiffening its free edge.

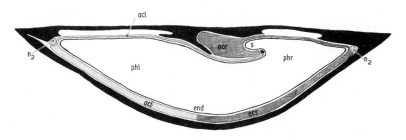

FIG. 27. Transverse section through *Placocystites*, with reconstructed soft parts, skeleton in black. Note the hook-shaped dorsal lamina, pointing downwards and rightwards, indicated by the small mid-dorsal process and associated structures: acl, acr = left and right anterior coeloms; end = position of endostyle; n_2 = nerve n_2; phl, phr = left and right pharynx; s = position of mucous rope, supported by dorsal lamina.

This fold of tissue was almost certainly a dorsal lamina. This is shown by its position at the dorsal meeting-place of right and left pharynxes and extending posteriorly from the presumed dorsal meeting-place of right and left peripharyngeal bands. The identification is strongly confined by the shape of the fold of tissue, curling rightwards and downwards like the dorsal lamina of all recent tunicates which possess one, and like the dorsal languets of those tunicates which have no lamina. The function of the dorsal lamina in living tunicates, as already mentioned under *Mitrocystella*, is to roll up the mucous rope (s in Fig. 27) and transport it backwards to the oesophagus. The hook-shaped section of the dorsal lamina in tunicates is adapted to rotate the rope anticlockwise when seen from behind (Werner & Werner, 1954: 79). The sense of rotation was presumably the same in mitrates, as it also is in amphioxus (Barrington, 1938: 302).

There is a rounded notch in the oblique ridge just left of the front end of the mid-dorsal process (op hyp in Fig. 23). By analogy with tunicates this represents the place where the ciliated organ connected with the pharynx.

The place where the oesophagus opened into the pharynx is indicated by a semicircular notch, with a rounded edge, in the right wall of the posterior coelom (oes op in Fig. 24). This notch is suitably situated to receive the mucous rope from the posterior end of the dorsal lamina. Its position well right of the mid-line is comparable with that of the oesophageal opening in salps.

A pair of nerves n_0 are indicated, in the first of the two polystyrene models of *Placocystites*, by two canals which leave the posterior coelom near the mid-line and pass forwards inside the ventral skeleton (Fig. 25). These nerves probably supplied the endostyle, since this would be the most important organ in the mid-ventral line of the pharynx. This same polystyrene model has a u-shaped process in the front wall of the posterior coelom, situated asymmetrically over the right canal n_0. I shall call this the retropharyngeal process.

The canals for n_0 do not show in the second polystyrene model, presumably having been filled with calcite after the animal's death. The retropharyngeal process in the second model is a complicated blade, fixed to the floor of the body in the same place as the retropharyngeal process of the first model, but much longer, and curved strongly rightwards (Fig. 26). There are two notches on each side of the blade near its base.

Combining information from both models, and assuming that the nerves n_0 supplied symmetrical right and left halves of the endostyle, then the retropharyngeal process was rooted near the posterior right corner of the endostyle and curved rightward toward the assumed position of the oesophageal opening. This suggests that the retropharyngeal process carried the retropharyngeal band on its front surface, for in that case the relations of the band would be exactly as in tunicates, where it is a posterior extension, leading to the oesophagus, of the right marginal band of the endostyle.

The pair of notches at the base of the retropharyngeal process in the second model (ph ep op in Fig. 26) probably represent openings between the posterior coelom and the right and left pharynx. These openings correspond to the pharyngoepicardial openings of *Ciona* which are likewise situated on either side of the retropharyngeal band and connect the right pharynx with the right epicardium, and the left pharynx with the left epicardium. In turn this indicates that the posterior coelom of *Placocystites* is homologous, as already mentioned, with the right and left epicardia of a tunicate.

The left and right atria of *Placocystites* were probably situated much as in *Mitrocystites* or *Mitrocystella*. The rectum would have opened into the left atrium. There is a notch in the left wall of the posterior coelom in *Placocystites* which indicates where the rectum passed out of the posterior coelom towards the left atrium.

The more anterior course of the oesophagus and rectum is indicated by the supra-alimentary components of the palmar nerves whose courses can be inferred from the skeleton. These components passed over the oesophagus, on the right, and the rectum on the left, in a way which implies that the oesophagus ran forwards into the right anterior coelom, and the rectum ran backwards out of it.

A reconstruction of the internal organs of the body of *Placocystites* is shown in Fig. 28. This drawing assumes that the asymmetrical condition of the peripharyngeal bands seen in certain salps is probably primitive. In these animals the endostyle has a right marginal band but no left marginal band. The right peripharyngeal band is an anterior extension of the right marginal band, and the retropharyngeal band is a posterior extension of it, while the left peripharyngeal band is a structure on its own (Garstang & Platt, 1928). The reconstruction does not attempt to show the division of the posterior coelom into right and left epicardia, nor the contents of the right anterior coelom apart from the non-pharyngeal gut. The right anterior coelom would in addition have contained the heart, the gonads, and the duct of the neural gland or hypophysis. The heart would be anterior to the gill slits as in an embryo myxinoid.

The brain and cranial nerves of *Placocystites* were very like those of *Mitrocystella* and will be described in detail elsewhere.

LAGYNOCYSTIS

I describe *Lagynocystis* on the basis of the only known species, which is *Lagynocystis pyramidalis* Barrande from the Lower Ordovician of Bohemia. *Lagynocystis* is important because it is probably closely related to the ancestry of amphioxus. I have described it in detail in Jefferies (1973).

The body of *Lagynocystis* was very elongate for a mitrate (Fig. 29a, b). It also showed a number of external asymmetries which were, so far as can be worked out, absent in the most primitive mitrate. These external asymmetries included a reduction in the right peripheral flange and the development of an anterior appendage left of the mid line. They would have the effect, if *Lagynocystis* swam forwards, of making the body rotate clockwise as seen from

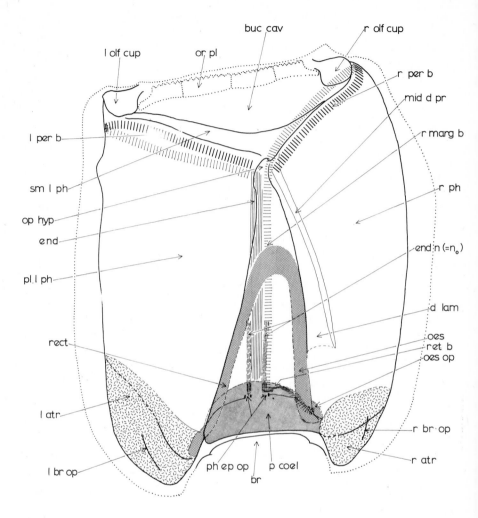

FIG. 28. *Placocystites*. Reconstruction of soft parts of body in dorsal aspect: br = brain; buc cav = buccal cavity; d lam = dorsal lamina; end = endostyle; end n (= n_0) = endostylar nerves (n_0); l atr = left atrium; l br op = left branchial opening; l olf cup = left olfactory cup; l per b = left peripharyngeal band; mid d pr = mid dorsal process; oes = oesophagus; oes op = oesophageal opening; op hyp = opening of hypophysis into pharynx; or pl = oral plates; p coel = posterior coelom, i.e. left & right epicardia; ph ep op = pharyngo-epicardial openings; pl l ph = pleated left pharynx; r atr = right atrium; r br op = right branchial opening; rect = rectum; ret b = retropharyngeal band; r marg b = right marginal band; r olf cup = right olfactory cup; r per b = right peripharyngeal band; r ph = right pharynx; sm l ph = smooth left pharynx.

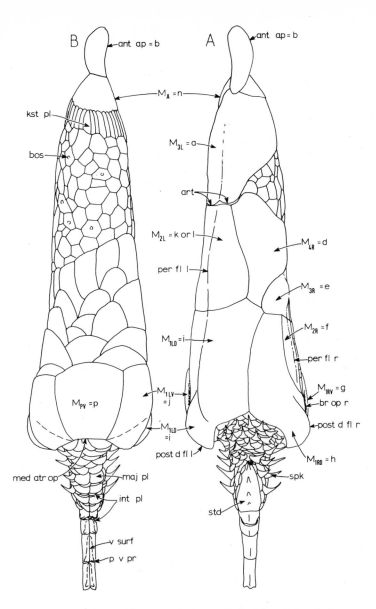

FIG. 29. *Lagynocystis pyramidalis* Barrande, Lower Ordovician, Bohemia. Reconstruction of A) dorsal and B) ventral aspects. Most of hind tail omitted: ant ap = b = anterior appendage; art = articulation posterior to anterior groups of plates; bos = bosses on plates beneath buccal cavity; br op r = right branchial opening; int pl = intercalary plate of fore, mid and proximal part of hind tail; kst pl = "keystone plate" (specialized oral); maj pl = major plate of fore tail; med atr op = median atrial opening; per fl l, r = right and left peripheral flanges; post d fl r, post d fl l = right and left postero-dorsal flanges; p v pr = postero-ventral process of hind-tail plate; spk = spike of fore tail; std = styloid; v surf = ventral surface of hind-tail plate; M_{PV} = p, M_A = n etc. = various marginal plates (two notations) (Reproduced with permission of the Royal Society of London).

behind, in the manner reported in larval amphioxus by Wickstead (1967).

The body openings included the mouth and right and left gill openings as in other mitrates and probably a median ventral gill opening as discussed below.

The interior of the body of *Lagynocystis* is aberrant among mitrates in a number of ways (Fig. 31a–c). Most important is the presence of a little symmetrical chamber in a posterior, ventral position. The front wall of this chamber is made up of about 40 calcite gill bars, separated by parallel-sided gill slits (Fig. 30). The

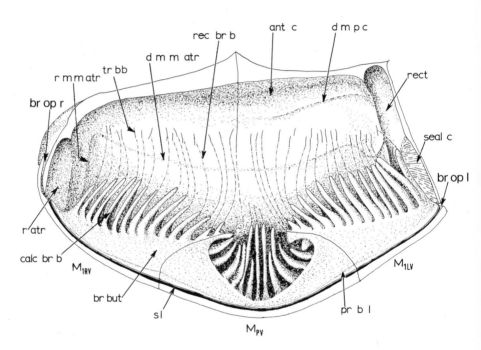

FIG. 30. *Lagynocystis.* Anterior aspect of the inside of the posterior part of the body, to show the calcitized branchial bars (calc br b) of the front wall of the median atrium. The soft dorsal portions of the gill bars were attached to the dorsal skeleton at a line of traces (tr bb), and this allows their soft parts to be reconstructed (rec br b). The dorsal and right margins of the median atrium (d m m atr, r m m atr) are also recorded on the skeleton. Note that the gill bars ended dorsally inside the posterior coelom (epicardia) whose dorsal margin is also visible (dmpc). Other structures: ant c = right anterior coelom; br but = branchial buttress; br op l, r = left and right branchial opening; pr b l = pre-branchial line; r atr = right atrium; rect = position of rectum; seal c = emplacement of cartilage sealing left branchial opening when the latter was shut; sl = soft layer in ventral skeleton. M_{1RV}, M_{PV}, M_{1LV} = posterior ventral marginal plates, as shown in Fig. 29b (Reproduced with permission of the Royal Society of London).

chamber does not make contact with either the right or the left gill opening (Fig. 31e). Furthermore the gill slits in its front wall converge backwards. The only way by which water could leave the chamber, therefore, was by means of a median ventral opening whose only possible position is just ventral to the tail. The chamber can therefore be interpreted as a median ventral atrium opening by a median ventral atriopore. The only other known animals with this arrangement of parts are living cephalochordates. There is evidence of a right atrium in *Lagynocystis*, as in other mitrates. A left atrium presumably also existed, but it must have been small, and there is no direct evidence of it.

The buccal cavity of *Lagynocystis* was very large for a mitrate. Its posterior margin, which was more posterior on the right than on the left, is defined by grooves on the internal cast and a change in the nature of the ventral integument.

The posterior coelom of *Lagynocystis* straddled the tops of the gill bars of the median atrium. This is important, for I have suggested above that the posterior coelom of mitrates is homologous with the epicardia of tunicates, and Berrill has argued that epicardia were primitively organs of excretion (Berrill, 1955: 103). The nephridia of amphioxus, each of which overlies a primary gill bar, could have arisen from the presumably excretory posterior coelom of *Lagynocystis* if the gill slits extended upwards to split the posterior coelom into one segment per gill bar.

The part of the body anterior to the posterior coelom and to the three atria, but posterior to the buccal cavity, was probably filled by a right and left pharynx and a right and left anterior coelom as in other mitrates. The oblique groove on the internal moulds is present but weak. It is likely that the right anterior coelom had been squashed upwards and forced outwards on both sides of the oblique groove by the growth of the ventral parts of the pharynx backwards to meet the median atrium. This lifting-up of the right anterior coelom would also lift up the proximal end of the rectum which seems actually to have run downwards towards the left gill opening. This upwards movement of the proximal part of the rectum also lifted the supra-alimentary component of the left palmar nerve and squashed it against the roof of the body.

A noteworthy asymmetry in amphioxus, interesting in this connection, is the position of the anus on the left side of the body.

The inferred order of appearance of the different groups of gill slits in *Lagynocystis* is interesting. The relations of the right pharynx to the oblique groove in all mitrates show that the left pharynx

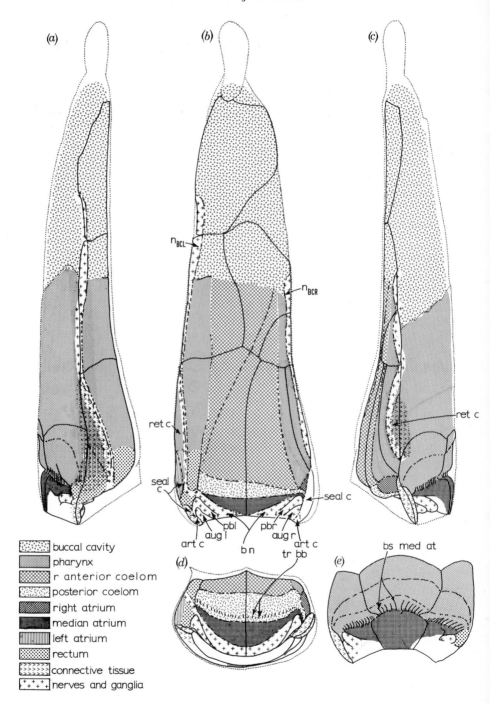

(a)

(b)

(c)

n_BCL

n_BCR

ret c

seal c

seal c

pbl

pbr

aug l

aug r

art c

b n

art c

tr bb

buccal cavity

pharynx

r anterior coelom

posterior coelom

right atrium

median atrium

left atrium

rectum

connective tissue

nerves and ganglia

(d)

(e)

bs med at

ret c

appeared earlier in ontogeny than the right pharynx. It is therefore likely that, in all mitrates, the gill slits leading into the left atrium appeared earlier in ontogeny than those leading into the right atrium. The relations of the gill slits of the median atrium to the growth lines of the plates in *Lagynocystis* show that this median group of gill slits was still being produced when the animal stopped growing or died. The likely order of appearance of the gill slits associated with the three different atria in the ontogeny of *Lagynocystis* is therefore: 1) left; 2) right; and 3) median. Similarly, in the observed ontogeny of amphioxus, the first gill slits to appear are morphologically left, then a group of right gill slits suddenly arises, and then further gill slits are added posterior to those already present (Willey, 1894).

The brain and cranial nerves of *Lagynocystis* were like those of other mitrates in basic plan, but modified, and simplified, by the presence of the median atrium (Fig. 32a, b). The brain was connected, as usual, to paired auditory and trigeminal ganglia. The medial part (optic) nerves of the brain had disappeared, together with the direct connection of the brain with the hypophysis. It is likely that nerves to the ventral wall of the body passed over the median atrium and down the gill bars in its front wall, much as in amphioxus. The nerves to the buccal cavity (n_{BCL} and n_{BCR}) are of special interest. These nerves correspond to the nerves n_2 of other mitrates, and probably to the maxillary trigeminal branches of fishes. The left nerve passes much further forward than the right one, which suggests that the mouth was predominantly innervated from the left. The same asymmetry recurs in amphioxus, where the right side of the oral hood is innervated from both the left and right sides, whereas the left side is innervated from the left only. The nervous system of *Lagynocystis*, although simple for a mitrate, was therefore much more complicated than that of amphioxus,

FIG. 31. *Lagynocystis.* Reconstruction of body chambers, based on natural internal moulds; a) left lateral, b) dorsal, c) right lateral, d) posterior, e) ventral aspect (of posterior part of body only). The oblique groove is visible as a line running from posterior left to anterior right in (b), but the right anterior coelom has been splayed out on both sides of it, probably because the ventral part of the pharynx extended backwards to meet the median atrium: art c = articular cartilage of right and left gill openings; aug l, r = left and right auditory ganglia; bs med at = branchial slits of median atrium; n_{BCL}, n_{BCR} = left and right nerves to buccal cavity; pbl, pbr = right and left pyriform bodies; ret c, seal c = retaining and sealing cartilages associated with right and left branchial openings; bn = nerves to body, from brain; tr bb = dorsal traces of branchial bars (Reproduced with permission of the Royal Society of London).

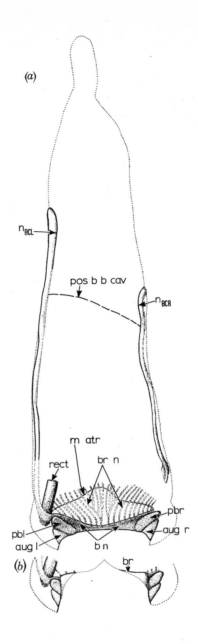

FIG. 32. *Lagynocystis*. Reconstruction of cranial nerves in dorsal aspect with brain (br) included in b, but omitted in a. Note that the left nerve to the buccal cavity (n_{BCL}) extends farther forward than the right one (n_{BCR}). The posterior boundary of the buccal cavity = poss b b cav. Other structures: aug l, aug r = left and right auditory ganglia; bn = body nerves, going directly out of brain; br n = branchial nerves, passing down the gill bars of the median atrium; m atr = median atrium; pbl, pbr = left and right pyriform bodies (trigeminal ganglia); rect = rectum (Reproduced with permission of the Royal Society of London).

especially in having a well developed brain with paired auditory and trigeminal ganglia.

The tail of *Lagynocystis* was regionated into fore, mid and hind parts as with other known mitrates. In the hind tail (Fig. 33) there is

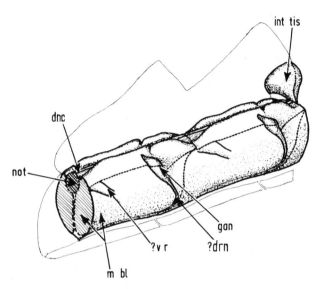

Fig. 33. *Lagynocystis*. Reconstructed anatomy of soft parts of approximately two segments of the hind tail. Skeleton imagined as transparent. Hard lines are based on direct skeletal evidence. Beaded lines indicate contacts between plates. dnc = dorsal nerve cord; ? drn = probable dorsal-root nerve; gan = segmental ganglion; int tis = interossicular tissue, probably ligament; m bl = muscle blocks; not = notochord; ?vr = probable ventral root, probably muscular (Reproduced with permission of the Royal Society of London).

evidence of segmental ganglia and also of ventral roots which, by analogy with amphioxus, were probably made of muscle rather than nerve (cf. Flood, 1966, 1975). The hind tail probably pulled the body backwards by pressing on ventral bearing surfaces, in which respect it was like the fore tail of *Reticulocarpos* and therefore probably primitive.

The reasons for thinking that *Lagynocystis* is specially related to amphioxus are therefore as follows: 1) the median atrium with its gill slits; 2) the deduction that the groups of gill slits appeared in ontogeny in the same asymmetrical order as in amphioxus; 3) the asymmetry of the nerve supply to the mouth region; 4) the plausible origin of the nephridia of amphioxus by the breaking up of a *Lagynocystis*-type posterior coelom by upward extension of the gill slits; 5) the suggestion that *Lagynocystis* rotated as it swam in the same manner and direction as larval amphioxus. *Lagynocystis* shows that amphioxus is by no means the ideal proto-vertebrate of the text books. Many of the most peculiar features of amphioxus are specializations which the ancestors of vertebrates never had.

CONCLUSIONS

The early history of the chordates was therefore probably as follows (Fig. 34). In late pre-Cambrian, or possibly earliest Cambrian, times a hemichordate resembling *Cephalodiscus* vacated its tube and took to crawling, right-side-down, on the sea floor. It lost the openings and tentacles of the originally right, but now ventral, side and acquired a calcite skeleton in which each plate was a single crystal.

From this form two groups evolved. One lost the remaining gill slit, elaborated the remaining, originally left tentacles and comprised the first echinoderms. The other descendant group elaborated the remaining gill slit, lost the originally left tentacles and comprised the first chordates. These latter developed a muscular, plated tail with a notochord, muscle blocks and probably a dorsal nerve cord, for pulling themselves backwards over the sea floor. Their most primitive known representative is the cornute *Ceratocystis*, from which other cornutes, such as *Cothurnocystis*, evolved. Primitive cornutes lived on sand.

Certain groups of cornutes took to a life on very soft mud relying on the weak strength of the mud to support themselves. The phylogenetically most important of such groups became small, flat and symmetrical as an adaptation to life on mud, and is represented by *Reticulocarpos*. From a close relative of this genus evolved the first mitrates which were adapted to stay up on soft mud by weight compensation ("buoyancy") and, most importantly, had developed new right gill slits, in addition to the left gill slits already present in the cornutes. The first mitrate, defined as the first chordate with right gill slits, would have appeared suddenly, as the child of cornute parents. It is possible that the earliest mitrates could swim, by wagging their tails.

The mitrates split into two groups. One of these, represented by *Lagynocystis*, developed a median ventral atrium and in a sense constituted the first cephalochordates. By loss of skeleton, by forward growth of the notochord and body muscles, and other changes, this group gave rise to living amphioxus and its relatives.

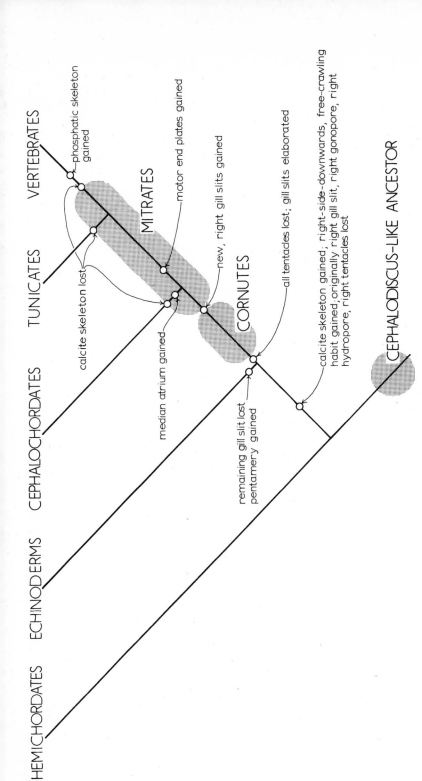

HEMICHORDATES ECHINODERMS CEPHALOCHORDATES TUNICATES VERTEBRATES

phosphatic skeleton
gained

MITRATES

motor end plates gained

calcite skeleton lost

new, right gill slits gained

median atrium gained

CORNUTES

all tentacles lost; gill slits elaborated

calcite skeleton gained, right-side-downwards, free-crawling
habit gained; originally right gill slit, right gonopore, right
hydropore, right tentacles lost

remaining gill slit lost
pentamery gained

CEPHALODISCUS-LIKE ANCESTOR

FIG. 34. The probable phylogenetic relationships of the deuterostome phyla.

All these animals retained the primitive, echinoderm-like method of innervating the muscle blocks by means of strips of muscle passing to the dorsal nerve cord.

The other group of mitrates acquired a new method of innervating the muscle blocks, by means of motor end plates like those of living vertebrates and tunicates. From an unknown member of this group of mitrates the tunicates probably arose. From some other member of it the vertebrates evolved. The mitrates most closely related to the vertebrates, and which in a sense *were* the first vertebrates, are probably the forms possessing a lateral line such as *Mitrocystites* or *Mitrocystella*, or the little primitive mitrate *Chinianocarpos*. The ancestor of modern vertebrates was probably some such mitrate which had taken to continuous swimming, and elaborated its sense organs and brain and lost its calcite skeleton as a result.

The vertebrate kidneys, necessary as a way of excreting the increased quantities of nitrogenous waste produced by continuous swimming, perhaps developed as an elaboration of the pre-existing left and right epicardia. Later these continuously swimming animals developed a phosphatic skeleton, which among chordates exists in vertebrates alone.

In my view, therefore, the calcichordates provide definite evidence, because of the happy fact of their calcitic skeletons, of how the earliest chordates really evolved. There are still wide areas of ignorance, but because of this new fossil evidence, the problem of chordate origins is no longer a speculative matter.

REFERENCES

Balinsky, B. I. (1960). *An introduction to embryology.* Philadelphia & London: Saunders.
Barrington, E. J. W. (1938). The digestive system of *Amphioxus (Branchiostoma) lanceolatus. Phil. Trans. R. Soc.* (B) **228**: 269–311.
Berrill, N. J. (1955). *The origin of vertebrates.* London: Oxford University Press.
Bone, Q. (1972). *The origin of Chordates.* Oxford Biology Readers (Ed. Head, J. J. & Lowenstein, O. E.). London: Oxford University Press.
Boolootian, R. A. & Campbell, J. L. (1964). A primitive heart in the echinoid *Strongylocentrotus parpuratus. Science, N.Y.* **145**: 173–175.
Carlisle, D. B. (1953). Origin of the pituitary body of chordates. *Nature, Lond.* **172**: 1098.
Carlisle, D. B. & Carlisle, R. C. (in press). Feeding mechanisms in tunicates. *Can. J. Zool.*
Caster, K. E. (1972). Review of Ubaghs (1970). *J. Paleont.* **45**: 919–921.

Chauvel, J. (1971). Les Echinodermes Carpoïdes du Paléozoique inférieur marocain. *Notes Serv. Géol. Maroc* **31**: 49–60.

Eaton, T. H. (1970). The stem-tail problem and the ancestry of chordates. *J. Paleont.* **44**: 969–979.

Fedotov, D. M. (1924). Zur Morphologie des axialen Organkomplexes der Echinodermen. *Z. wiss. Zool.* **123**: 209–304.

Flood, P. R. (1966). A peculiar mode of muscular innervation in amphioxus. *J. comp. Neurol.* **126**: 181–218.

Flood, P. R. (1975). Fine structure of the notochord of amphioxus. *Symp. zool. Soc. Lond.* No. 36: 81–104.

Garstang, W. & Platt, M. I. (1928). On the asymmetry and closure of the endostyle in *Cyclosalpa pinnata*. *Proc. Leeds phil. lit. Soc.* **1**: 325–334.

Gill, E. D. & Caster, K. E. (1960). Carpoid echinoderms from the Silurian and Devonian of Australia. *Bull. Am. Paleont.* **41** (185): 1–71.

Gislén, T. (1930). Affinities between the Echinodermata, Enteropneusta and Chordonia. *Zool. Bidr., Upps.* **12**: 199–304.

Godeaux, J. (1964). Que connaissons-nous de la glande neurale des Tuniciers? *Publs Univ. Elizabethville* **7**: 7: 51–64.

Grobben, K. (1924). Theoretische Erörterungen betreffend die phylogenetische Ableitung der Echinodermen. *Sber. Akad. Wiss. Wien* (Mat.—nat. Kl.) **132**: 262–290.

Jefferies, R. P. S. (1967). Some fossil chordates with echinoderm affinities. *Symp. zool. Soc. Lond.* No. 20: 163–208.

Jefferies, R. P. S. (1968). The Subphylum Calcichordata (Jefferies 1967)—primitive fossil chordates with echinoderm affinities. *Bull. Br. Mus. nat. Hist. (Geol.)* **16**: 243–339.

Jefferies, R. P. S. (1969). *Ceratocystis perneri* Jaekel—a middle Cambrian chordate with echinoderm affinities. *Palaeontology* **12**: 494–535.

Jefferies, R. P. S. (1971). Some comments on the origin of chordates. *J. Paleont.* **45**: 910–912.

Jefferies, R. P. S. (1973). The Ordovician fossil *Lagynocystis pyramidalis* (Barrande) and the ancestry of amphioxus. *Phil. Trans. R. Soc.* (B) **265**: 409–469.

Jefferies, R. P. S. & Prokop, R. J. (1972). A new calcichordate from the Ordovician of Bohemia and its anatomy, adaptations and relationships. *Biol. J. Linn. Soc.* **4**: 69–115.

Matsumoto, H. (1929). Outline of a classification of the Echinodermata. *Sci. Rep. Tôhuku Univ.* (Geol.) **13**: 27–33.

Millar, R. H. (1953). *Ciona. L.M.B.C. Mem. typ. Br. mar. Pl. Anim.* **35**: 1–84.

Millott, N. (1966). A possible function for the axial organ of echinoids. *Nature, Lond.* **209**: 594–596.

Millott, N. (1967). The axial organ of echinoids. Reinterpretation of its structure and function. *Symp. zool. Soc. Lond.* No. 20: 53–63.

Narasimhamurti, N. (1931). The development and function of the heart and pericardium in Echinodermata. *Proc. R. Soc.* (B) **109**: 471–486.

Pérès, J. M. (1943). Recherches sur le sang et les organes neuraux des Tuniciers. *Annls Inst. océanogr. Monaco* **21**: 229–359.

Robison, R. A. (1965). Middle Cambrian eocrinoids from Western North America. *J. Paleont.* **39**: 355–364.

Sprinkle, J. (1973). *Morphology and evolution of blastozoan echinoderms.* Cambridge, Mass.: Harvard Univ. (Special publications of the Museum of Comparative Zoology.)

Ubaghs, G. (1961). Sur la nature de l'organe appelé tige ou pedoncule chez les carpoïdes Cornuta et Mitrata. *C. r. hebd. Séanc. Acad. Sci. Paris* **253**: 2738–2740.

Ubaghs, G. (1967). Le genre *Ceratocystis* Jaekel (Echinodermata, Stylophora). *Paleont. Contr. Univ. Kans.* **22**: 1–16.

Ubaghs, G. (1968). Stylophora. In *Treatise on invertebrate palaeontology* pt. 5: Echinodermata 1: 3–60. Moore, R. C. (ed.). New York: Univ. of Kansas & Geological Society of America Inc.

Ubaghs, G. (1970). *Les echinodermes carpoides de l'ordovicien inférieur de la Montagne Noire (France)*. Paris: C.N.R.S. (Cahiers de Paléontologie).

Ubaghs, G. (1971). Diversité et specialisation des plus anciennes echinodermes que l'on connaisse. *Biol. Rev.* **46**: 157–200.

Werner, E. & Werner, B. (1954). Uber den Mechanismus des Nahrungserwerbs der Tunicaten, speziell der Ascidien. *Helgoländer Wiss. Meeresunters.* **5**: 57–92.

Wickstead, J. H. (1967). *Branchiostoma lanceolatum* larvae: some experiments on the effect of thiouracil on metamorphosis. *J. mar. biol. Ass. U.K.* **47**: 49–59.

Willey, A. (1894). *Amphioxus and the ancestry of the vertebrates*. New York & London: MacMillan.

Symp. zool. Soc. Lond. (1975) No. 36, 319–345.

THE POSSIBLE CONTRIBUTION OF NEMERTINES TO THE PROBLEM OF THE PHYLOGENY OF THE PROTOCHORDATES

E. N. WILLMER

Yew Garth, Grantchester, Cambridge, England

SYNOPSIS

Hubrecht in 1883 suggested that nemertine worms might possibly have given rise to the vertebrates. This idea was revived in 1960 and an examination of the more recent work on the physiology, histology, cytology and behaviour of nemertines adds weight to the general hypothesis. The possible conversion of the pharynx of a nemertine into a filter-feeding mechanism and of the proboscis into a notochord as an adaption for swimming in a fish-like manner, together with coincident changes in associated structures, could provide an explanation for such a highly specialized and peculiarly vertebrate structure as the pituitary body. It could also throw light on the relationships between the protochordates and the vertebrates. On this basis, it seems likely that the various groups of protochordates may have arisen from different nemertine-like creatures as these attempted to colonize the supernatant water for at least some part of their lives. Amphioxus is thus thought to have followed a parallel course of evolution to the vertebrates and to have started from a different nemertine. Other nemertines may have developed burrowing habits more directly and have evolved into such creatures as the enteropneusts.

INTRODUCTION

In a book published in 1970 I attempted to trace out in detail the possible origin of the various tissues and organs of vertebrates in relation to the types and properties of the cells that they contain (Willmer, 1970). In that book I particularly developed the idea, first promulgated by the Dutch embryologist Hubrecht (1883, 1887) some 90 years ago, that nemertines played a considerable part in this evolutionary process. In this paper it would be impossible to reconsider the whole of this problem, but there are certain aspects of it which I think are relevant to the discussions of this symposium and upon which observations made during the last few years have shed further light.

Table I shows a list of some of the major characters present in primitive vertebrates which are also present in some or all of the protochordates. They are sufficiently numerous and important to emphasize the close connexion between protochordates and vertebrates; and there are probably few of us who would doubt the reality of that connexion.

Table II, however, shows some features present in rather highly developed form even in primitive vertebrates which are

TABLE I

Characters common to vertebrates and protochordates

Elongated body, bilateral symmetry
Intestine, mouth, anus and liver (diverticulum)
Notochord
Tubular nervous system
Myotomes developing from somites
Gill slits
Serial excretory and/or gonadal ducts
Closed vascular system (? erythrocytes)
Heart
Ciliated and mucoid cells in skin
Coelomic cavities
Thyroid or iodine-binding tissue

TABLE II

Vertebrate characters not shared by protochordates

Two-chambered eyes with inverted retina
Pineal eyes
Adenohypophysis
Neurohypophysis
Oxyntic and chloride-secreting cells

nevertheless absent from, or only dubiously represented in the protochordates.

Two questions therefore arise from consideration of these two Tables. First, from what more primitive creatures did the protochordates and vertebrates derive these characters? Second, how did the vertebrates acquire the very complicated structures of their eyes and the pituitary body? I want first to examine this second question in some detail. This may seem an odd thing to do in a discussion on protochordates in which pituitary homologues are rather doubtfully represented, but I believe it will serve to illuminate the subject of my paper, namely the possible contribution of the nemertines to the phylogeny of the protochordates.

THE EVOLUTION OF NEMERTINES

Figure 1 illustrates a very synoptic view of the probable evolutionary progression which led up to the platyhelminthes and nemertines. I have used the suffix "oid" in each case in order to

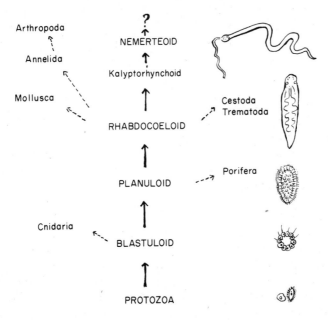

FIG. 1. Probable line of evolution that led up to the nemertines of the present day. The suffix "oid" indicates "-like", e.g. blastuloid = blastula-like. Note the everted proboscis of the nemerteoid.

emphasize the fact that the actual evolutionary path did not go through any organism now living but through organisms which had structures generally present in, and characteristic of the various groups described. For example, some rhabdocoel-like creatures, perhaps through some intermediate kalyptorhynch-like creature, evolved into something more like a nemertine.

Most of these flatworms are bottom-living or creeping in their habits. Other groups of organisms of course arose during the process of their evolution and some of these may have been able to remain afloat even as adults, like the medusae, others only as larvae, like the echinoderms. In general, however, size and weight probably kept the organisms with which we are concerned more or less anchored to the bottom.

In Cambrian times, therefore, or perhaps a little later, platyhelminthoid and nemerteoid creatures were probably abundant on the ocean floor, and even dominant there. The waters themselves were probably teeming with protozoa, small organisms and larval stages that were sufficiently active or buoyant to remain afloat, but probably contained relatively few larger creatures to feed on this abundant plankton. Thus there would be enormous selection pressure favouring the development of pelagic habits by rhabdocoeloids or nemerteoids. Similarly there were opportunities for any of the creatures of the ocean who could invade the fresh waters and vice-versa.

It was probably some advanced rhabdocoeloid or at least platyhelminthoid creature which acquired a strictly segmental organization and, by means of the semi-rigidity of the body so achieved, became able to swim in the sinusoidal manner characteristic of the annelids; and in the further development of this method they acquired parapodia and ultimately limbs and arthropodal organization.

The question relevant to our discussion, however, is "in what further ways did the nemerteoids evolve, if at all, or did their cumbersome method of feeding (and moving) by means of the proboscis preclude any further development?" Some may have retired into the substratum and modified their proboscis accordingly, i.e. towards more efficient feeding, digestion, burrowing or even all three. Indeed, it seems possible that pogonophores and enteropneusts could represent the outcome of this line of evolution if the proboscis were to develop as an evagination, somewhat after the fashion of an exogastrula, rather than as an invagination. The general similarity in structure and mode of life is certainly striking even down to such details as the curious parallelism between the so-called "vascular plug" of certain nemertines and the "glomerulus" of *Balanoglossus*. Rather than speculate along these lines, however, let us look at some of the many different ways in which present-day nemertines live and have evolved.

Cerebratulus is a very successful littoral nemertine in the north Atlantic which not only can swim by sinusoidal movements of its body in a vertical plane, like a number of other nemertines that have somehow solved the problem of staying afloat and have become pelagic, but has acquired a pharynx which is highly vascular and is used for respiration by alternate inflation and deflation with sea-water (Wilson, 1900). Some of the main features of *Cerebratulus* relevant to this discussion are illustrated in Fig. 2. Let us note that its proboscis opens at the anterior end through the rhynchodaeum, that it has a closed vascular system, that its intestine is formed into regular diverticula and its gonoducts are serially arranged between the diverticula.

Nemertopsis and *Prosadenoporus* (Fig. 3) are notable for possessing a pharynx which is extensively pouched (Bürger, 1895).

Malacobdella (Fig. 4) is an entocommensal in the mantle cavity of a lamellibranch mollusc. It feeds by filling its pharynx with the fluid from the mantle cavity and extracting the fine particles contained therein on a series of ciliated ridges. It has a proboscis which it only uses occasionally for the capture of larger prey and this opens, not anteriorly, but into the roof of the mouth, through which it can be ejected. It has a relatively simple intestine, without diverticula (Gibson & Jennings, 1969).

In *Lineus* (Fig. 5) attention should be paid to the presence of eyes, a complicated cephalic organ (which, incidentally, is widely distributed and very variable in structure among both rhabdocoels

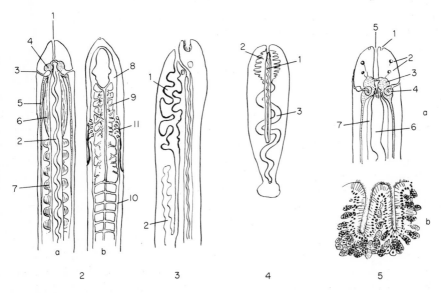

FIG. 2. Diagrams to show some important features of *Cerebratulus*. (a) Shows (1) rhynchodaeum, (2) proboscis in rhynchocoel, (3) cephalic organs, (4) cephalic ganglia and (5) nerve cords, (6) muscular pharynx and (7) intestine with diverticula. (b) Shows (8) the main blood vessels, with (9) rich supply to the pharynx, (10) serial vessels to the intestine. Nephridia are also shown (11).

FIG. 3. Longitudinal section through *Prosadenoporus* showing pouched pharynx (1) and forward diverticulum of the intestine (2), cf. amphioxus.

FIG. 4. Diagram of *Malacobdella* to show (1) the proboscis opening into the enlarged buccal cavity, (2) which has transverse ciliated folds. The intestine (3) is a simple tube.

FIG. 5. (a) Diagram of anterior end of *Lineus* showing (1) frontal organs, (2) two pairs of eyes, (3) cephalic ganglia and (4) cephalic organs, (5) rhynchodaeum, (6) proboscis and (7) rhynchocoel. (b) Diagram of pharyngeal epithelium. Note ciliated cells and at least three classes of glandular cells.

and nemertines), and a pharyngeal epithelium of extraordinary cytological complexity.

Most nemertines are marine, though some at least have some degree of euryhalinity, e.g. *Lineus*; a few live in fresh water, while *Geonemertes* is a terrestrial genus of which some species live at a height of 2000 ft (Coe, 1904; Hett, 1927; Pantin & Moore, 1969).

This brief survey gives some indication of the range of features and habitats displayed by existing nemertines. It is notable that no members of the group swim like the annelids with side-to-side progressive waves, nor like the vertebrates with side-to-side oscillation, though many, like *Pelagonemertes* and *Nectonemertes* can remain afloat and make progress by sinusoidal waves passing down the body in a vertical plane in which they may be assisted by fin-like extensions of the body (Coe, 1926).

Hubrecht (1883, 1887) and Jensen (1963) have suggested that, if the proboscis of a nemertine ceased to be used, the pressure of fluid in the rhynchocoel cavity could be sufficient to provide the semi-rigid rod (or notochord) which is now recognized as a necessary constituent for the successful development of side-to-side oscillations (R. B. Clark, 1964) and this formed the basis for the view that the notochord arose from the rhynchocoel of the nemertine, while the proboscis itself degenerated into an epithelial rudiment but remained functional as the anterior lobe of the pituitary. Since this idea has been latent for all these long years it is perhaps worth examining the situation afresh in the light of present knowledge.

Let us therefore try to envisage a nemerteoid with filter-feeding capabilities like those of *Malacobdella*, a respiratory habit like that of *Cerebratulus* and a pouched pharynx like that of *Prosadenoporus*. An embryonic modification of the pharynx could rather easily produce gill-slits in the manner seen in embryonic vertebrates, and thus improve both filter-feeding and respiration. This could lead to disuse of the proboscis and allow its modification into other structures. Ontogenetically the proboscis develops from a column of cells invaginating from the anterior end of the embryo and becoming surrounded by a split in the mesenchyme which develops into the rhynchocoel. The column of cells then hollows out and forms the characteristic tubular and epithelial structure surrounded by muscular elements. Figure 6 shows the similarity in proboscis and notochordal structures. The notochord of amphioxus differs from that of vertebrates in three ways. First, it is almost completely cellular; second, it extends to the anterior end of the

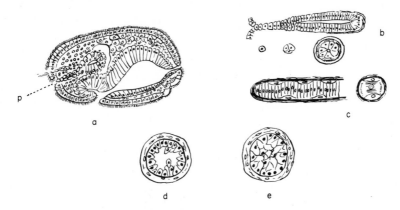

FIG. 6. Diagrams to illustrate: (a) Longitudinal section of nemertine embryo to show development of proboscis rudiment (p). (b) Longitudinal and transverse sections of proboscis rudiment. (c) Longitudinal and transverse sections of notochord of amphioxus. (d) Transverse section of regenerating proboscis of nemertine. (e) Transverse section of notochord of elasmobranch.

body; third, its cells are muscular (Welsch, 1968; Flood, Guthrie & Banks, 1969). These differences would be directly intelligible if the hypothetical nemerteoid from which amphioxus evolved was of the type whose proboscis opened at the anterior end of the body, if its development or differentiation occurred before the tube hollowed out and if the muscular elements of the proboscis were the cells utilized.

This idea receives some support from other considerations. In the first place notochords of different animals are by no means uniform in detailed structure (Fig. 7). In *Ciona* (Berrill & Sheldon, 1964) and *Tethyum* (Berrill, 1929) the notochord is a column of turgid cells. In amphioxus it is a column of cells rendered rigid by muscular contraction that is under neural control (Flood, 1970). In *Oikopleura* it is a hollow tube rendered turgid by secretion of fluid through an epithelium (Olsson, 1965) which has a curious similarity to that lining the rhynchocoel of some nemertines. In most vertebrates it is a tubular structure more or less filled with cells which are themselves rendered turgid by the secretion of a carbohydrate-containing fluid into vesicles, vacuoles or intercellular spaces. All these variations could have arisen by utilizing different aspects of the rather complex structure or potentialities of the developing nemertine proboscis or, as perhaps in *Oikopleura*, of the rhynchocoel in the manner suggested by Hubrecht. However that may be, these observations point towards different but

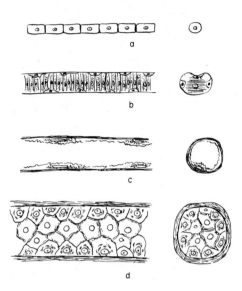

FIG. 7. Various forms of notochord. (a) Turgid cells in *Ciona* (after Berrill & Sheldon, 1964). (b) Muscular cells and Müller cells in Amphioxus (after Flood, 1970). (c) Turgid tube in *Oikopleura* (after Olsson, 1965). (d) Turgid cells in *Petromyzon* (after Schwarz, 1961).

parallel origins for the groups possessing the different types of notochord, since it is unlikely that once semirigidity had been developed by one method, e.g. muscular contraction, it would be replaced by another, e.g. cell turgidity.

The structure of the notochord of vertebrates suggests derivation from an originally tubular structure and it is significant that the notochord in them ends at some distance from the anterior end of the body. Thus, it could have developed from a proboscis which had progressed further in its differentiation than the single column of cells and which had already atrophied its connexion to the anterior end and had formed a secondary union with a diverticulum of the mouth as in *Malacobdella* (Figs 8 & 9). Moreover it used secretory cells as the source of turgidity.

It will be noted that this idea homologizes the vertebrate notochord with the proboscis itself and not with the rhynchocoel as Hubrecht, and later Jensen (1960, 1963) envisaged.

A major difficulty in accepting this hypothesis lies in the fact that the proboscis is essentially an epithelial (ectodermal) invagination from the anterior end of the body, while the notochord usually develops from the roof of the archenteron. However, in view of the plasticity of nemertine ontogeny and regenerative processes, and

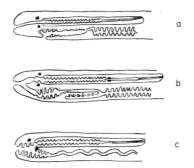

FIG. 8. Diagram to show the three ways in which the proboscis may be ejected. (a) *Lineus*, through separate rhynchodaeum. (b) *Amphiporus*, through common opening with the mouth. (c) *Malacobdella*, through pharynx and mouth.

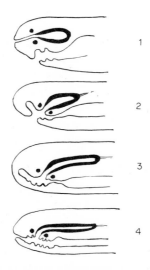

FIG. 9. Diagrams to show stages in the formation of the secondary opening of the proboscis as seen, for example, in the embryogenesis of *Malacobdella*. (1) Original development of the proboscis (black). (2) Closure of rhynchodaeum and secondary connexion with temporary mouth. Development of anterior diverticulum from mouth (3) Closure of mouth and further development of anterior diverticulum. (4) Opening of secondary and definitive mouth. Differentiation of ciliated ridges in mouth and pharynx. (After Hammarsten, 1918.)

incidentally the close similarity between notochordal development in tunicates and proboscis development in nemertines, it is difficult to know how much weight to attach to this difference, especially when both phylogenetic and cytological considerations are consistent with the idea that the proboscis itself may have originated as a definite structure by an embryological duplication of the gut, just as supernumerary notochords sometimes arise pathologically even in man.

The structure of the notochord and its sheath and the use made of existing embryonic rudiments seem now to favour such a view. A corollary of this argument of great interest in the present context is that it suggests that both amphioxus and the vertebrates could have arisen from nemerteoids, but their origins were different in that amphioxus arose from a nemerteoid whose proboscis opened anteriorly, while the vertebrates arose from a nemerteoid whose proboscis opened into the mouth cavity.

These ideas would do away with the anomaly that no nemertines apparently learned to swim with side to side oscillations. Vertebrates, cephalochordates and probably urochordates may be the products of those nemerteoids that did so. They also indicate that cephalochordates, urochordates and vertebrates are on parallel lines of evolution rather than on the same line. This is borne out by the absence of any true pituitary or of vertebrate-type eyes from amphioxus. The absence of the latter may be responsible for the fact that the amphioxus line has been less successful and only survives in members that have sought the protection of burrowing. The acquisition of swimming, to be really successful, must also involve the acquisition of efficient distance-receptors adequately integrated into the motor system.

In my book I have suggested how this may have come about especially by the development of eyes and their integration into a similarly expanding motor nervous system, and how this latter may have been related to the development of somites, but here I wish first to comment in more detail on the pituitary body because there is now more information available and the whole pattern has become much more clear and illuminating to the solution of the main problem.

THE ADENOHYPOPHYSIS

If we assume that vertebrates developed from nemerteoids which had the type of proboscis that opens into the mouth, like *Malacobdella*, then the notochord, derived from the proboscis rudiment, would end just below the cephalic ganglia. Moreover, the diverticulum from the buccal cavity, through which the proboscis was extruded, might still form in the embryo. This diverticulum could therefore be the anlage of Rathke's pocket from which the adenohypophysis of vertebrates develops. Figures 8, 9 show the variations in arrangement of the proboscis opening and they are strangely reminiscent of the variations in the hypophysial rudiments in primitive vertebrates. *Malacobdella* itself undergoes similar rearrangements to those of vertebrates during the course of its embryonic development (Hammarsten, 1918). It is therefore pertinent to examine the nature of the cells that this pouch might contain. During the last few years more or less detailed examinations have been made of the fore-gut of *Lineus* and *Malacobdella* (Jennings, 1962; Jennings & Gibson, 1969; Ling & Willmer, 1973; Gibson & Jennings, 1969) and some of the findings are very relevant to the problems that we are now considering.

Let us therefore examine the properties of the fore-gut of *Lineus* as a whole and consider its evolutionary potentialities. First, the whole of it picks up iodine far more avidly than does the rest of the worm. In ordinary sea-water most of this iodine appears to remain uncombined, but if the sea-water is concentrated, either by the addition of sodium chloride or by evaporation, the iodine is not only picked up still more rapidly but is also to a small extent combined into thyronine compounds (Balfour & Willmer, 1967). This is certainly consistent with the idea that nemertines and protochordates are related, though of course not by itself conclusive. Unfortunately we do not yet know which cells of the nemertine pharynx are responsible for which activity, because the pharynx, though lined with a ciliated epithelium, like the iodine-binding epithelium of the endostyle, is backed by a very complicated sub-epithelium of glandular cells showing five or six types or phases of activity.

In the living worm this complex epithelium must have a variety of functions. Powerful digestion goes on in the pharynx in an acid medium and some of the cells have a high content of carbonic anhydrase (Gibson & Jennings, 1969), thus resembling the oxyntic cells of the stomach and the chloride-secreting cells of fish gills. When *Lineus* is subjected to dilute sea-water, which it tolerates very well, the ciliated cells become temporarily engorged and regularly arranged, while most of the glandular cells become discharged and emaciated. In concentrated sea-water the reverse occurs and the glandular cells become packed with granules. One suspects therefore that this epithelium is very much concerned with the ionic balance of the body as a whole as well as with digestion, filter-feeding, respiration and so on. After a day or two in the abnormal medium the epithelium returns to normal.

It is perhaps of interest at this point to call attention to a curious parallel between the behaviour of the nemertine pharynx in fresh water and that of the endostyle of larval lampreys in the presence of thiourea or thiouracil. The endostyle is covered by ciliated cells, some of which specialize in collecting the iodine, and these overlie groups of glandular cells which produce a mucoid secretion with staining properties akin to that in the pharynx of *Lineus*. After treatment with the goitrogens the ciliated cells appear hyper-trophic while the mucoid cells first vacuolate and then shrink (Barrington & Sage, 1963a,b). How these observations should be interpreted remains to be discovered but the similarity between the reaction of the nemertine pharynx to water and that of the ammocoete's endostyle to thiourea is certainly morphologically striking and a study of the effects of alterations of salinity on the morphology of the endostyle could profitably be combined with a study of the effects of thiourea on the pharynx of *Lineus*. The former might be particularly profitable because it has been suggested that since lampreys have no power of storing calcium, e.g. in bone, yet can control its concentration in the blood with great accuracy, they do so by means of the pharyngeal epithelium and gills (Urist, 1963).

In *Lineus*, at least, the buccal cavity has the same sort of lining as the rest of the fore-gut with the complete gamut of cells. Thus the original Rathke's pocket could perhaps have contained the same classes of cells and this would indeed be suggested by the arrange-ment in *Malacobdella* (Gibson & Jennings, 1969). It would therefore seem reasonable to suppose that Rathke's pocket would initially have contained a variety of cells whose activities were integrated among themselves, with those in the rest of the pharynx and probably also with the skin and the excretory organs, and that many of their activities would be related to the ionic content of their surroundings. Initially the cells would be directly affected by their neighbours and by the contents of the pharyngeal lumen, but, after the pocket had become closed off, they would presumably be affected mainly by the tissue fluids or by their nerves. Thus these cells enclosed in the pouch could be used to monitor the contents of their immediate surroundings either directly or indirectly and still be able to react on the rest of the pharynx, the skin and the excretory organs.

One of the most striking functions of the adenohypophysis of fish is the secretion of prolactin which assists in the maintenance of the sodium level of the blood, especially in water of low salinity

(Ensor & Ball, 1968; Ball, 1969; Lam, 1972; Dharmamba, Mayer-Gostan, Maetz & Bern, 1973). This it does partly by its action on the chloride cells of the gills, though it also acts on the intestinal, renal and bladder epithelia (Utida *et al.*, 1972). In many of these actions it is opposed by cortisol (Utida *et al.*, 1972). It has been stated that sea-water fish exposed to fresh water show degranulation of the prolactin-secreting cells (Sage, 1968; Hopkins, 1969) though the evidence on this point is equivocal (Dharmamba & Nishioka, 1968; Emmart, Pickford & Wilhelmi, 1966). Furthermore the action of prolactin on amphibian tadpoles has been shown to be antagonistic to the action of thyroxin (Etkin & Gona, 1967; Hughes & Reier, 1972).

All these observations are pertinent in that the cells concerned would, on the present hypothesis, be regarded as derivatives of the pharyngeal membrane of the nemertine, whose iodine uptake and acid secretion from carbonic-anhydrase-containing cells perhaps foreshadow thyroid function and chloride-cell activity respectively. It is therefore very significant that the lining of the pharynx of *Lineus* contains acidophil cells with granules of very similar size and electron-density to those of the prolactin cells of vertebrate

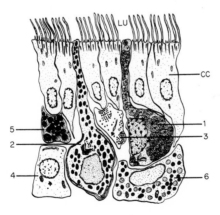

FIG. 10. Diagram showing the ciliated cells (CC) and the six types of glandular cells in the pharyngeal membrane of *Lineus*. LU = lumen. (From Ling & Willmer, 1973.)

pituitaries (Fig. 10) and that these degranulate when *Lineus* is placed in dilute sea-water.

To continue this line of thought, Table III shows a comparison between the cells of the pharynx of *Lineus* and those of the anterior

TABLE III

A comparison between the cells of the nemertine pharynx and those of the adenohypophysis in vertebrates

| | NEMERTINE PHARYNX | | | VERTEBRATE ADENOHYPOPHYSIS | |
Type	Staining	Granule size (nm)	Type	Staining	Granule size (nm)
1. Basophil	Acid mucopolysaccharide "Foamy" granules with 400 nm dense cores	1000	Gonadotroph	Basophil PAS positive	210
2. Acidophil	PAS negative	600	Mammotroph Somatotroph	Acidophil PAS negative	500 350
3. Basophil cytoplasm	Acidophil granules, PAS positive	120	Thyrotroph	Basophil PAS positive	130
4. Colourless			Chromophobe		
5. Very basophil granules		1000	(? Mast Cells)		(1000)
6. Granules of varying density		600			
7. Ciliated cells			Corticotroph Melanophore-stimulating Follicle		160 120

pituitary of vertebrates. The resemblance is sufficiently striking to encourage the belief that the pharyngeal cells contain the appropriate ingredients to permit their differentiation into adenohypophysial-like cells, if they became enclosed in Rathke's pocket and underwent, in the course of evolution, the consequent modifications and responses to changes in their new internal environment. In parenthesis it may be noted that if *amphioxus*, as we have supposed, evolved from a nemertine with a frontally opening proboscis, then it would not have developed a Rathke's pocket or an adenohypophysis of this type.

With regard to the other functions of the anterior pituitary there is much less evidence as yet. The thyrotroph cells, as the Table suggests, could possibly be derived from the Type 3 or brown-green cells of the nemertine (i.e. the cells that contain carbonic anhydrase) whose acidophil granules are comparable in size and distribution. The basophil cells of the nemertine pharynx contain granules that seem to have no direct counterpart in the adenohypophysis, but that does not necessarily mean that during evolution they could not have evolved into the gonadotrophs. They are at least basophil and the gonadotrophs have themselves evolved during the course of vertebrate evolution. Their granules within granules (Fig. 10) are peculiar; they could perhaps be interpreted as granules within endoplasmic reticulum, comparable with the granules of thyrotrophs in thyroidectomized animals. Curiously enough, somewhat similar but smaller structures are found in the glandular cells of the ammocoete endostyle (Fujita & Honma, 1968) and in the ultimobranchial bodies of reptiles, though not in such great numbers (N. B. Clark, 1972). The somatotrophs probably developed much later, even within the vertebrates, by modification of the prolactin cells as the growth hormone and prolactin have many common actions. Corticotrophs, "follicle cells" and melanophore-stimulating cells are even more problematical and the existence of cilia and microvilli on cells in the vertebrate pituitary requires explanation. It is therefore relevant that Rathke's pocket would originally be lined with ciliated epithelium and in the nemertine pharynx these cells are clearly altered by their environment in such a way that the ciliated cells flourish when the prolactin cells degranulate and vice versa. In view of the antagonism in teleosts between the actions of cortisol and prolactin on the water and salt balance (Utida *et al.*, 1972) this points strongly towards the corticotroph cells as derivatives of the ciliated cells.

To summarize this rather complicated discussion the following hypothesis is advanced. The pharyngeal membrane of the nemerteoid, in addition to its primary digestive function, was one of the main tissues concerned in regulating the internal environment. In this it was integrated with the skin, intestine and nephridia. When the proboscis evolved into the notochord the portion of the pharyngeal membrane within the secondary rhynchodaeum (and therefore of different embryonic origin from the proboscis itself) became closed off and evolved into the adenohypophysis whose function became that of controlling the activities of the other cells of the pharynx as they specialized in other directions to become chloride-secreting cells, oxyntic cells, thyroid cells, parathyroid cells, ultimobranchial cells, peptic cells and so on. The original interactions with skin, intestine and nephridia would probably also be preserved. The fundamental features of this pituitary control system may thus date back to the interactions of the cells in the nemertine pharyngeal membrane, though changes of environment, and further adaptations of both controller and controlled, are likely to have altered their modes of action very considerably during the course of evolution, as indeed one can see happening within the evolution of the vertebrates, at least with regard to the various hypophysial hormones.

THE NEUROHYPOPHYSIS

If this then was how the adenohypophysis arose, what was the origin of the neurohypophysis? Essentially the neurohypophysis is composed of modified endings of axons originating in the neurosecretory cells of the hypothalamus. How then can this be related to any progenitory tissue in the nemerteoid? First, it is notable that the proboscis is forcibly ejected and retracted by a combination of muscular movements of the proboscis itself and changes of pressure in the rhynchocoel caused by the body musculature (Fig. 11). The proboscis itself is well supplied with nerves. In addition to sensory fibres some are motor to the actual proboscis and others activate the retractor muscle by which the proboscis is drawn back into the rhynchocoel cavity. If, as has been supposed, the proboscis ceased to function as such, what then happened during the ensuing stages of evolution to the nerves and nerve cells which supplied its various muscles and glands? In *Lineus* the proboscis itself is sensitive to acetyl-choline and also to adrenalin, while the retractor muscle responds to oxytocin but not to either acetyl-choline or adrenalin (Willmer, 1970). Furthermore,

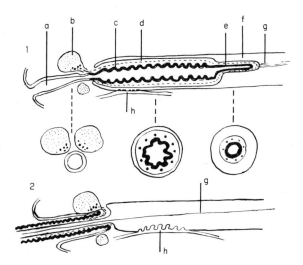

FIG. 11. Diagrams to show the relationships of the proboscis and associated tissues (1) when the proboscis is contained within the rhynchocoel and (2) when the proboscis is ejected. The central figures are transverse sections of (1) at the levels indicated by the dotted lines. (a) rhynchodaeum, (b) cephalic ganglia with neurosecretory cells, (c) anterior proboscis, (d) proboscis nerve (dotted), (e) posterior proboscis, (f) rhynchocoel, (g) retractor muscle, (h) dorsal vessel with vascular plugs or villi, collapsed in (1) and dilated in (2).

in addition to characteristic cholinergic and aminergic vesicles in the proboscis nerve (Ling, 1971) neurosecretory granules are present in some of the ganglion cells (Lechenault, 1963; Bianchi, 1969) and in the nerves along the proboscis as far as the junction with the rather peculiar retractor muscle (Ling, 1971). While it has not yet been shown that oxytocin (or a near relative) is actually used in the nemertine, the evidence is very suggestive that the neurohypophysis is the evolutionary result of atrophy of the nerves to the proboscis and its retractor. These nerves enter the proboscis at its anterior end and are therefore morphologically in the appropriate place for the pituitary at the junction between the supposed notochord and Rathke's pocket.

A further suggestion may be of some interest. When the proboscis is retracted it has to fit into a cavity that is already filled with fluid. This is presumably accomplished fairly easily by relaxation of the muscles of the body, but such relaxation could well impede further movements on the part of the worm as a whole. If, however, fluid was encouraged to leave the rhynchocoel cavity the position could be relieved. *Lineus* has a sort of valve system connecting the rhynchocoel with the dorsal and lateral blood vessels by which pressure changes can be accommodated (Ling, 1971), and *Geonemertes* also has so-called vascular plugs which perhaps function in a similar way (Pantin & Moore, 1969; Moore, 1973). In the longer term, however, an increase in permeability of the wall of the rhynchocoel, or of the vascular plugs, could be advantageous in restoring equilibrium. Anyone who has pondered over the "whys and wherefores" of anatomical structure and physiological function must have wondered why the neurohypophysis is placed where it is, why it is composed of modified nerve cells and processes, why its associated neurosecretory cells produce oxytocin that primarily activates certain types of muscle, and vasopressin that causes changes in permeability. If the nemertine produced a neuropeptide with both these properties, or even two separate neuropeptides in order to retract and rehouse the proboscis, the answers to these queries would perhaps be forthcoming. It may be pertinent to notice that the endothelium over the posterior part of the proboscis shows innumerable pinocytotic vacuoles; so also does the retractor muscle itself after treatment with oxytocin (Ling, 1971). Perhaps some fluid is transferred from the rhynchocoel to the proboscis. It still remains, however, to explain the persistence of the neural remnants after the primary purpose of their secretions became obsolete. The answer may be that certain originally secondary effects of the neuropeptides became of primary importance in vertebrates, in the regulation of water excretion, muscular contraction and vascular tone.

Though it would not be appropriate now to discuss how the neurohypophysis relates to the rest of the nervous system, since this has been fully discussed elsewhere (Willmer, 1970) it can be briefly suggested that the nervous system as it existed in nemerteoids probably corresponded in a general way to the autonomic system in the vertebrates and that the main central nervous system of the latter is an essentially new structure which has been grafted on to the existing ganglia and nerves in relation to the coordination of the acquired swimming movements and the increased sensory input necessitated thereby. If this is the correct view, then the cerebral ganglia of nemertines are comparable with the hypothalamus of vertebrates and the neurosecretory cells of the latter find their homologues in those of the nemertine ganglia.

EYES

We may now turn to a brief consideration of the origin of the vertebrate eye. This very elaborate and perfect organ is present in its fully developed form in all living vertebrates, the only exceptions being explicable in terms of atrophy, as in cave-dwellers or among the inhabitants of the deep sea. Primitive fossil forms, like *Jamoytius* and many heterostracan fish indicate the presence of large eyes, though, of course, little of their detailed structure can be discerned. There is nothing really comparable in amphioxus though there are two sets of light-sensitive cells. The so-called eye in the vesicle of larval ascidians has little in common with the paired eyes of vertebrates which possess lens, anterior and posterior chambers, inverted retina with receptors, bipolar cells, ganglion cells and a pigment layer. If the suggestion made earlier that the nemerteoid gave rise to the vertebrate more or less directly is to be upheld it is necessary to enquire whether among nemertines there are any structures which could be fore-runners of the vertebrate eye and yet not be developed in a similar manner in the protochordates.

Many existing nemertines have eye-spots often buried in the dermis at the anterior end. These generally consist of cups of pigmented cells containing melanin into which receptor cells are invaginated in much the same way as they are in the eyes of rhabdocoels. In the latter the receptor cells have innumerable microvilli adjacent to the pigment cells and from their other pole, they send fibres back to the cerebral ganglia. These eyes therefore have little in common with those of vertebrates apart from the inverted receptor cells and the pigment layer. Although *Lineus* generally has at least two of these eye-spots on either side of its head they can be removed without affecting the sensivity of the worm to the direction of incident light (Gontcharoff, 1953, 1956; Ling, 1969). Thus although it would not be impossible that these structures could be elaborated during the course of evolution, they do not seem to offer much more than the eye-spots in ascidian larvae.

In *Geonemertes* there is, indeed, a variation of this structure that is of considerable significance. The eyes of these creatures are vesicular and the vesicle is filled with a transparent jelly (Schröder, 1918). The outer surface is composed of a cap of transparent cells while the inner hemisphere has pigmented cells between which the processes of receptor cells penetrate to terminate in the substance of the gel. This is precisely the arrangement that is found in the eye-spot of an ascidian larva (Barnes, 1971) and the pineal eyes of a cyclostome (Dendy, 1907), and is but little different from that in the pineal eye of lizards (Eakin & Westfall, 1959, 1960) (Fig. 12). The nuclei of the sensory cells mingled with those of the interstitial

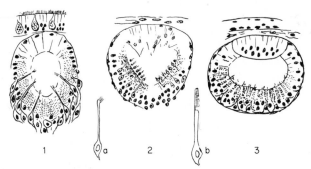

FIG. 12. Diagrams of (1) Eye-spots of *Geonemertes*, (2) Pineal eye of lamprey, (3) Pineal eye of lizard. (a) Receptor cells of *Geonemertes*. (b) Receptor cells of lizard. (After Schröder, 1918; Dendy, 1907; Eakin & Westfall, 1959.)

pigment cells form a substantial retina but bipolar cells are not represented and ganglion cells are few. From the functions that are now generally assigned to the pineal eye, namely determining light from dark, measuring total radiation and day length, producing melatonin and stimulating melanophores, it may perhaps be inferred that these elementary eye-spots in nemertines are concerned with similar functions since these could be of extreme importance to animals living in positions where they would be exposed both to radiation and to predators. Certainly, land nemertines are well supplied with eyes and the littoral *Lineus* changes colour in response to illumination.

Perhaps even more interesting is the observation (Gontcharoff & Lechenault, 1958) that when *Lineus lacteus* is deprived of its head and thus of its eyes, cephalic ganglia and cephalic organs, it responds by rapid development of the gonads and spawning, a response which could correlate with the now recognized function of the mammalian and avian pineals in inhibiting the development and function of the gonads. It is perhaps rather significant that *L. lacteus* is a very pale worm which, according to Bürger (1895), has very numerous eyes which are connected to the cephalic ganglia by quite large nerves, so that, although the authors connect the effect on the gonads with the removal of the cephalic organ, their observations are equally consistent with removal of the ganglia and eyes. Indeed they are better explained in that way, since other species of *Lineus*, with similar cephalic organs but less well-endowed with eyes, do not spawn in the same way when beheaded, though in animals of appropriate age the gonads are released from inhibition by removal of the cephalic ganglia, while the cephalic organs have no effect (Bierne, 1966).

Thus there is a strong case for considering the eye-spots of nemerteoids, and in particular the type found in *Geonemertes,* as being the fore-runners of the pineal system, and it is not inconceivable that such an organ could be modified into the type of visual organ which developed into the lateral eyes of vertebrates. There is, however, another possibility for the latter which should be considered seriously and which in some ways has more to recommend it.

From the rhabdocoels upwards into the nemertines there is a structure known as the cephalic pit or cerebral organ which has among its innumerable and extensive variations certain cytological, histological and embryogenic features which could correlate with similar features in the lateral eyes of vertebrates (Fig. 13). This

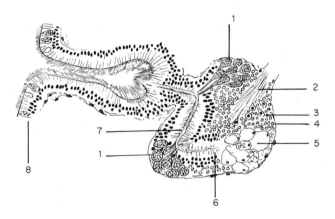

FIG. 13. Diagram of cephalic canal and organ 1. Secretory cells, 2. Nerve to brain, 3. Nerve cells, 4. Bipolar cells, 5. Vesicular cells, 6. Termination of canal, 7. Canal partitioned by flagella, 8. Superficial epithelium. (After Ling, 1969.)

organ is essentially an invagination of the ectoderm into a canal lined with cells that are characterized by the possession of long flagella and microvilli. No two species seem to have the organ developed in the same way and, though it is always connected with the brain, the position and nature of this connexion is variable. In some species the canal is distended into a vesicle, in others it is narrow and bent twice at right angles (Ling, 1969, 1970). Two groups of secretory cells pour their mucoid secretions into the canal.

The function of this organ is still unknown, but in *Lineus* the secretory cells alter their activity in response to the salt

concentration of the medium in which the worm is immersed and
also in response to light (Willmer, 1970; Ling, 1969, 1970).
Although *Lineus* is unaffected in its immediate reaction to light by
removal of its "eyes", light shone on the cephalic ganglia and
therefore probably on the cephalic organ does cause directional
movement (Ling, 1969). Removal of the organ impairs the capacity
of the worm to osmoregulate (Lechenault, 1965).

The ciliated cells of the inner part of the canal are backed by
bipolar cells and there are other nerve cells which send processes to
the cephalic ganglia so that the arrangement of receptor cells,
bipolar cells and ganglion cells of the vertebrate retina seems to be
foreshadowed. Embryologically, in those nemertines that develop
by means of placodes (Fig. 14), each cephalic organ appears as a

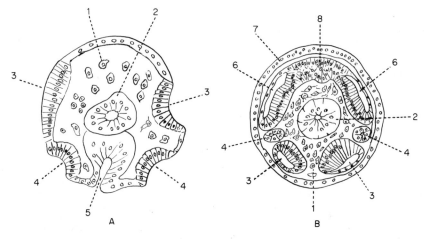

FIG. 14. Diagrams to show the development of placodes and the formation of "amnia". A.
First invaginations. B. Formation of "amnia". 1. Mesenchyme, 2. Mid-gut, 3. Trunk placode,
4. Cephalic-organ placode, 5. Fore-gut, 6. Cephalic placode, 7. Larval ectoderm, 8. Proboscis
rudiment. (After Schmidt, 1934.)

separate vesicle or placode comparable in its manner of formation
with the optic vesicle of the vertebrates. Whatever this organ is
doing in the nemertines it clearly has the required versatility and
rather special combination of structural features that would be
necessary for any fore-runner of the vertebrate eye.

Among the problems facing a nemerteoid learning to swim,
directional sensitivity to light and ability to use such sensitivity
quickly would obviously be important. The cephalic organ is

sometimes, as in *Lineus,* very closely integrated with the brain and thus its information may already be partly processed in the required direction for producing a quick response. It is conceivable that a shore-living nemerteoid might have gained more detritus for its filter-feeding mechanism if it invaded estuarine waters where depth, turbidity and salinity might all be more variable. Under those conditions stimuli of decreased salinity and increased light might both require the same response from the worm so that the cephalic organ, initially developed for the detection of decreased salinity but incidentally sensitive to light, might develop its sensitivity to the latter at the expense of its sensitivity to the lack of salt, which as we have seen may have been transferred to other tissues, i.e. the pharyngeal complex and, later, the pituitary. This may sound fantastic, but it should be remembered that the receptors of vertebrate eyes are themselves extremely sensitive to changes of sodium and potassium, and exposure to light involves ionic movements across the membranes of the receptor processes (Sillman, Ito & Tomita, 1969a,b; Cavaggioni, Sorbi & Turini, 1973). Unfortunately we do not yet know anything of the presence or absence of visual pigments in the eyes or cephalic organs of nemertines.

One thing is important to bear in mind in connexion with the evolution of eyes. Once directional sensitivity was established there would be a premium on some sort of image formation and of form vision, which of course means a neural mechanism to interpret the image. As soon as that was established the value of perfecting it would be enormous. In consequence, evolution may have progressed extremely rapidly as soon as an elementary optical system was integrated with an efficient motor system. This I believe to be the cause of the hiatus in the pre-history of the vertebrates and to explain why such apparently intermediate creatures as *Jamoytius* are so extremely rare. Evolution took a leap forward when the nemerteoid creature learned to swim efficiently. The fact that no nemertines swim like vertebrates is no longer surprising; had they done so they would have become vertebrates or perished by the way, unless they escaped from the competition by burrowing like amphioxus or returning to a sedentary life with their newly acquired filter-feeding mechanism to keep them going, like the ascidians.

SUMMARY

Figure 15 summarizes the point of view which it has been the purpose of this paper to promulgate. The very versatile nemer-

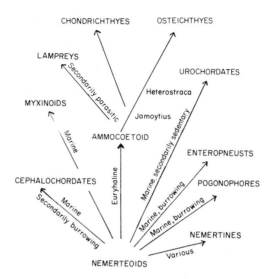

FIG. 15. The suggested relationships between nemertines, protochordates and verteb-rates.

teoids probably made several attempts to swim or to colonize new niches and many new features developed in consequence. The really successful swimmers went on evolving till both the pelagic nemertines of to-day and the vertebrates resulted. Others fell by the wayside and either perished or successfully adopted some other mode of life. It is these last that constitute the uro- and cephalo-chordates. The pterobranchs, enteropneusts, pogo-nophores and such creatures as *Lagynocystis* (Jefferies, 1973) may well be derivatives of nemerteoids that never got off the ground, though, as already mentioned, their embryonic development may pose serious problems for this hypothesis. However, Barraclough Fell (1948) has, in a similar situation, warned us that phylogenetic arguments based on embryonic forms are not always reliable. Even so it is possible that the ancestors of the echinoderms may have jumped for a time on to the swimming and filter-feeding band wagon. There is certainly a bizarre, if superficial, resemblance between some early echinoderms (Ubaghs, 1971), nemertines, enteropneusts and early heterostracan fishes (Halstead, 1973), but such resemblances mean little unless they can be substantiated by such physiological, microstructural and ecological considerations as have formed the backbone of this paper.

References

Balfour, W. E. & Willmer, E. N. (1967). Iodine accumulation in the nemertine *Lineus ruber. J. exp. Biol.* **46**: 551–556.

Ball, J. N. (1969). Prolactin and osmoregulation in teleost fishes: a review. *Gen. comp. Endocr. Suppl.* **2**: 10–25.

Barnes, S. N. (1971). Fine structure of the photoreceptor and cerebral ganglion of the tadpole larva of *Amaroucium constellatum* (Verrill) Subphylum Urochordata, Class Ascidiacea. *Z. Zellforsch. mikrosk. Anat.* **117**: 1–16.

Barrington, E. J. W. & Sage, M. (1963a). On the responses of the glandular tracts and associated regions of the endostyle of the larval lamprey to goitrogens and thyroxine. *Gen. comp. Endocr.* **3**: 153–165.

Barrington, E. J. W. & Sage, M. (1963b). On the responses of the iodine-binding regions of the endostyle of the larval lamprey to goitrogens and thyroxine. *Gen. comp. Endocr.* **3**: 669–679.

Berrill, N. J. (1929). Studies on tunicate development. I. General physiology of development of simple ascidians. *Phil. Trans. R. Soc.* (B) **218**: 37–78.

Berrill, N. J. & Sheldon, H. (1964). The fine structure of the connections between muscle cells in ascidian tadpole larva. *J. Cell Biol.* **23**: 664–669.

Bianchi, S. (1969). On the neurosecretory system of *Cerebratulus marginatus* (Heteronemertini). *Gen. comp. Endocr.* **12**: 541–548.

Bierne, J. (1966). Localisation dans les ganglions cérébroides du centre régulateur de la maturation sexuelle chez la femelle de *Lineus ruber* (Müller) (Hétéronémertes). *C. r. hebd. Séanc. Acad. Sci. Paris* **262**: 1572–1575.

Bürger, O. (1895). Die Nemertinen des Golfes von Neapel. *Fauna Flora Golf Neapel* **22**: 1–743.

Cavaggioni, A., Sorbi, R. T. & Turini, S. (1973). Efflux of potassium from isolated rod outer segments; a photic effect. *J. Physiol.* **232**: 609–620.

Clark, N. B. (1972). Calcium regulation in reptiles. *Gen. comp. Endocr. Suppl.* **3**: 430–440.

Clark, R. B. (1964). *Dynamics in metazoan evolution.* London and New York: Oxford University Press.

Coe, W. R. (1904). The anatomy and development of the terrestrial nemertean (*Geonemertes agricola*) of Bermuda. *Proc. Boston Soc. nat. Hist.* **31**: 531–570.

Coe, W. R. (1926). The pelagic nemerteans. *Mem. Mus. comp. Zool. Harvard* **49**: 9–244.

Dendy, A. (1907). On the parietal sense-organs and associated structures in the New Zealand lamprey (*Geotria australis*). *Q. Jl microsc. Sci.* **51**: 1–30.

Dharmamba, M. & Nishioka, R. S. (1968). Responses of 'prolactin secreting cells' of *Tilapia mossambica* to environmental salinity. *Gen. comp. Endocr.* **10**: 409–420.

Dharmamba, M., Mayer-Gostan, N., Maetz, J. & Bern, H. A. (1973). Effect of prolactin on sodium movement in *Tilapia mossambica* adapted to sea water. *Gen. comp. Endocr.* **21**: 179–187.

Eakin, R. M. & Westfall, J. A. (1959). Fine structure of the retina in the reptilian third eye. *J. biophys. biochem. Cytol.* **6**: 133–134.

Eakin, R. M. & Westfall, J. A. (1960). Further observations on the fine structure of the parietal eye of lizards. *J. biophys. biochem. Cytol.* **8**: 483–499.

Emmart, E. W., Pickford, G. E. & Wilhelmi, A. E. (1966). Localization of prolactin within the pituitary of a cyprinodont fish, *Fundulus heterolitus* (Linnaeus) by specific fluorescent antiovine prolactin globulin. *Gen. comp. Endocr.* **7**: 571–583.

Ensor, D. M. & Ball, J. N. (1968). A bioassay for fish prolactin (Paralactin). *Gen. comp. Endocr.* **11**: 104–110.

Etkin, W. & Gona, A. G. (1967). Antagonism between prolactin and thyroid hormone in amphibian development. *J. exp. Zool.* **165**: 249–258.

Fell, H. Barraclough (1948). Echinoderm embryology and the origin of chordates. *Biol. Rev.* **23**: 81–107.

Flood, P. R. (1970). The connection between spinal cord and notochord in amphioxus (*Branchiostoma lanceolatum*). *Z. Zellforsch. mikrosk. Anat.* **103**: 115–128.

Flood, P. R., Guthrie, D. M. & Banks, J. R. (1969). Paramyosin muscle in the notochord of amphioxus. *Nature, Lond.* **222**: 87–88.

Fujita, H. & Honma, Y. (1968). Some observations on the fine structure of the endostyle of larval lampreys, ammocoetes of *Lampetra japonica*. *Gen. comp. Endocr.* **11**: 111–131.

Gibson, R. & Jennings, J. B. (1969). Observations on the diet, feeding mechanisms, digestion and food reserves of the entocommensal rhynchocoelan, *Malacobdella grossa*. *J. mar. biol. Ass. U.K.* **49**: 17–32.

Gontcharoff, M. (1953). Le phototropisme chez *Lineus ruber* et *Lineus sanguineus* au cours de la régénération des yeux. *Annls Sci. nat. (Zool.)* (II) **15**: 369–372.

Gontcharoff, M. (1956). Le phototropisme chez *Lineus ruber* et *Lineus sanguineus* au cours de la régénération des yeux. *Int. Congr. Zool.* **14**: 208.

Gontcharoff, M. & Lechenault, H. (1958). Sur le déterminisme de la ponte chez *Lineus lacteus*. *C. r. hebd. Séanc. Acad. Sci. Paris* **246**: 1929–1930.

Halstead, L. B. (1973). The heterostracan fishes. *Biol. Rev.* **48**: 279–332.

Hammarsten, O. (1918). Beitrag zur Embryonalentwicklung der *Malacobdella grossa*. *Arb. zootom. Inst. Univ. Stockh.* No. 1.

Hett, M. L. (1927). On some land nemertens from Upolu Island (Samoa) with notes on the genus *Geonemertes*. *Proc. zool. Soc. Lond.* **1927**: 987–997.

Hopkins, C. R. (1969). The fine structural localization of acid phosphatase in the prolactin cell of the teleost pituitary following the stimulation and inhibition of secretory activity. *Tissue Cell* **1**: 653–671.

Hubrecht, A. A. W. (1883). On the ancestral form of the chordate. *Q. Jl microsc. Sci.* **23**: 349–368.

Hubrecht, A. A. W. (1887). The relation of the Nemertea to the Vertebrata. *Q. Jl microsc. Sci.* **27**: 605–644.

Hughes, A. & Reier, P. (1972). A preliminary study on the effects of bovine prolactin on embryos of *Eleutherodactylus ricordii*. *Gen. comp. Endocr.* **19**: 304–312.

Jefferies, R. P. S. (1973). The Ordovician fossil *Lagynocystis pyramidalis* (Barrande) and the ancestry of Amphioxus. *Phil. Trans. R. Soc.* (B.) **265**: 409–469.

Jennings, J. B. (1962). A histochemical study of digestion and digestive enzymes in the rhynchocoelan *Lineus ruber* (O. F. Müller). *Biol. Bull. mar. biol. Lab. Woods Hole* **122**: 63–72.

Jennings, J. B. & Gibson, R. (1969). Observations on the nutrition of seven species of rhynchocoelan worms. *Biol. Bull. mar. biol. Lab. Woods Hole* **136**: 405–433.

Jensen, D. D. (1960). Hoplonemertines, myxinoids and deuterostome origins. *Nature, Lond.* **188**: 649–650.

Jensen, D. D. (1963). Hoplonemertines, myxinoids and vertebrate origins. In *The*

Lower Metazoa: Comparative biology and phylogeny: 113–126. (Eds.) Dougherty, E. C., Brown, Z. N., Hanson, E. D. and Hartman, W. D., Berkely, L. A.: University Calif. Press.

Lam, T. J. (1972). Prolactin and hydromineral regulation in fishes. *Gen. comp. Endocr.* Suppl. **3**: 328–338.

Lechenault, H. (1963). Sur l'existence de cellules neurosécrétrices chez les Hoplonémertes. Caractéristiques histochemiques de la neurosécrétion chez les Némertes. *C. r. hebd. Séanc. Acad. Sci. Paris* **256**: 3201–3203.

Lechenault, H. (1965). Neurosécrétion et osmorégulation chez les Lineidae (Hétéronémertes). *C. r. hebd. Séanc. Acad. Sci. Paris* **261**: 4868–4871.

Ling, E. A. (1969). The structure and function of the cephalic organ of a nemertine, *Lineus ruber. Tissue Cell* **1**: 503–524.

Ling, E. A. (1970). Further investigations on the structure and function of cephalic organs of a nemertine, *Lineus ruber. Tissue Cell* **2**: 569–588.

Ling, E. A. (1971). The proboscis apparatus of the nemertine *Lineus ruber. Phil. Trans. R. Soc.* (B.) **262**: 1–22.

Ling, E. A. & Willmer, E. N. (1973). The structure of the foregut of a nemertine, *Lineus ruber. Tissue Cell* **5**: 381–392.

Moore, J. (1973). Land nemertines of New Zealand. *Zool. J. Linn. Soc.* **52**: 293–313.

Olsson, R. (1965). Comparative morphology and physiology of the *Oikopleura* notochord. *Israel J. Zool.* **14**: 213–220.

Pantin, C. F. A. & Moore, J. (1969). The genus *Geonemertes. Bull. Br. Mus. (nat. Hist.) Zool.* **18**: 263–310.

Sage, M. (1968). Responses to osmotic stimuli of *Xiphophorus* prolactin cells in organ culture. *Gen. comp. Endocr.* **10**: 70–74.

Schmidt, G. A. (1934). Ein zweiter Entwicklungstypus von *Lineus Gesseriensis-ruber* (O. F. Müll.) Nemertini. *Zool. Jb.* (Abt. Anat. Ontog. Tiere) **58**: 607–660.

Schröder, O. (1918). Beiträge zur Kenntnis von *Geonemertes palaensis* Semper. *Abh. senckenb. naturforsch. Ges.* **35**: 155–175.

Schwarz, W. (1961). Elektronenmikroskopische Untersuchungen an den Chordazellen von *Petromyzon. Z. Zellforsch. mikrosk. Anat.* **55**: 597–609.

Sillman, A. J., Ito, M. & Tomita, T. (1969a). Studies on the mass receptor potential of the isolated frog retina. I. General properties of the response. *Vision Res.* **2**: 1435–1442.

Sillman, A. J., Ito, M. & Tomita, T. (1969b). Studies on the mass receptor potential of the isolated frog retina. II. On the basis of the ionic mechanism. *Vision Res.* **2**: 1443–1451.

Ubaghs, G. (1971). Diversité et spécialisation des plus anciens Échinodermes que l'on connaisse. *Biol. Rev.* **46**: 157–200.

Urist, M. R. (1963). The regulation of calcium and other ions in the serums of hagfish and lampreys, *Ann. N. Y. Acad. Sci.* **109**: 294–311.

Utida, S., Hirano, T., Oide, H., Ando, M., Johnson, D. W. & Bern, H. A. (1972). Hormonal control of the intestine and urinary bladder in teleost osmoregulation. *Gen. comp. Endocr.* Suppl. **3**: 317–327.

Welsch, U. (1968). Über den Feinbau der Chorda dorsalis von *Branchiostoma lanceolatum. Z. Zellforsch. mikrosk. Anat.* **87**: 69–81.

Willmer, E. N. (1970). *Cytology and evolution* 2nd Ed. New York: Academic Press.

Wilson, C. B. (1900). Habits and early development of *Cerebratulus lacteus* (Verrill). *Q. Jl microsc. Sci.* **43**: 97–198.

AUTHOR INDEX

Numbers in italics refer to pages in the References at the end of each article.

Wissig, S. L., 39, *41*
Wolken, J. J., 13, *15*

Y

Yarsagil, G. M., 66, *78*
Young, B. A., 170, *176*

Yudkin, W. H., 114, 115, 118, *126, 127*

Z

Zappi, E., 132, *158*
Zeller, C., 196, *212*

SUBJECT INDEX

Roman numbers indicate text pages, *italic* numbers indicate Figs and numbers in parentheses e.g. (118) indicate tables

A

Amphioxus (see Branchiostoma)
Amphiphorus, *327*
Aplysia, 56
Apical macula (eye spot), 69, 75
Arenicola marina, 109, (121)
Aristotle's lantern, kinasein muscles of, 122–123, (122, 123)
Ascending fast response, 57, 62, 63
  slow response, 57, 67
Ascidia mentula, 117, (118)
Ascidiacea, 100
Ascidians,
  iodine in tissues of, 129–130
  iodine uptake in organs of, 151, 152
  phylogenetic relationship to vertebrates, 131, 132
Ascidiella, 134, 135, 145
Ascidiella aspersa, 230, (118)
Atelecyclus septemdentatus, (116)
ATP, as energy store, 106
Atriopore, 54, 66, 67, 189
Atrium,
  muscles of, 50
  receptors of, 76–77
    responses from, 76–77
  wall of, 225
Atubaria, 2
Audouinia tentaculata, (121)

B

Balanoglossus, 192, 322
B. clavigerus, (114)
B. salmoneus, 116, (114)
Botryllus schlosseri, (118)
Branchiostoma
  adults, distribution, 185, 187, 188, 189, 204, *187*
  adults, establishment of populations, 185–189, *186*

Branchiostoma—contd.
  diagnostic features of populations, 205
  difference factor, in population analysis, 206, 207, 209, *208*, (207)
  distribution and current patterns, 180, 185, 187, 188, 209, 210, *187*
  growth, comparison of species, 228–230, *229*
  notochord, differences with vertebrate, 324–325
  odour of, 192
  planktonic larva, 180–182, 218, 230
    behaviour, 181, 182
    currents, effect on distribution, 183, 184, 185, 186, 209
    delay in metamorphosis, 183
    distribution, 182–184, *182*, (184)
  metamorphosis, 185, 218
  settlement, 184–185, 214, *186, 187*
  substrate preference, 189–191, 192, 209, *191*
    behaviour and substrate, 189, *191*
Branchiostoma belcheri, 181, 182, 183, 213, 228, 229, 230, 231, 232, *182, 229*, (184, 207)
B. californiense, 207, (207)
B. capense, 207, *208*, (187, 207)
B. caribaeum, 228, 229, 230
  growth, and distribution, 230, 231
B. elongatum, 207, (207)
B. gambiense, 183, *208*, (184)
B. lanceolatum, 81, 160, 182, 199, 201, 204, 206, 207, 213, 214, 218, 219, 223, 225, 226, 228, *180, 200, 201, 203, 208, 215*, 230, 231, 232, (184, 188, 216), *217, 218, 221, 224, 227, 229*
  atrial system of, 50–51
  brain of, 68–74, *69, 73, 74*